Current Issues in Asian Tourism: Volume II

Edited by
C. Michael Hall and Chris Cooper

Routledge
Taylor & Francis Group
LONDON AND NEW YORK

First published 2021
by Routledge
2 Park Square, Milton Park, Abingdon, Oxon, OX14 4RN

and by Routledge
52 Vanderbilt Avenue, New York, NY 10017

Routledge is an imprint of the Taylor & Francis Group, an informa business

British Library Cataloguing-in-Publication Data
A catalogue record for this book is available from the British Library

ISBN13: 978-0-367-67963-7 (hbk)
ISBN13: 978-1-003-13356-8 (ebk)

Typeset in Times New Roman MT Std
by codeMantra

Publisher's Note
The publisher accepts responsibility for any inconsistencies that may have arisen during the conversion of this book from journal articles to book chapters, namely the inclusion of journal terminology.

Contents

Citation Information

The following chapters were originally published in the *Current Issues in Tourism*, volume 23, issue 1 (2020) and volume 22, issue 5 (2019). When citing this material, please use the original page numbering for each article, as follows:

Chapter 2
Tourist arrivals in Korea: Hallyu as a pull factor
SungKyu Lim and Evangelos Giouvris
Current Issues in Tourism, volume 23, issue 1 (2020) pp. 99–130

Chapter 3
Local and international perspectives of the influence of creative experiences of Chinese traditional culture on revisit intentions
Chih-Hsing Liu
Current Issues in Tourism, volume 23, issue 1 (2020) pp. 17–35

Chapter 4
A study of segmentation of cannabis-oriented tourists from China based on motivation
Jun Wen, Fang Meng, Tianyu Ying and Yaniv Belhassen
Current Issues in Tourism, volume 23, issue 1 (2020) pp. 36–51

Chapter 5
A journey of hope: an institutional perspective of Japanese outbound reproductive tourism
I-Chieh Michelle Yang
Current Issues in Tourism, volume 23, issue 1 (2020) pp. 52–67

Chapter 6
Tourist behaviour through consumption in Melaka World Heritage Site
Nurul Diyana Md Khairi, Hairul Nizam Ismail and Syed Muhammad Rafy Syed Jaafar
Current Issues in Tourism, volume 22, issue 5 (2019) pp. 582–600

Chapter 7
Perceived pollution and inbound tourism for Shanghai: a panel VAR approach
Xu Xu and Markum Reed
Current Issues in Tourism, volume 22, issue 5 (2019) pp. 601–614

Chapter 8
Profiling the elite middle-age Chinese outbound travellers: a 3rd wave?
Jigang Bao, Xin Jin and David Weaver
Current Issues in Tourism, volume 22, issue 5 (2019) pp. 561–574

Chapter 9
Structural relationships among mega-event experiences, emotional responses, and satisfaction: focused on the 2014 Incheon Asian Games
HakJun Song, MinCheol Kim and Yunseon Choe
Current Issues in Tourism, volume 22, issue 5 (2019) pp. 575–581

Chapter 10
When destination attractiveness shifts in response to climate change: tourists' adaptation intention in Taiwan's Kenting National Park
Wei-Ching Wang, Chung-Hsien Lin, Wen-Bor Lu and Su-Hsin Lee
Current Issues in Tourism, volume 22, issue 5 (2019) pp. 522–543

Chapter 11
Understanding the role of local food in sustaining Chinese destinations
Qian Chen and Rong Huang
Current Issues in Tourism, volume 22, issue 5 (2019) pp. 544–560

Chapter 12
How to improve tourism energy efficiency to achieve sustainable tourism: evidence from China
Lamei He, Jianping Zha and Hui Ann Loo
Current Issues in Tourism, volume 23, issue 1 (2020) pp. 1–16

Chapter 13
The interaction effect of tourism and foreign direct investment on urban–rural income disparity in China: a comparison between autonomous regions and other provinces
Joon Ho Kim and Kyung Ho Kang
Current Issues in Tourism, volume 23, issue 1 (2020) pp. 68–81

Chapter 14
Towards an understanding of Chinese tourist photography: evidence from the UK
Mohan Li, Richard Sharpley and Sean Gammon
Current Issues in Tourism, volume 22, issue 5 (2019) pp. 505–521

Chapter 15
Exploring qualitative applications of social media data for place-based assessments in destination planning
Allie McCreary, Erin Seekamp, Mae Davenport and Jordan W. Smith
Current Issues in Tourism, volume 23, issue 1 (2020) pp. 82–98

For any permission-related enquiries please visit:
http://www.tandfonline.com/page/help/permissions

Contributors

Jigang Bao School of Tourism Management, Sun Yat-sen University, Guangzhou, People's Republic of China.

Yaniv Belhassen The Program of Hotel & Tourism Management, Ben-Gurion University of the Negev, Beer Sheva, Israel.

Qian Chen School of Tourism and Culinary Science (Food Science & Engineering), Yangzhou University, People's Republic of China.

Yunseon Choe Department of Tourism Management, Kyung Hee University, Seoul, Republic of Korea.

Chris Cooper School of Events, Tourism and Hospitality Management, Leeds Beckett University, UK.

Mae Davenport Forest Resources, University of Minnesota, St. Paul, USA;

Sean Gammon School of Management, University of Central Lancashire, Preston, UK.

Evangelos Giouvris SOM, Royal Holloway, UK.

C. Michael Hall Department of Management, Marketing and Entrepreneurship, University of Canterbury, Christchurch, New Zealand; Department of Service Management and Service Studies, Lund University, Helsingborg, Sweden; and Geography Research Unit, University of Oulu, Finland.

Lamei He School of Tourism, Sichuan University, Chengdu, People's Republic of China.

Rong Huang School of Tourism and Hospitality, Plymouth University, UK.

Hairul Nizam Ismail Faculty of Built Environment, Universiti Teknologi Malaysia, Skudai, Malaysia.

Syed Muhammad Rafy Syed Jaafar Faculty of Built Environment, Universiti Teknologi Malaysia, Skudai, Malaysia.

Xin Jin Department of Tourism, Sport and Hotel Management, Griffith University, Southport, Australia.

Kyung Ho Kang College of Hotel and Tourism Management, Kyung Hee University, South Korea.

Joon Ho Kim Department of Korean Language Education, Changshin University, Changwon-si, South Korea.

MinCheol Kim Department of Sports Industry, Chosun University, Gwangju, Republic of Korea.

Su-Hsin Lee Department of Geogrpahy, National Taiwan Normal University, Taipei, Taiwan.

Mohan Li School of Tourism Management, Sun Yat-sen University, Guangzhou, People's Republic of China.

SungKyu Lim SOM, Royal Holloway, UK.

Chung-Hsien Lin Department of Leisure and Recreation, National Formosa University, Taiwan.

Chih-Hsing Liu Department of Tourism Management, National Kaohsiung University of Science and Technology, Kaohsiung, Taiwan.

Hui Ann Loo School of Tourism, Sichuan University, Chengdu, People's Republic of China.

Wen-Bor Lu Department of Art Design and Creative Industry, Nanfang College of Sun Yat-Sen University, Guangzhou, *People's Republic of China.*

Allie McCreary School of Kinesiology, Recreation & Sport, Western Kentucky University, Bowling Green, USA.

Nurul Diyana Md Khairi Faculty of Built Environment, Universiti Teknologi Malaysia, Skudai, Malaysia.

Fang Meng School of Hotel, Restaurant, and Tourism Management, University of South Carolina, Columbia, USA.

I-Chieh Michelle Yang Department of Marketing, Monash University, Malaysia.

Markum Reed School of Business, Henderson State University, Arkadelphia, USA.

Erin Seekamp Parks, Recreation and Tourism Management, North Carolina State University, Raleigh, USA.

Richard Sharpley School of Management, University of Central Lancashire, Preston, UK.

Jordan W. Smith Environment and Society & Director, Institute of Outdoor Recreation and Tourism, Utah State University, Logan, USA.

HakJun Song Department of Hotel & Convention Management, Pai Chai University, Daejeon, Republic of Korea.

Wei-Ching Wang Business School, Nanfang College of Sun Yat-Sen University, Guangzhou, People's Republic of China.

David Weaver Department of Tourism, Sport and Hotel Management, Griffith University, Southport, Australia. Dongbei University of Finance and Economics, Dalian, People's Republic of China.

Jun Wen School of Business and Law, Edith Cowan University, Joondalup, Australia.

Xu Xu School of Business, Henderson State University, Arkadelphia, USA.

Tianyu Ying Department of Tourism and Hotel Management, School of Management, Hangzhou, People's Republic of China.

Jianping Zha School of Tourism, Sichuan University, Chengdu, People's Republic of China.

Marketing, Consumption and Demand for Asian Tourism

Introduction to Current Issues in Asia: COVID-19 and beyond

C. Michael Hall and Chris Cooper

Asia is now central to the global tourism economy. Even before COVID-19 Asia had been for many years the region that was providing new international tourism generating markets as it transformed from being primarily a destination economy to an increasingly more balanced tourism region with strong domestic, intra-regional travel and outbound tourism (Singh, 2009; Winter, 2009; Hall & Page, 2016; Cooper & Hall, 2020; Dolezal et al., 2020). Although in global terms Asia has been the region most affected by COVID-19 with a drop of 51% in international arrivals in the January–April 2020 period (this compares with −44% for Europe and −36% for the Americas for the same period) (UNWTO, 2020a), it is also the region that is expected to help regenerate global tourism growth once the impacts of COVID-19 begin to abate (Hall et al., 2020a).

The impacts of COVID-19 are undoubtedly substantial. Because the diffusion of the virus was closely connected to the international aviation and tourism system, tourism was one of the first sectors to feel its effects (Falk & Hagsten, 2020). The near-complete lockdown of destinations imposed in May 2020, in response to the pandemic, led to a global 98% decline in international tourist numbers in the month when compared to the previous year (UNWTO, 2020b). This followed a fall of 97% of international arrivals in April and a 55% decline in March (UNWTO, 2020a). At a global level the UNWTO reported a 56% year-on-year drop in tourist arrivals between January and May. This translates into a fall of 300 million tourists and US$320 billion lost in international tourism receipts – more than three times the loss during the global economic and financial crisis of 2009 (UNWTO, 2020b). The longer that the restrictions on mobility that have been put into place to manage the pandemic last, then the worse the international figures will get.

By the middle of 2020 the overall decline in international arrivals in Asia showed that northeast Asia was the worst affected with a decline of 88% international arrival for the year to date compared with 2019, and southeast Asia with a 64% decline. South Asia was the least affected with a 55% fall (UNWTO, 2020c). Nevertheless, the tourism sectors in all Asian countries have been affected with corresponding disproportional effects on the lives and well-being of some of the poorest and the most vulnerable members of society, particularly workers in the large tourism informal sector (UN Economic and Social Commission for Asia and Pacific (ESCAP), 2020). In the case of Malaysia, for example, which was promoting its Visit Malaysia 2020 (VM2020) campaign, a total of 170,084 hotel room bookings during the period 11 January to 16 March 2020 were cancelled as a result of COVID-19, causing a loss of revenue amounting to RM68,190,364 (Foo et al., 2020) and corresponding impacts on those working in the sector. Foo et al. (2020) report that as of April 2020 with respect to a sample of hotel workers in Kuala Lumpur, 3% had been laid off, 20% had been asked to take unpaid leave and 16% workers had faced pay cuts.

However, the international tourism figures, as significant as they are painting a rather partial picture for tourism in Asia given the substantial attention given by governments to the promotion of domestic tourism as a response to COVID-19 international travel restrictions. Indeed, the growth of domestic tourism has been influenced not just by international restrictions but by the effects of domestic measures to curtail the disease. For example, although Asia accounted for almost 100% of reported daily cases in mid-February 2020, by the beginning of July the figure had fallen to 14%, which were almost entirely outside of China. In contrast, at this time the USA and Canada accounted for 29% of cases and the rest of the Americas 30% (World Health Organization and John Hopkins University in McKinsey & Co, 2020a).

China's rise as tourism generation region and as a destination in its own right reflects the overall importance of Asia for international tourism (Hall & Page, 2016). However, China is increasingly becoming a bellwether for the global tourism industry. As Hall and Page (2016, p. 3) commented, "such is the wider economic importance of Asia, and of the Chinese economy in particular, changes in the region and their effects now have major repercussions on confidence in the global economy". As a result of early action to restrict the spread of COVID-19, China is one of the first countries in which the tourism economy has begun to rebound. Several indicators illustrate the factors in the external environment that help create a context in which tourists are more likely to travel and book holidays. In an international survey of business confidence conducted in July 2020, more respondents in Greater China regarded business conditions as better (47% of respondents), and 45% as worse than six months previously, in all other regions conditions were regarded as considerably worse (McKinsey & Co, 2020a). Although Chinese consumer sentiment has fluctuated during the pandemic it has gradually become more positive. With the proportion optimistic that China's economy will rebound within the next 2–3 months and grow just as strong or stronger than before COVID-19 growing from 42% in a survey conducted in late February to 52% in mid-June (McKinsey & Co, 2020b). Such positive shifts in sentiment have influenced travel behaviours. As of June 2020, domestic airline passenger levels in China had continued to improve, and sat at half of pre-crisis levels. By late August domestic arrivals at Chinese airports were reported to have reached 86% of 2019 levels (BBC, 2020b). In the same period hotel occupancy rates had returned to 60% of 2019 levels although there was considerable variation between different levels and types of accommodation (McKinsey & Co, 2020c). For example, Marriott had reopened all its 350 outlets across China at the end of May reporting a 40% occupancy rate, a substantial improvement from the low of 7% in late January 2020, when China had its peak of COVID-19 cases. Furthermore, the Chinese occupancy figures for late May were twice that of Marriott's occupancy levels in the US at the same time (BBC News, 2020a).

Interestingly, the most positive travel sentiment at the time in China was with families with children as opposed to single persons. In addition, positive sentiment toward undertaking tourist-related travel was continuing to rise together with gradual improvements in perceptions of safety, with the greatest improvement in positive sentiment coming from the Wuhan region, in which the coronavirus was first reported (McKinsey & Co, 2020c). However, safety is a paramount issue in the minds of most travellers and will likely remain so for the foreseeable future. An example of this was that Chinese travel to Beijing in August was still 24.8% behind the same period in 2019, because of the impact of restrictions as a result of the city's second coronavirus outbreak in early June and the associated perceptual issues that this created (BBC, 2020b). In order to meet consumer concerns over safety, hotel chains have introduced cleaning and safety programmes. For example, Hilton reopened all of its 255 hotels in China in late May 2020 and introduced a CleanStay initiative to protect guest and employees (BBC, 2020a).

As the industry begins to bounce back, one of the clear implications of COVID-19 for tourism is the enormous turbulence that has been caused within the tourism system. Although the pandemic is worldwide in scope, it is experienced locally, and mediated via the role of governments and the critical role of the media (Chen et al., 2020), with a profound effect on consumer behaviour. Writing in a tourism and services context, Hall et al. (2020b) argue that COVID-19 non-pharmaceutical intervention measures (NPIs) restrict individual mobility and personal contact (physical or social distancing) (Anderson et al., 2020), and include the closure of certain business functions, transport routes and restrictions on congregations of people, i.e. events, the lodging sector and restaurants. This displaces consumption in five main ways, all of which have implications for tourism:

Where consumption occurs (*spatial aspects of consumption*): Restrictions on mobility, including restricting access to destinations, and/or enforced physical distancing constraints, the type of business or setting in which face-to-face consumption can occur (Dube et al., 2020), as well as limiting purchasing to areas close to the consumer's residence (Foo et al., 2020). Hall et al. (2020b) refer to this as a form of 'enforced localisation' and which reflects the switch from international outbound travel to domestic tourism in many countries as a result of lack of access to international destinations, although due to restrictions this process can also occur within countries (Foo et al., 2020).

When consumption occurs (*temporal aspects*): COVID-19 and NPIs have affected the availability of access to different products in time. Although in retailing such impacts are usually related to purchasing being brought forward in time, i.e. so-called "panic buying", in tourism they are usually pushed back as a result of risk perceptions or personal financial concerns, or because the product that would have been purchased is seasonal and therefore, consumption has to be delayed to the next time it becomes available again (Bae & Chang, 2020).

What and why consumption occurs: NPIs, such as quarantine and mobility restrictions in accessing destinations, have psychosocial effects with respect to what products are consumed as a result of different sets of motivations and needs that emerge as a result of COVID-19. This will potentially influence destination and product perceptions and particularly the rationale to engage in travel. Given the social impacts of the pandemic this may be especially significant for visiting friends and relations tourism in particular, as well as for those destinations to which consumers have a strong place attachment.

How consumption occurs: Restrictions in physical contact favour some forms of purchasing relations over others. In the case of tourism this potentially only serves to reinforce the growth of e-tourism and virtual and extended reality (Kwok & Koh, 2020; Yu et al., 2020). However, it may have other impacts on travel consumption as well. For example, threat of the spread of COVID-19 by touching surfaces has influenced how payment is made by encouraging contactless purchasing rather than use of cash, as well as to whether purchases are conducted in person or online (Pal & Bhadada, 2020).

The impacts of COVID-19 are clearly substantial but they are also very uneven as a product of the diffusion of the disease, the quality of health and welfare systems in place and the policy responses to the pandemic. Undoubtedly, the recovery will also be uneven both within and between countries. Shao et al. (2020) note, for example, how the focus of policy measures to assist tourism's recovery from COVID-19 varies by region within China. However, critical to tourism's rebound will be a firm research base from which destination managers and marketers, policy makers and the wider industry, including educators, can draw. Although the research in this collection of papers in the second volume on *Current Issues in Asian tourism* was conducted prior to the advent of COVID-19 the insights and knowledge gained will remain relevant for tourism's future in the region.

The works in this volume are collected into three main themes. The first theme is that of marketing consumption and demand for Asian tourism and features a range of papers from different countries in the region. The continued importance of the Korean cultural economy is the focus of Lim and Giouvris' chapter on hallyu tourism, while more traditional forms of culture are examined by Khairi et al. at the Melaka World Heritage Site in Malaysia and Liu's study on the creative experiences provided to tourists from Chinese traditional culture. More specialised and niche forms of tourism are discussed in Wen et al.'s chapter on cannabis-oriented tourists from China and Yang's study of Japanese outbound reproductive tourism. These seemingly disparate chapters are united by their common identification of regulatory structures on different types of behaviour to be a pull or push factor in relation to international tourism. Given its above discussed role as a metric of Asian and international tourism in general, Chinese outbound tourism will serve a critical role in the post-COVID-19 era. Bao et al.'s chapter on elite middle-age Chinese outbound travellers therefore provides invaluable insights on the future direction of this market.

The second theme of the book is that of sustainability. Wang et al. discuss the important issue of shifts in the relative attractiveness of destinations as a result of climate change using the example of Taiwan's Kenting National Park. This study also resonates with the earlier chapter by Xu and Reed on perceived pollution and inbound tourism in Shanghai. Responses to climate change are seen in He et al.'s chapter on energy efficiency in tourism, while broader interpretations of sustainability and sustainable tourism can be seen in Chen and Huan's chapter on local food in Chinese destinations and Kim and Kang's chapter on the effect of tourism and foreign direct investment on urban–rural income disparity in China. The latter chapter highlights an important area of future research for the wider Asian region given the rapid urbanisation that is occurring.

The collection is then rounded out with two chapters that focus on research methods. Li and Sharpley examine Chinese tourist photography while McCreary et al. explore qualitative applications of social media data for destination planning.

This volume provides a further contribution to the growing literature on tourism in Asia and provides a useful snapshot of some of the main research concerns and the methods by which they are evaluated. Although COVID-19 will undoubtedly shape much of the tourism agenda in the short-term, research such as that contained in this volume will not only provide a firm base from which to better understand contemporary tourism processes in the region but will also provide insights to help develop an appreciation of the key themes and issues in the long-term future of Asian tourism.

References

Anderson, R. M., Heesterbeek, H., Klinkenberg, D., & Hollingsworth, T. D. (2020). How will country-based mitigation measures influence the course of the COVID-19 epidemic? *The Lancet*, *395*(10228), 931–934. https://doi.org/10.1016/S0140-6736(20)30567-5.

Bae, S. Y., & Chang, P. J. (2020). The effect of coronavirus disease-19 (COVID-19) risk perception on behavioural intention towards 'untact' tourism in South Korea during the first wave of the pandemic (March 2020). *Current Issues in Tourism*, https://doi.org/10.1080/13683500.2020.1798895.

BBC (2020a). Coronavirus: Marriott reopens all hotels in China as travel rebounds. BBC News, 2 June, https://www.bbc.com/news/business-52874716.

BBC (2020b). Coronavirus: Flights within China to 'fully recover' next month. BBC News, 27 August, https://www.bbc.com/news/business-53927980.

Chen, H., Huang, X., & Li, Z. (2020). A content analysis of Chinese news coverage on COVID-19 and tourism. *Current Issues in Tourism*, https://doi.org/10.1080/13683500.2020.1763269.

Cooper, C. P., & Hall, C. M. (Eds.). (2020). *Current issues in Asian tourism*, Vol. 1. Routledge.

Dolezal, C., Trupp, A., & Bui, H. T. (Eds.). (2020). *Tourism and development in Southeast Asia*. Routledge.

Dube, K., Nhamo, G., & Chikodzi, D. (2020). COVID-19 cripples global restaurant and hospitality industry. *Current Issues in Tourism*, https://doi.org/10.1080/13683500.2020.1773416.

Falk, M. T., & Hagsten, E. (2020). The unwanted free rider: Covid-19. *Current Issues in Tourism*, https://doi.org/10.1080/13683500.2020.1769575.

Foo, L. P., Chin, M. Y., Tan, K. L., & Phuah, K. T. (2020). The impact of COVID-19 on tourism industry in Malaysia. *Current Issues in Tourism*, https://doi.org/10.1080/13683500.2020.1777951.

Hall, C. M., & Page, S. J. (Eds.). (2016). *The Routledge handbook of tourism in Asia*. Routledge.

Hall, C. M., Scott, D., & Gössling, S. (2020a). Pandemics, transformations and tourism: Be careful what you wish for. *Tourism Geographies*, *22*(3), 577–598. https://doi.org/10.1080/14616688.2020.1759131.

Hall, C. M., Prayag, G., Fieger, P., & Dyason, D. (2020b). Beyond panic buying: Consumption displacement and COVID-19. *Journal of Service Management*, https://doi.org/10.1108/JOSM-05-2020-0151.

Kwok, A. O., & Koh, S. G. (2020). COVID-19 and Extended Reality (XR). *Current Issues in Tourism*, https://doi.org/10.1080/13683500.2020.1798896.

McKinsey & Company (2020a). COVID-19: Briefing materials, Global health and crisis response, 6 July, https://www.mckinsey.com/~/media/McKinsey/Business%20Functions/Risk/Our%20Insights/COVID%2019%20Implications%20for%20business/COVID%2019%20July%2023/COVID-19-Facts-and-Insights-July-23.pdf

McKinsey & Company (2020b). Survey: Chinese consumer sentiment during the coronavirus crisis, 2 July, https://www.mckinsey.com/business-functions/marketing-and-sales/our-insights/survey-chinese-consumer-sentiment-during-the-coronavirus-crisis?

McKinsey & Company (2020c). *China's travel recovery gains steam: How families are planning their summer vacations*. McKinsey & Company.

Pal, R., & Bhadada, S. K. (2020). Cash, currency and COVID-19. *Postgraduate Medical Journal*, http://dx.doi.org/10.1136/postgradmedj-2020-138006.

Shao, Y., Hu, Z., Luo, M., Huo, T., & Zhao, Q. (2020). What is the policy focus for tourism recovery after the outbreak of COVID-19? A co-word analysis. *Current Issues in Tourism*, https://doi.org/10.1080/13683500.2020.1806798.

Singh, S. (Ed.). (2009). *Domestic tourism in Asia: Diversity and divergence*. Earthscan.

UN Economic and Social Commission for Asia and Pacific (ESCAP) (2020). *Policy responses to COVID-19: Addressing the impact of the pandemic on tourism in Asia-Pacific small island developing states*. ESCAP.

UNWTO (2020a). New data shows impact of COVID-19 on tourism as UNWTO calls for responsible restart of the sector. UNTourism News 14, 22 June, https://www.unwto.org/news/new-data-shows-impact-of-covid-19-on-tourism

UNWTO (2020b). Impact of COVID-19 on global tourism made clear as UNWTO counts the cost of standstill. UNTourism News 15, 28 July, https://www.unwto.org/news/impact-of-covid-19-on-global-tourism-made-clear-as-unwto-counts-the-cost-of-standstill

UNWTO (2020c). International tourism and COVID-19, 14 September, https://www.unwto.org/international-tourism-and-covid-19

Winter, T. (2009). Asian tourism and the retreat of Anglo-Western centrism in tourism theory. *Current Issues in Tourism*, *12*(1), 21–31. https://doi.org/10.1080/13683500802220695.

Yu, M., Li, Z., Yu, Z., He, J., & Zhou, J. (2020). Communication related health crisis on social media: A case of COVID-19 outbreak. *Current Issues in Tourism*, https://doi.org/10.1080/13683500.2020.1752632.

Tourist arrivals in Korea: Hallyu as a pull factor

SungKyu Lim and Evangelos Giouvris

Econometric modelling on tourist arrivals has concentrated mostly on forecasting. We abstract and assess the impact on arrivals to Korea from China, Hong Kong, Indonesia, Japan, Taiwan, Singapore and Malaysia using a number of macroeconomics variables (unemployment, inflation, GDP and exchange rate) sentiment, mood and most importantly 'Hallyu' (Korean cultural wave) which is quantified for the very first time. Effectively, we test how each variable 'measures up' against all other variables. Our analysis is twofold. Firstly, we use impulse response functions to assess the risk, magnitude and duration that each shock has on short-term arrivals from each country separately. Secondly, we concentrate on systemic effects and use causality tests to examine the effect of each variable on arrivals for all countries simultaneously. We find that the effect of our variables is not uniform but most importantly that 'Hallyu' cannot be ignored.

1. Introduction

Tourism is a very important industry for some countries since it contributes quite a lot to their GDP. Korea is a highly industrialized country which is mostly known as a car (Hyundai, Kia and SsangYong) and electronics manufacturer (Samsung, LG and Daewoo) and less as a tourist destination. However, the last few years, Korea has attracted attention and is now a very popular tourist destination among neighbouring countries (Kim & Ryoo, 2007; "Soap, Sparkle and Pop," 2014). This recent emergence of Korea as a popular tourist destination can be attributed among other factors to the effect that Korean culture (movies, music, language, fashion, animation, books, etc.) as a tourism pull factor has on neighbouring countries and beyond. This infatuation with Korean culture is greatly documented in the local and international press and it has been dubbed the 'Korean wave' or 'Hallyu' in Korean. This made local press such as *The Korean Times* (May 6, 2009) to feature an article entitled 'Hallyu becoming global cultural icon' to capture the effect of this phenomenon among others. Hallyu has also helped to attract foreign investment (Yoon, 2014) and has reached as far as Israel and Palestine (Otmazgin & Lyan, 2013). Even though nowadays Korea is considered an established tourist destination, there is still a huge variability in tourist arrivals on a yearly level, let alone on a monthly level that makes modelling future arrivals especially difficult. In order to make the modelling process easier and identify the most important tourism determinants, firstly we investigate idiosyncratic short-term effects of (i) macroeconomic, (ii) mood, (iii) sentiment and (iv) Hallyu shocks on tourist arrivals from each country. Secondly, we concentrate on systemic effects and investigate causalities between the same variables and arrivals in an attempt to help authorities to manage their long-term strategy. Variability in tourist arrivals is an accepted global phenomenon. Variability whether short term or long term can have an adverse effect on planning, policy-making and regulation for the government but also on budgeting for tourist operators who make their bookings and close their deals a year in advance (Uysal & O'Leary, 1986). Shocks on the variables mentioned above and their effect on arrivals (magnitude, rate of decay, etc.) are of tremendous significance to tourist operators.

Taking variability in tourist arrivals as a fact of life and responding to the econometric modelling challenge, researchers in the area of tourism have embarked on a 'crusade' to model arrivals in the best way possible. We also respond to this challenge and assess the impact on arrivals/visitors in Korea from China, Hong Kong, Indonesia, Japan, Taiwan, Singapore and Malaysia. Those countries represent more than 60% of total visitors to Korea (see Figure 1).

In the two paragraphs that follow, we present the past literature. We place emphasis on econometric techniques and explanatory variables used in the modelling process and how we differ. A number of techniques have been used in the past to forecast tourism demand. Law (2000) uses back-propagation and neural networks for Taiwan and Hong -Kong; Chu (1998, 2004, 2008, 2009) looks into nine Asia-Pacific region countries using cubic polynomial and ARMA techniques. Emphasis is on forecasting tourism demand. We, on the other hand, innovate by investigating (i) the effects of shocks (macroeconomic, sentiment, mood and Hallyu) on arrivals and (ii) causalities between our variables and arrivals. In order to achieve this, we use VAR, impulse response functions (IRFs) and perform Dumitrescu–Hurlin (D-H) panel Granger causality tests. Analytically IRFs are used to study the short-term effect, magnitude, duration and decay (within a period of 12 months) of a shock in our variables on arrivals from each country. This is helpful for both Korean authorities and operators to adjust their short-term strategy if unexpected events (shocks) occur. The emphasis here is on idiosyncratic shocks from individual countries. D-H panel causality tests concentrate on systemic effects. Such tests will help the Korean authorities and operators to broaden their outlook and determine their long-term strategy by indicating which variables have an impact on the industry as a whole from all neighbouring countries.

In an attempt to improve econometric modelling, researchers have added a number of variables over time besides macroeconomic indicators. Obviously, it makes sense for

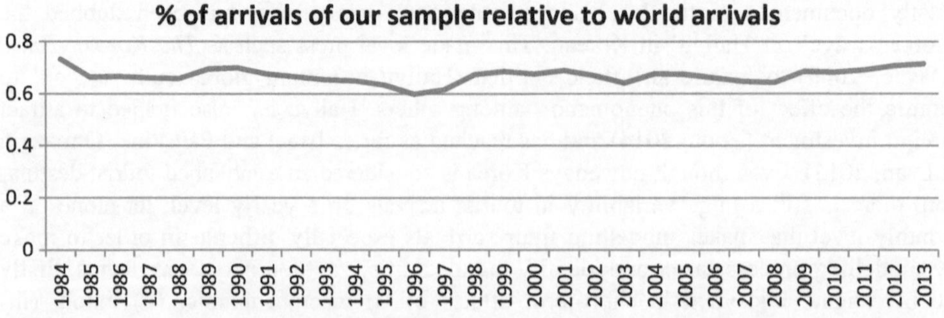

Figure 1. % of arrivals of our sample countries relative to world arrivals. This figure presents percentage changes in tourist arrivals/visitors from our sample countries relative to world arrivals/visitors.

individuals to consider exchange rate and consumer price index when deciding where to go. Smithers (2016), a reporter for *The Guardian*, indicates that tourism booms in the UK after post referendum fall in pound. Kah and Lee (2013) show that a depreciation of the Japanese Yen has made it more expensive for Japanese tourists to travel to Korea. Song, Li, Witt, and Athanasopoulos (2010) and Song, Li, Witt, and Fei (2010) employ GDP of the country of origin and CPI in the destination country (Hong Kong) relative to CPI in the country of origin adjusted by the exchange rate. Gounopoulos, Petmezas, and Santamaria (2012) use unemployment of the country of origin, relative CPI ($CPI_{GREECE}/CPI_{ORIGIN}$) and confidence index of the country of origin in order to forecast short-term arrivals in Greece. More recently, Dragouni, Filis, Gavriilidis, and Santamaria (2016) use confidence indexes to capture sentiment and financial indexes such as S&P500 to capture mood.[1]

This paper considers all variables discussed above and innovates by adding 'Hallyu' using a variety of techniques. 'Hallyu' is quantified by obtaining expenditure on Korean cultural products such as movies, TV series, music, animation and books adjusted by number of tourists. The higher the spending, the stronger the 'Hallyu'. We call this new variable cultural consumption per visitor or CCV. Effectively, CCV is a proxy for 'Hallyu' which is a tourism pull factor. We are not aware of a similar phenomenon such as 'Hallyu' occurring in any other country and we believe that it is exactly this characteristic that makes this study unique and innovative. We have not come across a study that investigates the impact of macroeconomic variables, sentiment (captured by confidence index, CCI) and mood (captured by financial indexes, FIN) on Korean tourism let alone including 'Hallyu' as a pull factor. We believe that we add substantial value by quantifying 'Hallyu' to the best of our ability and data available to us and examine how it compares against established macroeconomic variables, sentiment (captured by confidence index, CCI) and mood (captured by financial indexes, FIN). Essentially, we are testing the hypothesis if cultural factors such as Hallyu have an effect on tourist arrivals considering established factors as well. This is achieved by using IRFs and D-H (2012) panel data analysis following Galariotis and Giouvris (2015) and Bilen, Yilanci, and Eryüzlü (2015). D-H non-causality test for heterogeneous panel models with fixed effects allows for heterogeneity in the regression model and the causal relationships. This technique will help us identify if there is a unidirectional or bi-directional relationship between our variables and arrivals.

In this paragraph, we provide a glimpse of our results. IRFs show that Hallyu has an impact on short-term arrivals regardless of country. On the contrary, unemployment has a time-varying effect and it is more country specific. D-H confirms a unidirectional effect from UNE to arrivals. In a nutshell, these are the two variables that stand out in terms of consistent results obtained by both techniques. It is in the interest of Korean authorities to reinforce Hallyu and manipulate EXR and RCPI to achieve a higher number of arrivals.

To the best of our knowledge, research in assessing factors that could have an impact on tourist arrivals in Korea is scarce. We aim at closing this literature gap; however, the most important contribution of this paper is the introduction of a new time-series variable that captures 'Hallyu' and the effect that Korean culture has on tourism among other variables. The first part of our analysis is a longitudinal/time-series study. The second part is a panel study. On one hand, our IRFs allow readers to visualize and assess idiosyncratic short-term effects and their magnitude on arrivals over time. On the other hand, D-H (2012) panel data analysis allows a systemic outlook and provides an overall picture of causality between visitors and all other variables.[2]

This paper is structured as follows. Section 2 discusses the relevant literature and each variable separately. We draw the reader's attention to common reference points between

'Hallyu' and film-induced tourism. Section 3 discusses our sample; we present and analyse descriptive statistics as well as correlations between our variables including orthogonalization. Section 4 concentrates on methodology and in Section 5 we discuss our findings. Section 6 concludes.

2. Related literature

In this section, we discuss 'Hallyu', mood (captured by financial indexes, FIN), sentiment (captured by confidence index, CCI) and the macroeconomic variables that we include in our model. As you can see, we dedicate a large part of this section on 'Hallyu'. We believe that it is important that readers get a very good understanding of 'Hallyu'-induced tourism. Before proceeding with our variables discussion, we would like to remind readers of the hypothesis of this study briefly mentioned in the introduction. It will help with putting all variables in perspective. Essentially, in this study, we are testing the hypothesis if cultural factors such as Hallyu have an effect on tourist arrivals considering established factors as well. For example, is Hallyu strong enough to counterbalance a higher price level in Korea or a more expensive Won that will normally put off potential tourists? Is the desire, to ride the Korean wave (be part of the Hallyu wave), strong enough so as to put well-established factors in the second place? How do all variables measure up against each other? We start our literature review by discussing Hallyu first.

2.1. *Hallyu-cultural consumption per visitor (CCV) as a tourism pull factor: capitalizing on the Korean wave*

Korean popular culture has attracted a lot of attention the last decade. By 'popular culture', we mean TV dramas, movies, pop songs, fashion and books/magazines. Neighbouring countries such as China, Taiwan, Hong Kong, Japan, Singapore, Malaysia and Indonesia appear to 'consume' Korean popular culture at an increasing rate. Korean culture is gaining ground and even overtaking Japanese and/or Hollywood popular culture in Asia (Visser, 2002). Newspapers and magazines have dubbed this infatuation with Korean popular culture the 'Korean wave' or Hallyu (Han ryu) in Korean. One of the most recent and well-known manifestations of 'Hallyu' is rapper Psy's performance of *Gangnam style* which became an instant sensation. In addition, Lee and Bai (2016) interview 24 fans of Korean pop star Ahn Jae-wook and find that pop culture as an autonomous agent delivers a more powerful and stronger influence on the destination image than previously recognized in the literature. Specifically, the Korean wave started back in 1997 with the airing of Korean television dramas such as *What is love all about?* (and then again in 1998 as a result of popular demand achieving the second highest rating in Chinese TV, Heo 2002), *Stars in my heart* which became a big hit in both China and Taiwan and *Shiri*, a blockbuster which earned US$14 million in Japan only (Kim, 2000).

A more recent hit was the *Winter sonata* which was first aired by the Nippon Hoso Kyoaki broadcast company in 2002 and re-aired again three times. This TV drama was not confined within Asia but reached as far as Egypt, Ghana and the USA (Kim, Agrusa, Lee, & Chon, 2007). Following the broadcast of the *Winter sonata* in neighbouring countries and other countries further away increased tourist arrivals in Korea by 35.5% (Korean National Tourism Office, 2006). In the words of Park Young Su, assistant bureau chief at the Korea National Tourism Organization: 'thanks to the success of shows like "*Autumn In My Heart*" and "*Winter Sonata*" we have had 130,000 tourists coming to visit the locations where the dramas were filmed'. In addition, a report by the Hyundai Economic Research Institute (2004), the economic impact of *Winter sonata* reached US$1 billion in Korea and US$2 billion in Japan. KBS News (2005) reported that the sales of video tapes, DVDs, CDs and photos exceeded US$370 million. Besides the impact that Korean movies/TV shows have on tourist arrivals in the country as indicated above, there seems to be an increasing demand for buying remake rights for older Korean horror films such as *A tale of two sisters* by major Hollywood film-makers, namely DreamWorks SKG in an attempt to capitalize and ride the 'Korean wave'. Specifically, Heo (2002) reports that DreamWorks SKG paid US$2 million for the remake rights of *A tale of two sisters* which is twice the amount paid for *The ring*, a Japanese horror movie. Of course, Korean pop music videos did not fall back but contributed to the Korean wave as well. We are not going to go into details as we did before for the movie/TV drama series wave, but it will suffice to say that in South-East Asia there are two pop music movements, K-pop which stands for Korean pop music and J-pop which stands for Japanese pop music.

Film-induced tourism is a well-established phenomenon. Korea is not the only country which has benefitted from it. Actually, the effect in Korea was so strong so that a new term was established. It is called 'Hallyu' tourism. Other countries which benefitted from film-induced tourism but without branding it were Australia, which according to Tooke and Baker (1996) showed an annual increase of 20.5% from 1981 to 1988 for a number of reasons including the shooting of movies such as *Mad max*, *The man from snowy river* and *Crocodile dundee*. The USA, which is the biggest film/TV series-producing country in the world, also benefited from film-induced tourism. Riley, Baker, and Van Doren (1998) analysed 12 film locations in the USA and conducted a before/after movie release study. They found that visitation increased by 43%, 5 years after the movie was released compared to pre-release numbers. A similar before/after TV show broadcast (*Balamory*) study is that of Connell (2005a, 2005b). Connell reported a 43.3% increase in average per capita spending 5 months after the broadcast compared to 1 year before. Associated Press of London reports that 1/5 of the 28 million tourists who visit Britain do so after watching a movie or TV series shot in the UK. They report that the income generated as a result of film-induced tourism is in the region of US$2.8 billion (Associated Press, 2005, p. 1). The effect of movies, TV series or music on tourism is clear; however, all of the studies above are static in nature and they only report an increase in tourist arrivals and income. They are not longitudinal/time-series studies and cannot investigate if cultural consumption (proxied by money spent on movies, TV series, CDs, DVDs, music, animation and cartoons adjusted per visitor) can predict future tourist arrivals among other competing variables (mainly macroeconomic variables such as inflation, exchange rate, unemployment, GDP or even sentiment (captured by confidence index, CCI) and mood (captured by financial indexes, FIN)). In addition because of their static nature, all previous studies although very useful cannot investigate how a shock/change in cultural consumption or any other variable can affect arrivals in the short/long run and how long it will take to return to pre-shock levels. Admittedly, it is difficult to quantify cultural consumption or 'Hallyu' over a long period of time in order to investigate the effects that cultural consumption has on tourist arrivals; however, we have managed to identify a good proxy for 'Hallyu'. This is the amount of money spent on movies, TV series and music. The Korea National Tourism Organization has data on money spent on movies music, TV series, etc. for each country in our sample. Obviously, the more money potential tourists spend on movies, TV series and music, the more acquainted they get with Korean culture and the more likely they are to visit the country due to the positive impression delivered by cultural products leading to a higher number of tourists and even more spending.

Obviously, there could be a bi-directional feedback but this needs to be investigated. According to *The Korean Times* (May 6, 2009) in an article entitled 'Hallyu becoming global cultural icon', a local polling agency, namely 'Net Intelligence & Research' conducted a public survey in China and Japan in 2004 and 2006. It showed that 60% of Chinese who have been watching Korean TV dramas have a positive image of the country, while 42.5% of Japanese said that their impression of Korea has changed in a positive manner (Kang, 2009). Korean TV dramas also had a positive impact on Korean cuisine and how it is perceived internationally. 이상미 (2015) investigates the impact of Hallyu on Korean food consumption. The author identifies Korean food's function and recognition which is observed in films and dramas. The study finds that Hallyu is highly influential and increases purchase intention which includes cooking Korean food, visiting Korean restaurants and recommending Korean food to other people.

Hallyu had an impact on other areas of the economy such as fashion (Lee, 2014) and cosmetic industry (Park, 2014). Specifically, exports of local cosmetic products increased by 20.9% in 2013 (*The Korea Times*, 2014). KOTRA (대한무역투자진흥공사)/KOFICE (한국문화산업교류재단) (2016) reports that the consumption of cultural products and tourism spending in 2015 is $7 billion which indicates an increase of 2.4% from 2014 (title: 한류의 경제적 효과에 관한 연구). There has also been observed an increase in Korean medical tourism. According to the Korea Health Industry Development Institute, 'Mi Clinic' in Myeong Dong, Seoul, had 9366 foreign patients and 'Jaseng Hospital of Oriental Medicine' had 1500 foreign patients in 2012. Foreign tourists spent $115 million on medical services in 2011 (Park, 2013). 이형룡, 김유진, 김정매 (2013) investigate the impact of Hallyu on tourists' intentions from China and Japan to visit Korea for medical treatments. According to이형룡, 김유진, 김정매, the most important consumer of cultural products such as music, film and TV drama is women. The study shows that the perceived image of Korean culture obtained from music, films and TV dramas has a statistically significant positive impact on their intention to visit Korea for medical treatments (title: 한류문화컨텐츠에 대한 이미지가 의료서비스 참여도에 미치는 영향-일본, 중국 여성관광객을 대상으로-, pp. 129–144).

The effect of Hallyu was so strong that the Korean government began thinking of ways to capitalize on the Hallyu effect in an attempt to further enhance South Korea's exports, international presence and diplomatic leverage among other objectives. For this reason, the government established a presidential council on nation branding in 2009. The aspiration of the council was to raise South Korea's position by a massive 18 places over 4 years in the Anholt-GFK Roper Nation Brands Index.[3] The presidential council was dismantled in 2013 as Korea achieved a nation brand above the OECD average, overtaking countries such as Spain, Finland and Ireland. As of 2016, Korea is not at the top 10 list which is an incentive to resurrect the council and promote Korea to achieve a place at the top 10 list. Despite not being at the top 10, there were huge benefits. The Export Import Bank of Korea (한국수출입은행, 2012) investigates cultural exports and consumption across 92 countries between 2001 and 2011. Their study (entitled: 한류수출 파급효과 분석 및 금융지원 방안) shows that cultural product exports can stimulate and enhance the overall volume of exports. They emphasize that cultural consumption is not just an important factor for exports but it is also related to the country brand image which provides a stimulus for maximizing the economic effect. Additionally, the Korea Chamber of Commerce and Industry: 대한상공회의소 (2012), cited in The Export and Import Bank of Korea (한국수출입은행, 2012), conducted a market-based survey which included 300 mid-sized manufacturing companies. The survey showed that 51.9% of the companies observed a positive direct impact on their profits due to the expansion of the Hallyu phenomenon.

Hallyu did not have only positive results as most readers would expect. The effect was so strong that it actually ignited an anti-Hallyu sentiment in some neighbouring countries. This is definitively an indication of its success. Specifically, Taiwan's national broadcasting regulator ordered Gala TV (GTV) station to cut down on broadcasting Korean dramas. The reason for this order was that broadcasting Korean dramas inhibit the development of local content/dramas. The overall percentage of domestic productions broadcasted should not be less than 20% ('Taiwan orders TV station to reduce S Korean programs', *The Korea Herald*, January 3, 2015). Of course, the Japanese anti-Hallyu sentiment was not as soft as in Taiwan. Japan and Korea are well-known antagonists and their rivalry has started many years ago at least since the Second World War. Anti-Hallyu feelings in Japan were expressed through rallies, protests and chanting of hatred messages. Korean film promoters in Japan report having problems signing new deals in Japan but this is not confined only in the film industry. It appears that it affects all businessmen selling cultural products in Japan. Korean restaurants, cosmetic and gift shops are also affected. Actually, every single aspect of Korean business appears to be affected even Japanese websites that promote Korean culture. For instance, WK, the biggest Japanese website that promotes Korean culture, reported a 25% drop in visitors as a result of the anti-Hallyu sentiment. The author of the article, Park Si-Soo, believes that this is a result of the sluggish Japanese economy (Park, 2013).

To summarize Hallyu has been met with mixed feelings. Korean officials report an increase in visitors and spending (Korean National Tourism Office, 2006) as a result of it while at the same time in neighbouring countries, we observe anti-Hallyu protests (for example, in Japan) or measures against Hallyu (Korean TV show prohibitions in Taiwan). At the moment, all research in Hallyu is static in nature and employs a 'before Hallyu/after Hallyu' methodology similar to Tooke and Baker (1996) and Riley et al. (1998). There is no research that looks into the dynamics of Halluy over a longer period of time (since data have become available) so as to actually ascertain if there is a real effect on tourism using causality testing and consequently to the economy of Korea. Most importantly, there is no research on how Hallyu affects tourism in the short run. Oh (2005) employing a Granger causality approach shows that the Korean economy did not grow as a result of tourism at least for the period under examination (1975–2001). We innovate by investigating (ever since expenditure on movies, music, books, animation and fashion has become available) if Hallyu (i) is driving tourism in the long run (using causality) and consequently contributes to the growth of the Korean economy and (ii) short-run shocks and their effect on arrivals (using IRFs). Currently, the contribution of tourism to GDP is 5.1% despite a continuous downward trend. If the effect of Hallyu is statistically significant considering all other tourism-inducing variables described below, then Hallyu is contributing to the Korean economy through tourism as well. If this is found to be the case, then Hallyu must be reinforced.[4] Investigation of the effect of Hallyu on tourism and consequently to the Korean economy is achieved by looking at shocks on inbound tourism from each neighbouring country separately and as a totality emphasizing systemic effects employing a D-H causality approach. Looking into the effect of all variables for all countries simultaneously will help Korean authorities determine which variables have the strongest effect and in what direction (single or bi-directional) and pursue a basic strategy for their tourism industry. Our individual regressions and IRFs allow us to concentrate on specific countries and short-run effects from those countries. IRFs would be helpful to operators who need to adjust their short-term strategies. We believe that by looking at both individual countries and all countries simultaneously gives this study an extra advantage.

2.2. *Sentiment (captured by confidence index, CCI) and mood (captured by financial indexes, FIN)*

Sentiment (captured by consumer confidence, CCI) and mood (captured by financial indexes, FIN) are variables that capture the 'feeling' of the respondents about the economic condition in a country. Essentially, those variables reflect expectations and aspirations. Better sentiment or being in a good mood means that consumers feel very confident and are enthusiastic about the future economic conditions in their country and increase their spending. Better sentiment and happy mood could potentially translate into more holidays in foreign countries. The distinction between sentiment and mood is at best hazy and the two terms are used interchangeably. Dragouni et al. (2016) show that those two terms capture different dimensions of the human psycho-synthesis and use two proxies to capture sentiment and mood. They use Michigan's index of consumer sentiment to capture sentiment and S&P500 to capture mood. Yap and Allen (2011) find that the consumer sentiment index has significant impact when visiting friends and relatives but not on holiday tourism. Croce (2016) finds that tourism confidence index (TCI) provides a meaningful indication about tourist arrivals and improves forecast accuracy; however, accuracy gains vary greatly across regions and can hardly be generalized. Gounopoulos et al. (2012) find that the consumer confidence has no bearing on future tourists' arrivals. Given the ambiguity as to the effect of sentiment (captured by consumer confidence, CCI) in predicting/determining tourist arrivals, we decided to include it in our study in an attempt to provide further evidence. We have not come across a study that uses a financial index such as S&P500 to capture mood except Dragouni et al. (2016) so we decided to add mood as well by incorporating a financial index for every country in our sample.

2.3. *Unemployment and GDP*

Unemployment and GDP are hardcore macroeconomic indexes that we are considering in our analysis. Unemployment and GDP[5] are good indicators of the current state of the economy of each country in our sample. Kim, Lee, and Mjelde (2016) find that per capita GDP among other variables (relative prices and exchange rate) is a significant determinant of Japanese inbound tourism from Korea. Unemployment also appears to have an effect on the psychological well-being and satisfaction of consumers (Winkelmann, 2014). A negative psychological condition implies that fewer people will consider holidays abroad.

2.4. *Relative consumer price index*

Relative consumer price index (RCPI) which is defined as CPI of destination country (Korea) relative to CPI of the country of origin shows how expensive Korea is in relation to the other countries. Past research shows that tourists from Japan, the USA, the UK and Hong Kong consider the current price level in Korea relative to their country to make a decision (see Lee, 2013). Similarly, Korean tourists planning a visit to Japan also consider the price level in Japan relative to Korea (Kim et al., 2016). The price level in Hong Kong is a very important determinant of expenditure incurred by visitors from Australia, the UK and the USA relative to their countries (Song, Li, Witt, & Fei, 2010; Song, Li, Witt, &

Athanasopoulos, 2010). It is only normal that potential visitors consider how expensive goods and services are in the country they plan to visit. If it is very expensive, then they may decide to cancel their trip. If they decide to visit despite the high prices observed, then they keep their spending to a minimum as indicated above. Research also shows that price competitiveness is time varying for tourists from certain countries. Chang, Khamkaew, and McAleer (2012) show that price competitiveness is important for tourism demand for Japan, Korea and Hong Kong in the long run, and for Hong Kong and Taiwan in the short run for Thai tourists visiting those countries.

2.5. *Exchange rate*

The prevailing exchange rate is an important tourism demand determinant. An overvalued local currency will deter tourists visiting the country in question or reduce the number of days they spend in their chosen destination. The press is 'littered' by articles on the effect of the exchange rate on tourist demand. Smithers (2016), a reporter for *The Guardian*, indicates that tourism booms in the UK after post referendum fall in pound. Kah and Lee (2013) show that a depreciation of the Japanese Yen has made it more expensive for Japanese tourists to travel to Korea. Kim et al. (2016) find that the exchange rate among other variables (relative prices, and GDP) is a significant determinant of Japanese inbound tourism from Korea. In addition, Lee (2013) with reference to Korean tourism reports that the coefficients of relative prices and exchange rates were significant and elastic.

3. Data

3.1. *Sample*

Our database starts in 2007 and ends in 2013. We use monthly observations for our VAR IRF and D-H (2012) panel data analysis. We would love to use a much bigger database but cultural consumption per visitor (CCV) is not available before 2007 so this variable determines the starting point of our dataset. Also, there is no data after 2013. All of our variables unemployment, exchange rate, relative consumer price index, sentiment (captured by confidence index, CCI), GDP and mood (captured by financial indexes, FIN) for all our countries were downloaded from Datastream. Expenditure on Korean cultural products (movies, TV series and music) and tourist arrivals from China, Hong Kong, Indonesia, Japan, Taiwan, Singapore and Malaysia was obtained following communication from the Korean National tourism office. The reader at this point might be thinking what made us choose those specific countries to be our sample. Data availability was a very important factor but as you can see in Figure 1, the countries selected make up on average more than 60% of total arrivals in Korea. The data used in this graph are much longer than the data used in our VAR IRFs and D-H analysis and as you can see the percentage of arrivals from our sample countries remains quite high and is relatively stable over a number of years. The lowest percentage (60%) occurs in 1996. The lowest percentage achieved in our VAR impulse response sample and D-H (2001–2013) is 65% and occurs in 2007–2008.

3.2. *Descriptive statistics*

3.2.1. *Tourist arrivals: short overview*

Tourist arrivals show a constant increase from 1984 on a yearly basis (Figure 2, lower graph). There appears to be a single decrease in 2003. The effect is quite short-lived, just

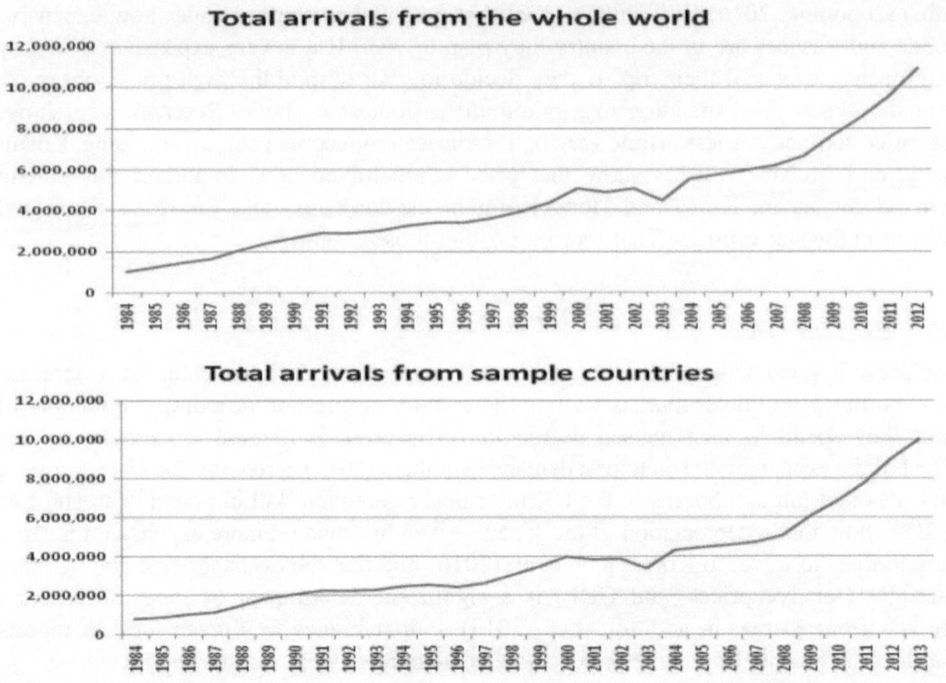

Figure 2. Total arrivals (in absolute values) from the whole world and from sample countries.

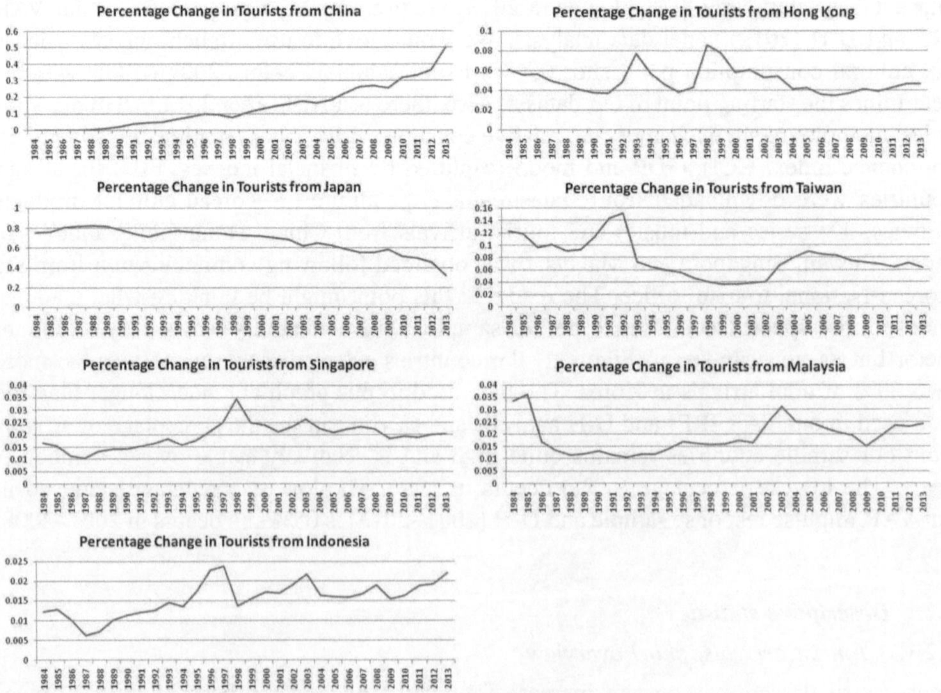

Figure 3. Percentage changes in tourist arrivals/visitors from individual countries.

for a year. Now looking at each country separately (Figure 3), we can clearly see that the percentage of Japanese tourists relative to arrivals from the rest of the countries in our sample is constantly reducing over time. This contrasts with a continuous increase observed in Chinese arrivals. Hong Kong, Singapore and Indonesia exhibit peaks and troughs.

3.2.2. *Cultural consumption per visitor (CCV)*

Cultural consumption per visitor (CCV): the average change in CCV presented in Table 1 is positive for all countries in the sample. Singapore shows the highest value and standard deviation followed by Indonesia. Positive average percentage changes over time indicate that 'Hallyu' has a strong effect on cultural consumption per visitor.

3.2.3. *Sentiment captured by consumer confidence index (CCI) and mood captured by financial indexes (FIN)*

The mean percentage change in consumer confidence index (CCI) is negative for China, Hong Kong and Japan which perhaps could translate into reduced arrivals from those countries. Indonesia and Taiwan show positive mean changes which could translate into higher tourist arrivals to Korea. CCI is not available for Singapore and Malaysia. Returns of financial indices are positive but small.

3.2.4. *Relative consumer price index (RCPI)*

Relative CPI (CPI in Korea/CPI in the country of origin) is calculated in accordance with Gounopoulos et al. (2012), Song, Li, Witt, and Athanasopoulos (2010), and Song, Li, Witt, and Fei (2010). A low relative CPI means that prices in Korea are lower compared to the country of origin. We obtain negative mean percentage changes in relative CPIs for all countries in the sample except Japan and Taiwan (see Table 1). This means that the level of inflation is higher in those countries (China, Hong Kong, Indonesia, Singapore and Malaysia) and Korea could be an attractive destination. We obtain the highest CPI for Japan but Japan is well known for its low level of inflation so relative CPI appears high.

3.2.5. *Unemployment (UNE)*

The mean percentage change in unemployment presented in Table 1 is negative for all countries in our sample except Malaysia and Taiwan but even for those countries for which we observe a positive value, the mean percentage change is very small.

3.2.6. *Exchange rate (EXR) and GDP*

The mean change in the exchange rate is positive for all countries in the sample except Indonesia and Malaysia. GDP is negative for four out of the seven countries in our sample.

3.3. *Correlations between variables*

Table 2 presents correlations between all our variables for all countries in our sample. The first number in each cell of the table represents correlation and right underneath we present *p*-values. Some correlations are statistically significant (highlighted in bold), and for this reason, we orthogonalize all our variables. We decided to orthogonalize all our variables

Table 1. Descriptive statistics.

	UNE	CCI	CPI	CCV	EXR	GDP	FIN
China							
Mean	−0.0221	−0.0587	−0.0261	3.9801	0.3909	−0.0374	0.0098
SD	0.7692	2.0400	0.6150	41.4469	3.0664	0.2715	0.0925
Min	−2.5018	−5.1756	−2.1686	−50.0650	−7.7605	−0.6000	−0.2463
Max	4.6736	8.0321	1.6439	348.1944	16.8831	0.6000	0.2745
Hong Kong							
Mean	−0.4493	−0.2726	−0.0392	7.2482	0.0780	−0.0407	0.0064
SD	3.5156	4.4657	1.0808	50.4298	3.1651	0.5635	0.0677
Min	−8.5714	−21.6000	−3.1507	−49.2708	−7.9611	−1.2700	−0.2247
Max	12.1951	19.4978	2.7946	354.0173	16.4565	1.3650	0.1707
Japan							
Mean	−0.1305	−0.0685	0.2214	5.2527	0.5018	0.0022	0.0018
SD	3.2469	3.9823	0.3609	38.5805	4.3486	0.5127	0.0645
Min	−6.8182	−12.3596	−0.4703	−63.4932	−8.8713	−1.2575	−0.2159
Max	10.0000	12.3636	1.2898	301.2085	24.5075	1.5275	0.2103
Taiwan							
Mean	0.0645	0.1252	0.1238	6.3420	0.3096	−0.0298	0.0043
SD	1.9115	3.3277	0.8446	49.8987	2.5598	0.8005	0.0628
Min	−3.8674	−10.4848	−1.8958	−87.0698	−6.7912	−1.7000	−0.1728
Max	8.6207	10.8821	2.2831	445.0708	12.2866	2.5700	0.1925
Indonesia							
Mean	−0.5266	0.3437	−0.2447	15.7332	−0.3864	0.0037	0.0159
SD	1.8042	3.2643	0.5507	132.7061	2.5595	0.1799	0.0716
Min	−7.8148	−7.2588	−2.7231	−77.2520	−10.6059	−0.5875	−0.2525
Max	4.6921	8.7698	0.7053	1247.0860	8.6304	0.3750	0.1918

Singapore						
Mean	−0.0910	−0.0361	39.5511	0.3519	−0.0331	0.0045
SD	6.2776	0.6450	281.7608	2.2699	1.0048	0.0619
Min	−30.3030	−1.5933	−51.3435	−5.7053	−2.0550	−0.2650
Max	18.5185	1.8330	2606.2430	11.5163	2.8975	0.2113
Malaysia						
Mean	0.3371	−0.0058	5.8592	−0.0708	0.0154	0.0082
SD	9.0237	0.3225	56.8088	1.6935	0.7815	0.0276
Min	−20.5882	−0.6920	−41.6717	−3.3716	−0.9475	−0.0642
Max	17.8571	0.6972	264.8596	3.5556	2.7725	0.0705

Notes: The table shows mean value, standard deviation, minimum value and maximum value for all variables. UNE indicates the unemployment rate, CCI is for the consumer confidence index (sentiment), CPI is the relative consumer price index, CCV is the cultural consumption per visitor, EXR is the exchange rate, GDP is the gross domestic product, and FIN is the financial indexes returns and captures mood.

Table 2. Correlations before and after orthogonalization for all countries.

	China before orthogonilization							China after orthogonilization						
	UNE	CCI	CPI	CCV	EXR	GDP	FIN	UNE	CCI	CPI	CCV	EXR	GDP	FIN
UNE	1							1						
CCI	0.08	1						−2.8E−16	1					
	(0.74)							(1)						
CPI	0.00	−0.02	1					1.3E−16	−4E−16	1				
	(0.99)	(0.77)						(1)	(1)					
CCV	**−0.22**	0.01	0.05	1				5.4E−16	8.9E−17	−1.7E−16	1			
	(0.02)	(0.89)	(0.59)					(1)	(1)	(1)				
EXR	**0.35**	−0.08	**0.31**	−0.10	1			−9E−16	3.4E−16	−1.9E−16	−7.5E−18	1		
	(0)	(0.42)	**(0)**	(0.31)				(1)	(1)	(1)	(1)			
GDP	−0.10	**0.20**	−0.05	0.04	**−0.48**	1		−8.7E−18	1.9E−16	1.1E−16	−1.47E−16	−2.4E−16	1	
	(0.29)	**(0.04)**	(0.59)	(0.65)	**(0)**			(1)	(1)	(1)	(1)	(1)		
FIN	**−0.20**	**0.10**	−0.15	−0.07	**−0.38**	**0.35**	1	−5.4E−17	1.8E−16	2.7E−16	−1.25E−18	3.8E−16	−1.4E−16	1
	(0.04)	**(0.31)**	(0.12)	(0.44)	**(0)**	**(0)**		(1)	(1)	(1)	(1)	(1)	(1)	
Japan before orthogonilization								Japan after orthogonilization						
	UNE	CCI	CPI	CCV	EXR	GDP	FIN	UNE	CCI	CPI	CCV	EXR	GDP	FIN
UNE	1							1						
CCI	−0.02	1						−0.02	1					
	(0.81)							(0.00)						
CPI	0.02	0.08	1					−0.00	−0.00	1				
	(0.82)	(0.42)						(0.95)	(0.95)					
CCV	−0.15	0.15	**0.24**	1				−0.00	−0.00	−0.00	1			
	(0.12)	(0.14)	**(0.01)**					(0.97)	(0.97)	(0.99)				
EXR	−0.04	**−0.25**	0.11	−0.15	1			−0.00	−0.00	−0.00	−0.00	1		
	(0.54)	**(0.01)**	(0.25)	(0.10)				(0.98)	(0.98)	(0.99)	(0.99)			
GDP	−0.07	**0.20**	−0.14	**0.20**	**−0.22**	1		−0.00	−0.00	−0.00	−0.00	−0.00	1	
	(0.45)	**(0.04)**	(0.15)	**(0.04)**	**(0.03)**			(0.93)	(0.93)	(0.98)	(0.99)	(0.99)		
FIN	−0.07	**0.37**	−0.11	0.13	**−0.57**	**0.22**	1	0.02	0.02	0.07	0.00	0.00	0.00	1
	(0.44)	**(0)**	(0.25)	(0.17)	**(0)**	**(0.03)**		(0.79)	(0.80)	(0.94)	(0.97)	(0.98)	(0.92)	

	Taiwan before orthogonilization							Taiwan after orthogonilization						
	UNE	CCI	CPI	CCV	EXR	GDP	FIN	UNE	CCI	CPI	CCV	EXR	GDP	FIN
UNE	1							1						
CCI	**–0.23**	1						–3.2E–35	1					
	(0.02)							(1)						
CPI	0.14	0.09	1					2.8E–15	2.4E–16	1				
	(0.15)	(0.35)						(1)	(1)					
CCV	0.01	0.03	**0.19**	1				–1E–17	4.9E–17	–1.6E–16	1			
	(0.85)	(0.75)	**(0.05)**					(1)	(1)	(1)				
EXR	**0.20**	–0.12	**0.39**	0.05	1			–5.9E–17	–1.1E–17	–5.1E–16	6.4E–17	1		
	(0.04)	(0.21)	**(0)**	(0.58)				(1)	(1)	(1)	(1)			
GDP	**–0.30**	**0.39**	0.03	0.08	**–0.29**	**1**		2E–16	–2.3E–17	–3.7E–16	1.6E–15	5.3E–16	1	
	(0)	**(0)**	(0.74)	(0.42)	**(0)**			(1)	(1)	(1)	(1)	(1)		
FIN	**–0.22**	**0.24**	–0.10	0.11	**–0.35**	**0.31**	1	–8.6E–16	–1E–16	2.3E–16	2.1E–16	–8.4E–16	–1.1E–16	1
	(0.02)	**(0.01)**	(0.32)	(0.28)	**(0)**	**(0)**		(1)	(1)	(1)	(1)	(1)	(1)	

Table 2. Continued.

	Hong Kong before orthogonilization							Hong Kong after orthogonilization						
	UNE	CCI	CPI	CCV	EXR	GDP	FIN	UNE	CCI	CPI	CCV	EXR	GDP	FIN
UNE	1							1						
CCI	**0.19**	1						3.2E–16	1					
	(0.06)							(1)						
CPI	–0.02	–0.04	1					2E–17	–2.1E–16	1				
	(0.79)	(0.64)						(1)	(1)					
CCV	**0.35**	0.08	0.16	1				1.2E–16	4.7E–16	1.2E–16	1			
	(0)	(0.39)	(0.10)					(1)	(1)	(1)				
EXR	**0.20**	–0.03	**0.37**	0.03	1			–3.2E–17	–1.3E–16	3.2E–17	8.6E–17	1		
	(0.04)	(0.72)	**(0)**	(0.74)				(1)	(1)	(1)	(1)			
GDP	**–0.38**	0.06	–0.03	–0.04	–0.40	1		–5E–16	1.8E–16	3.4E–16	9.2E–17	–2.1E–16	1	
	(0)	(0.50)	(0.73)	(0.67)	(0)			(1)	(1)	(1)	(1)	(1)		
FIN	–0.03	–0.04	**–0.22**	–0.05	–0.59	0.27	1	1.3E–16	–3.5E–16	1.5E–16	–4.8E–17	–3.7E–16	1.4E–16	1
	(0.74)	(0.65)	**(0.02)**	(0.61)	(0)	(0)		(1)	(1)	(1)	(1)	(1)	(1)	

	Singapore before orthogonilization							Singapore after orthogonilization						
	UNE	CCI	CPI	CCV	EXR	GDP	FIN	UNE	CCI	CPI	CCV	EXR	GDP	FIN
UNE	1							1						
CCI		1							1					
CPI	0.09		1					6E–16		1				
	(0.34)							(1)						
CCV	–0.03		–0.03	1				6.5E–19		7E–17	1			
	(0.75)		(0.75)					(1)		(1)				
EXR	–0.09		0.06	**0.23**	1			1.4E–16		–1.8E–16	1.5E–16	1		
	(0.33)		(0.54)	**(0.02)**				(1)		(1)	(1)			
GDP	**–0.17**		–0.03	**0.24**	0.03	1		1.1E–16		–3.2E–17	–3.9E–17	–2.6E–16	1	
	(0.09)		(0.75)	**(0.01)**	(0.75)			(1)		(1)	(1)	(1)		
FIN	–0.14		0.11	–0.10	–0.10	**0.18**	1	2.8E–16		2.5E–16	1.9E–17	–3.3E–16	–2.6E–16	1
	(0.15)		(0.28)	(0.32)	(0.30)	**(0.07)**		(1)		(1)	(1)	(1)	(1)	

Table 2. Continued.

	Indonesia before orthogonilization							Indonesia after orthogonilization						
	UNE	CCI	CPI	CCV	EXR	GDP	FIN	UNE	CCI	CPI	CCV	EXR	GDP	FIN
UNE	1							1						
CCI	–0.08	1						–1.6E–16	1					
	(0.43)							(1)						
CPI	–0.06	**0.23**	1					–1.7E–16	2.2E–16	1				
	(0.52)	**(0.02)**						(1)	(1)					
CCV	–0.10	–0.03	0.04	1				–3.3E–17	5.1E–17	–1.1E–16	1			
	(0.32)	(0.71)	(0.67)					(1)	(1)	(1)				
EXR	–0.11	0.03	**0.17**	–0.06	1			–2.7E–16	1E–16	1.3E–16	3E–17	1		
	(0.25)	(0.74)	**(0.09)**	(0.53)				(1)	(1)	(1)	(1)			
GDP	–0.06	–0.15	–0.14	0.02	0.09	1		–1E–16	1.4E–16	3.2E–16	–5.3E–17	–3.1E–16	1	
	(0.53)	(0.14)	(0.17)	(0.77)	(0.36)			(1)	(1)	(1)	(1)	(1)		
FIN	–0.05	0.08	0.10	0.02	0.04	–0.02	1	–1.1E–16	7.2E–16	4.6E–16	–3.1E–17	3.1E–16	2.5E–16	1
	(0.56)	(0.43)	(0.33)	(0.79)	(0.64)	(0.81)		(1)	(1)	(1)	(1)	(1)	(1)	
Malaysia before orthogonilization								Malaysia after orthogonilization						
	UNE	CCI	CPI	CCV	EXR	GDP	FIN	UNE	CCI	CPI	CCV	EXR	GDP	FIN
UNE	1							1						
CCI		1							1					
CPI	**0.24**		1					–2.5E16		1				
	(0.09)							(1)						
CCV	0.06		0.16	1				–5.2E–18		–1.6E–16	1			
	(0.57)		(0.25)					(1)		(1)				
EXR	0.00		**0.31**	1.9E–5	1			–4.4E–15		–5.5E–16	–3.5E–16	1		
	(0.98)		**(0.02)**	(0.99)				(1)		(1)	(1)			
GDP	0.15		–0.01	–0.10	–0.14	1		4E–17		–4.7E–16	1.7E–16	–2.4E–16	1	
	(0.30)		(0.92)	(0.46)	(0.34)			(1)		(1)	(1)	(1)		
FIN	0.03		–0.07	–0.20	0.05	–0.00	1	–9E–17		–3.2E–16	–5.5E–17	–3E–16	–8.3E–17	1
	(0.80)		(0.61)	(0.15)	(0.70)	(0.95)		(1)		(1)	(1)	91)	(1)	

Notes: The first and second part represent correlations before and after orthogonalization, respectively. The first number in each cell of the tables represents correlation and right underneath we present p-values. Significant correlations are in bold. $p = 1$ means no correlation.

and

$$VV^{\mathrm{T}} = \Omega. \tag{6}$$

After successful transformation of the orthogonalized innovation, we replace u_t with $e_t V$ and we rewrite the VAR as a moving average:

$$Y_t = \sum_{n=0}^{\infty} A_n V e_{t-n}. \tag{7}$$

By setting $B_n = A_n V$, our VAR becomes

$$Y_t = \sum_{n=0}^{\infty} B_n e_{t-n}, \tag{8}$$

where B_n are the impact multipliers.

For the Dumitrescu–Hurlin (2012) panel test, the regression is given below:

$$Y_{i,t} = \alpha_i + \sum_{k=1}^{K} \gamma_i^{(k)} Y_{i,t-k} + \sum_{k=1}^{K} \beta_i^{(k)} X_{i,t-k} + \varepsilon_{i,t}, \tag{9}$$

where $Y_{i,t}$ captures arrivals in Korea as before and $X_{i,\ t-k}$ are our macroeconomic variables and cultural consumption. $K \in N$ and $\beta_i = (\beta_i^{(1)}, \ldots, \beta_i^{(K)})'$. The individual effects α_i are supposed to be fixed in the time dimension. Initial conditions $(y_{i,-K}, \ldots, y_{i,0})$ and$(x_{i,-K}, \ldots, x_{i,0})$ of both individual processes $y_{i,t}$ and $x_{i,t}$ are given and observable. We assume that lag orders K are identical for all cross-section units of the panel and the panel is balanced. Besides, we allow the autoregressive parameters $\gamma_i^{(k)}$and the regression coefficients slopes $\beta_i^{(k)}$ to differ across groups. However, contrary to Weinhold (1996) and Nair-Reichert and Weinhold (2001), parameters $\gamma_i^{(k)}$ and $\beta_i^{(k)}$ are constant in time. It is important to note that our model is not a random coefficient model as in Swamy (1970), it is a fixed coefficient model with fixed individual effects (Dumitrescu & Hurlin, 2012, p. 1451).

5. Empirical results

5.1. *The impact of macroeconomic shocks on future tourism arrivals*

Table 4 presents results of the VAR model. Each equation is dedicated to a specific country. The number in brackets represents the *p*-value. The output we obtain is quite big so we present coefficients significant only at 10%. Relative consumer price index (ΔRCPI) and unemployment (ΔUNE) are the only two variables which are significant for all countries in the sample. Each of the rest of our variables is insignificant only once. For example, cultural consumption per visitor (ΔCCV) is insignificant in Hong Kong only. The only variable which is insignificant twice is ΔGDP (in Hong Kong and Japan). All variables carry mixed signs and there is sign reversal (from positive to negative and again from negative to positive) at different lags for all countries in their sample. This is because our variables are in differences (Δ), not levels. Also, correlation feedbacks in addition to fluctuations of estimations at different lags could create further complications making interpretation of coefficients a bit difficult (see Gounopoulos et al., 2012).

5.2. *IRF results and discussion*

Unemployment (ΔUNE) affects arrivals to Korea in a negative way primarily from Hong Kong followed by Japan and Singapore. The reaction appears to be imminent and occurs within the first month. For the rest of the countries (Indonesia, Taiwan, China and Malaysia), there appears to be an initial positive reaction (arrivals increase despite unemployment) but this is short-lived. Specifically, in the case of Indonesia and Taiwan, there is a quick steep downwards correction with a lag of 1 month. For China and Malaysia, it takes longer to correct. We observe a lag of 2–3 months. Gounopoulos et al. (2012) obtain similar results for US arrivals to Greece following a shock in unemployment. The position of the confidence bands suggests that the impulse response is reliable and robust. Also, the impulse response reduces to zero in less than 7 months for all countries. The process is stationary.

Exchange rate (ΔEXR) affects arrivals to Korea in a negative way for visitors from Indonesia, Japan and Malaysia. The reaction is imminent. With reference to Japan, Kah and Lee (2013) show that a depreciation of the Japanese Yen has made it more expensive for Japanese tourists to travel to Korea. It is worthwhile to note the response from Malaysia which is quite strong (steep fall in arrivals). For the rest of the countries, the reaction is positive. It is worth noting a spike in the case of Hong Kong while for China and Taiwan the reaction is sluggish and this is visually captured by a straight line following the initial positive reaction. Overall, the reaction to exchange rate shocks is mixed. The position of the confidence bands suggests that the impulse response is reliable and robust. Also, the impulse response reduces to zero in less than 7 months for all countries. The process is stationary.

Relative consumer price index (ΔCPI) affects arrivals to Korea in a negative way for visitors from Singapore and Malaysia. Recovery occurs after 2 months. In the case of Hong Kong, the response is sluggish. The line is flat. For China, Indonesia, Japan and Taiwan, the reaction is initially positive but corrects within 1–2 months. The decline is more pronounced for Japan and Taiwan. The position of the confidence bands suggests that the impulse response is reliable and robust. Also, the impulse response reduces to zero in less than 7 months for all countries. The process is stationary.

ΔCCV appears to have a positive effect on arrivals to Korea for all countries in the sample. The reaction is imminent and strong (steep increase) for China, Hong Kong, Japan, Taiwan and Malaysia. The response from Indonesia and Singapore although positive is sluggish and the effect dies out much faster than the rest of the countries. Given that it takes quite some time for potential tourists to get acquainted with Korean culture through movies, music, etc., the response appears to be quite fast and consistent for all countries in the sample. The position of the confidence bands suggests that the impulse response is reliable and robust. Also, the impulse response reduces to zero in less than 7 months for all cases. The process is stationary.

Sentiment captured by the consumer confidence index (ΔCCI) affects arrivals to Korea in a positive way mainly from Japan even though the response is short-lived. The response from Indonesia is quite flat while from China, Hong Kong and Taiwan is initially negative and correction occurs in 2–3 months. The reason we observe this pattern could be that visitors from China, Hong Kong and Taiwan consider Korea to be an inferior destination. This means that the more confident they feel about economic conditions in their country, the more likely they are to travel further abroad and not confine themselves within their region. Unfortunately, there is no data available for Singapore and Malaysia. The position of the confidence bands suggests that the impulse response is reliable and robust. Also, the impulse response reduces to zero in less than 7 months in all cases. The process is stationary.

ΔGDP affects arrivals to Korea in a positive way from China, Hong Kong, Indonesia, Japan and Malaysia. It is worth noticing that the response from China and Indonesia is weak and sluggish (almost a flat line). ΔGDP affects arrivals to Korea in a negative way from Taiwan and Singapore. The position of the confidence bands suggests that the impulse response is reliable and robust. Also, the impulse response reduces to zero in less than 7 months in all cases. The process is stationary.

Financial indexes returns (FIN) which are a proxy for mood affect arrivals to Korea in a positive way from China, Hong Kong and Japan. The response from Hong Kong is quite weak (flat line), while the response from Japan is steeper. The response from Taiwan, Singapore and Malaysia is negative, the stronger one obtained from Malaysia (steep decrease) while for Taiwan and Singapore the response is considerably less intense. The position of the confidence bands suggests that the impulse response is reliable and robust. Also, the impulse response reduces to zero in less than 7 months in all cases. The process is stationary.

5.2.1. *Discussion of IRFs results and implications*

In the previous paragraphs, we merely presented results from the IRFs. In this section, we provide explanations for the results above and discuss how Korean authorities could take advantage of those findings. The IRF graphs show that it is mainly cultural consumption (CCV) that affects short-term arrivals to Korea (results are positive and consistent for all countries in the sample). Normally, we would expect the effect of CCV to be less intense and appear in our graphs with a time lag. We would also expect the whole effect to be short-lived but it appears that it dies out very slowly. It takes at least 4 months before the response line gets to zero. GDP and unemployment (UNE) come second in our hierarchy in terms of consistent results. We would expect unemployment and GDP to have an even bigger impact perhaps comparable to that of CCV. The results obtained are neither weak nor a surprise really. Gounopoulos et al. (2012) find that there is a serious lag in intensity with which unemployment affects tourist arrivals in Greece from European countries. In our case, Japan and Singapore are known to have very low levels of unemployment, while China and Hong Kong are developing fast, especially China which is considered to be the power train of the world economy. These are factors that dampen the effect of unemployment. Also, an increase in GDP takes quite some time to be 'felt' by consumers. Even though unemployment IRFs are not as consistent as CCV IRFs, it is worth noticing that the effect of unemployment is more persistent compared to that of GDP. The effect of unemployment dies out much more slowly (it takes more than 5 months for the effect to die out), while the effect of GDP lasts for about 4 months. At this point, it is worth reminding our readers that our variables are orthogonalized so we are observing the 'net' effect of unemployment and GDP. Obviously, unemployment has a long-lasting effect because it is a more 'personal' issue and directly affects potential tourists' decision to go on holidays. Financial indexes returns (mood), exchange rates and the price index come third in our hierarchy. The results we obtain from IRFs are mixed. Financial indexes returns can be very volatile and unpredictable and this can make potential tourists to change their decisions really fast affecting the outcome from IRFs. A possible reason that RCPI and EXR are not having a more consistent and stronger impact could be that tourists choose 'all-inclusive' packages at fixed prices; therefore, some of the higher cost/higher inflation and/or more expensive Korean Won are absorbed by other stakeholders. Gounopoulos et al. (2012) find that there is an asymmetry and a serious lag in intensity with which RCPI affects tourist arrivals in Greece from European countries. Also, Chang et al. (2012) find that price competitiveness is important for tourism demand for Japan, Korea and Hong Kong in the long run, and for Hong Kong and Taiwan in the short run for Thai tourists visiting those countries confirming a time-varying effect. Finally, CCI (sentiment) appears to produce the least consistent results even though the error bands are tight and the shock dies in about 6 months which indicates good stability of the model. Gounopoulos et al. (2012) find that shocks to consumer confidence after undertaking IRF analysis have no bearing on future tourists' arrivals and their system appears to have widening error bands. Croce (2016) even though finds that tourism confidence index (TCI) provides a meaningful indication about tourist arrivals and improves forecast accuracy stresses that accuracy gains vary greatly across regions and can hardly be generalized. If our readers wonder why CCI (sentiment) performs so inconsistently compared to financial indexes returns, we need to remind them that financial indexes returns capture mood, while CCI captures sentiment. They are two different aspects of the human psycho-synthesis according to Dragouni et al. (2016). In addition, all our variables are orthogonalized, so CCI (sentiment) and financial indexes returns (mood) capture two entirely different aspects of human behaviour which justifies their individual performance.

Considering the results from the IRFs, one might wonder how the Korean Tourism Organization could take advantage of the findings above. As stated, CCV is the only variable which produces consistent results for all countries followed by GDP and unemployment. Given that it is impossible for Korean authorities to control unemployment and GDP in other countries, the only option they have is to reinforce the Hallyu phenomenon in order to increase tourist arrivals from neighbouring countries. In the literature review, we mentioned the establishment of a presidential council on national branding which aimed at raising Korea's position in the Anholt-GFK Roper nation brands index by 18 places. Even though this council does not exist anymore perhaps Korean authorities should consider 'resurrecting' it. The role of the resurrected national branding council would be to reinforce Hallyu. Tooke and Baker (1996) suggest that a subsidy might be justified to encourage film companies to choose locations which display attractive national landscapes. Taking onboard this suggestion, the Korean national branding council could introduce a subsidy system that would encompass all areas of Hallyu (movies, music, language, fashion, etc., and not just film companies as suggested by Tooke and Baker) so as to magnify the effect. This could compensate for a reduction in economic activity observed (specifically exports which is the basis of the Korean economy) as a result of newly introduced protectionism policies following US elections and Brexit (Allen, 2016; Jones, 2016). Of course, Korean authorities could also manipulate the value of their currency in order to attract visitors from abroad but our IRFs present mixed results as to the effect of EXR on short-term tourist arrivals. Perhaps this could be their long-term strategy if causality analysis which follows in the next section shows that EXR is Granger causing visitors.

To summarize the three variables that stand out in terms of consistency are CCV, GDP and unemployment. The rest of the variables present mixed results with consumer confidence (CCI) that captures sentiment being the most inconsistent. Mood proxied by financial indexes returns performs considerably better. Given the inability of the Korean authorities to control GDP and unemployment in other countries, their best option is to reinforce the Hallyu phenomenon in every possible way to increase arrivals. This could be achieved by introducing a new national branding council.

5.3. *Dumitrescu–Hurlin panel causality tests and discussion*

Having performed individual analysis for each country, the next step is to look at all countries together and test for Granger causality in a panel data setting. This will broaden Korean authorities' understanding of the systemic effect of all variables on arrivals. The null hypothesis is that visitors (ΔVISI) *do not* homogenously cause ΔCCI, ΔCCV, ΔRCPI, ΔEXR, ΔUNE, ΔGDP and FIN. The alternative hypothesis states that ΔCCI, ΔCCV, ΔRCPI, ΔEXR, ΔUNE, ΔGDP and FIN *do not* homogenously cause visitors (VISI). Results are presented in Table 5.

There appears to be a two-way relationship between cultural consumption (ΔCCV) or Hallyu and visitors (ΔVISI). This can be explained in the following way. Tourists who have already visited Korea will probably spend more money in Korean cultural products which will perhaps increase their desire to visit Korea again in the future. Those who have never visited Korea in the past will get to familiarize themselves with the country through its cultural products and perhaps visit the country in the future to further enhance their understanding of the country and 'quench their thirst' for Korea. This relationship can be conceptualized as a circle where one variable feeds into the other. Our IRFs indicated a consistent positive reaction to ΔVISI as a result of a shock in ΔCCV. Even though IRFs are designed to measure the impact of a shock its duration and rate of decay, they also provide support to the results obtained here.

There appears to be a two-way relationship between relative consumer price index (RCPI) and visitors. Obviously, higher (lower) prices in Korea compared to the country of origin will make it a less (more) attractive destination and reduce (increase) arrivals. This is easy to conceptualize. Lee (2013) finds that the relative CPI is a significant variable for Korean inbound tourism from Japan, the USA, the UK and Hong Kong. Chang et al. (2012) find that price competitiveness is important for tourism demand for Korea in the

Table 5. Pairwise Dumitrescu–Hurlin panel causality tests.

Pairwise Dumitrescu–Hurlin panel causality tests			
Null hypothesis	*W*-statistics	*Z* bar-statistics	Probability
VISI does not homogeneously cause CCI	2.1921	0.1554	0.8765
CCI does not homogeneously cause VISI	2.3998	0.3771	0.7061
VISI does not homogeneously cause CCV	6.1732	4.4053	0.0000
CCV does not homogeneously cause VISI	4.4217	2.5355	0.0112
VISI does not homogeneously cause RCPI	5.2781	3.4496	0.0006
RCPI does not homogeneously cause VISI	5.5348	3.7237	0.0002
VISI does not homogeneously cause EXR	1.4047	−0.6850	0.4933
EXR does not homogeneously cause VISI	1.8451	1.9201	0.0548
VISI does not homogeneously cause UNE	1.5377	−0.5431	0.5871
UNE does not homogeneously cause VISI	4.2172	2.3172	0.0205
VISI does not homogeneously cause GDP	2.5131	0.4981	0.6184
GDP does not homogeneously cause VISI	0.8295	−1.2991	0.1939
VISI does not homogeneously cause FIN	2.9245	0.9372	0.3486
FIN does not homogeneously cause VISI	1.6467	−0.4267	0.6696

Notes: The table shows results of the Dumitrescu–Hurlin (2012) panel Granger causality tests. ΔVISI is the change in total tourism arrivals/visitors to Korea, ΔUNE is the change in unemployment rate, ΔRCPI is the change in relative consumer price index (Korea/country of origin), ΔCCI is changes in consumer confidence index (sentiment), ΔEXR is changes in the exchange rate, ΔCCV is changes in cultural consumption per visitor, ΔGDP is changes in gross domestic product, and FIN is financial indexes returns (mood). All explanatory variables are orthogonalized.

long run for Thai tourists. Those two studies support our findings of RCPI Granger causing visitors. Of course, our results also show that arrivals/visitors Granger cause CPI as well. It makes sense that a greater number of tourists will push prices up since a greater number of people will be chasing a fixed amount of goods and services. Demand exceeds supply and this pushes prices up.

The effect that EXR has on arrivals/visitors is significant. The probability we obtain is 0.0548. The exchange rate can affect tourist arrivals even though our IRFs results are mixed regarding the effect and intensity. Kah and Lee (2013) show that a depreciation of the Japanese Yen has made it more expensive for Japanese tourists to travel to Korea. Visitors do not homogenously Granger cause EXR. Even though it makes sense to believe that demand for a specific currency will push the price up for that currency, the impact on price of a currency that demand from tourists has is reduced. The price of each currency is determined by other more important factors that relate mostly to industrial production and purchase of materials including oil and of course investment.

The relationship between arrivals and unemployment (UNE) is much simpler to explain. Arrivals/visitors do not homogenously Granger cause unemployment and this makes perfect sense since UNE is unemployment in the country of visitors, not Korea. The probability is 0.81. However, the alternative 'UNE does not homogenously cause VISI' is rejected indicating that unemployment in the country of visitors does affect arrivals to Korea from that specific country. The probability is 0.02. Overall, the results obtained here provide support to results obtained by looking at individual countries.

Now looking at the three remaining variables: (i) GDP, (ii) sentiment captured by consumer confidence index (CCI) and (iii) mood captured by financial indexes returns (FIN), our causality tests show that there is no causality in any direction. Our IRFs for the effect of GDP on tourist arrivals indicated a short-term impact that dies out quite fast. IRFs results for financial indexes returns provided a mixed short-term impact while for CCI were inconsistent. Of course, IRFs and causality tests are designed to study different effects, so one technique does not necessarily imply similar results for the other.

To summarize results obtained from both IRFs and causality tests are along the same line. The effect of some variables such as CCV and UNE is indisputable while for others such as CCI (sentiment) is very weak. Causality for relative consumer price index (RCPI) is bi-directional while for exchange rate (EXR) is unidirectional. The exchange rate affects arrivals. Manipulation of EXR and RCPI in order to attract more visitors would require intervention in international FX markets and changes in the mix of macroeconomic policies by the Korean government. Perhaps it would be easier to manipulate CCV by reinforcing the phenomenon in order to increase arrivals.

6. Conclusion

Tooke and Baker (1996) look into the effect of movies in tourist arrivals adopting a static before/after approach. In their conclusion, they state that research could make possible the construction of an operational model to estimate the effects (in terms of both visitor numbers and spending and on inbound tourism) of deciding to use a particular location for filming. Such a model could be used to calculate the potential value and the costs of a movie or a television series to the proposed location and to the country. This suggestion inspired us to look into econometric models (both overtime and in the cross section, not static) that investigate the effect of Hallyu (which encompasses movies, music, language, fashion, etc., and not just films) and other variables on arrivals.

Firstly, we investigate the impact that random shocks have on short-term arrivals in Korea in terms of intensity and duration employing IRFs. We use a number of macroeconomic variables (GDP, UNE, EXR, RCPI) as well as variables that capture mood (financial indexes returns, FIN), sentiment (consumer confidence index, CCI) and most importantly a new variable that captures the effect that 'Hallyu' has on tourist arrivals in Korea over time. 'Hallyu' can be seen as a pull factor. All our variables are orthogonalized. To the best of our knowledge, this is the very first study that quantifies the effect of 'Hallyu' to be used in a time series (using IRFs) and panel data (using D-H) study. The IRF graphs allow us to visualize the effect of each variable on arrivals within 12 months and assess the extent to which each variable from each country poses a risk for Korean tourism (Figure 4).

Secondly, we broaden our outlook and investigate causalities between our variables and tourist arrivals/visitors employing the D-H (2012) non-homogenous Granger causality panel data approach. This allows us to obtain a more general picture of possible systemic

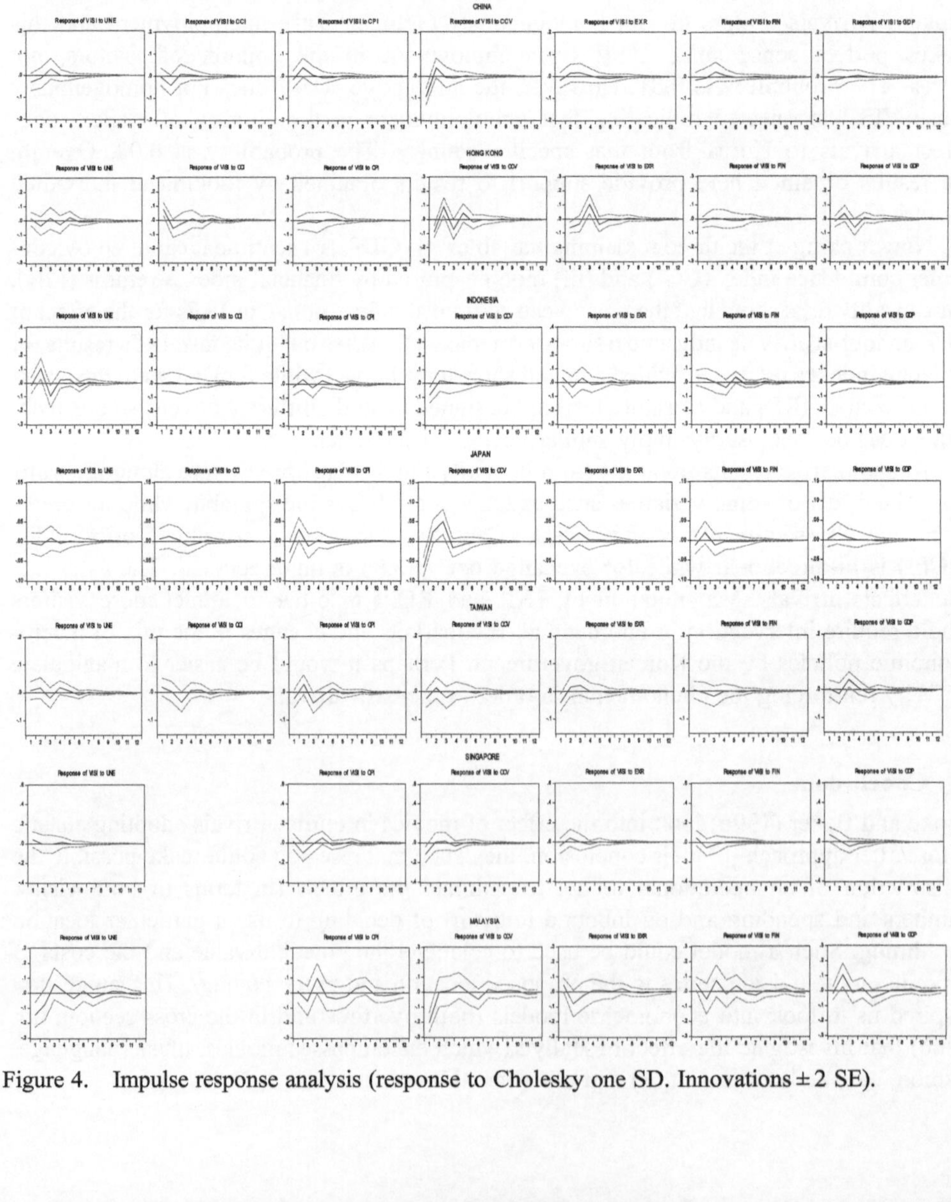

Figure 4. Impulse response analysis (response to Cholesky one SD. Innovations ± 2 SE).

interactions between visitors and our variables. A good understanding of possible systemic interactions would help Korean authorities to determine their long-term strategy and policies.

Thirdly, the 'Hallyu' phenomenon adds extra value to this study because it has never been examined in conjunction with other established macroeconomic variables as well as sentiment (consumer confidence index, CCI) and mood (financial indexes returns, FIN) which have recently been introduced in the tourism literature as distinct aspects of the human psycho-synthesis (Dragouni et al., 2016). Effectively, we conduct a 'horse race' between all our variables. We are not aware of a similar phenomenon such as 'Hallyu' occurring in any other country; therefore, it is very difficult for this study to be replicated and we believe that it is exactly this characteristic that makes it unique. We believe that we add substantial value by quantifying 'Hallyu' to the best of our ability and data available to us and examine how it compares against established determinants.

Both techniques provide similar results. The 'Hallyu' phenomenon cannot be ignored. This is confirmed by both IRFs and D-H. There is a significant short-term effect as indicated by IRFs but there is also bi-directional causality as shown by D-H. Even though we were not surprised by this finding, we thought that its effect would not be so strong because it was competing against well-established macroeconomic variables/determinants as well as sentiment (captured by confidence index CCI) and mood (captured by financial indexes returns or FIN). Some variables from specific countries appear to have a more intense impact on arrivals compared to others. For example, the effects of unemployment on short-term arrivals are more country specific and have a varying effect while Hallyu appears to have an impact on short-term arrivals regardless of the country. The D-H causality test confirms a unidirectional effect from UNE to visitors for all countries. In contrast, the effect of CCI (sentiment) is quite limited as shown by both IRFs and D-H. Gounopoulos et al. (2012) also find that the consumer confidence index has no bearing on tourist arrivals in Greece. Croce (2016) with reference to tourism confidence index (TCI) stresses that accuracy gains vary greatly across regions and can hardly be generalized. Mood captured by financial indexes returns also has a limited effect confirmed by both IRFs and D-H. GDP appears to have an impact in the short run. The effect is short-lived. There is no causality from GDP to arrivals. In short, there are two variables that stand out in terms of consistency, namely CCV and unemployment. All our variables are orthogonalized, so we are observing the 'net effect' of each variable.

Generally speaking, policy-makers should consider variables both at a country level (idiosyncratic) and overall (systemic) before deciding on policies to be implemented; however, some general conclusions can be drawn. Firstly, variables that capture 'Hallyu' pull-type phenomena (if possible to be quantified) should be included in quantitative analysis because they can add value and prove useful in assessing the impact and risk that popular destinations face. This is not going to be an easy task for researchers in the area of tourism. Secondly, Korean authorities should reinforce the Hallyu phenomenon by 'resurrecting' the national branding council as discussed in the literature review even though it has achieved its primary objective. Tooke and Baker (1996) suggest that a subsidy might be justified to encourage film companies to choose locations which display attractive national landscapes. Taking onboard this suggestion, the Korean national branding council could introduce a subsidy system that would encompass all areas of Hallyu (movies, music, language, fashion, etc., and not only for film companies as suggested by Tooke and Baker) so as to magnify the effect. The national branding council could be solely responsible for initiating and managing the subsidy scheme. If they do not take action they may observe a further decrease in the contribution that tourism has on their GDP given the existing downward trend. Korea is a highly industrialized country and its economy relies heavily on exports (cars, electronics, etc.) and it is highly vulnerable to protectionism, tariffs and quotas. The last year following Brexit and US elections there has been lots of discussion of reintroducing protectionism. This could have a serious impact on the Korean economy which is export led (Allen, 2016; Jones, 2016). Korea is already suffering reduced exports as a result of protectionism. Currently, there are 184 restrictions on Korean products, up from 159 a year earlier, according to figures from the Ministry of Trade, Industry and Energy (Kim et al., 2016). In order to reduce risk and keep other avenues open, the Korean government should reinforce Hallyu and consequently tourism in case protectionism becomes harsher.

Unemployment in visitor countries is also an important tourism determinant but the Korean government has no way of controlling the effect of this. EXR and RCPI as indicated by causality tests have an effect but manipulating the EXR could be difficult, especially if it requires constant intervention in the FX market. Inflation and consequently RCPI is easier to tame but this would require fiscal and monetary interventions. It seems that any improvement in tourist arrivals which does not require changes in economic policy (monetary, fiscal and intervention in the FX market) depends on Hallyu. Korean authorities need to ride the Korean wave[6] and take advantage of the present momentum.

Disclosure statement

No potential conflict of interest was reported by the authors.

Notes

1. In their paper, they make a distinction between mood and sentiment, as we do.
2. For example, if there is a one-way causal relationship running from exchange rate to visitors, then it is in the interest of the Korean authorities to keep the Korean Won as low as possible to reinforce tourism. Country-specific regressions or IRFs may provide different results for the effect of a cheap Won on visitors from different countries but causalities for all countries will help Korean authorities determine and pursue a basic strategy.
3. The index is based on six dimensions one of which is tourism.
4. Perhaps the Korean authorities should consider resurrecting the presidential council on nation branding which does not exist anymore.
5. GDP is available quarterly. In order to include it in our sample which is monthly, we divide changes in GDP equally between the 3 months in each period.
6. We are not aware of Hallyu-type phenomena in other countries but if present somewhere else besides Korea, new research could assess the impact of such phenomena on each country separately as well as simultaneously following the techniques applied here.

References in English

Allen, K. (2016, November 28). Protectionism and trade disputes threaten world growth, says OECD. *The Guardian*. Retrieved February 26, 2017, from https://www.theguardian.com/business/2016/nov/28/protectionism-trade-wars-world-growth-oecd-donald-trump

Associated Press London. (2005). Britain's film industry rewinds. *Taipei Times*.

Bilen, M., Yilanci, V., & Eryüzlü, H. (2015). Tourism development and economic growth: A panel Granger causality analysis in the frequency domain. *Current Issues in Tourism*, *20*(1), 27–32.

Chang, C. L., Khamkaew, T., & McAleer, M. (2012). *Estimating price effects in an almost ideal demand model of outbound Thai tourism to East Asia* (Working Paper). National Chung Hsing University, Maejo University and Erasmus University.

Chu, F. L. (1998). Forecasting tourism demand in Asian-Pacific countries. *Annals of Tourism Research*, *25*, 597–615.

Chu, F. L. (2004). Forecasting tourism demand: A cubic polynomial approach. *Tourism Research*, *25*, 209–218.

Chu, F. L. (2008). Analyzing and forecasting tourism demand with ARAR algorithm. *Tourism Management*, *29*, 1185–1196.

Chu, F. L. (2009). Forecasting tourism demand with ARMA-based methods. *Tourism Management*, *30*, 740–751.

Connell, J. (2005a). Toddlers, tourism and Tobermory: Destination marketing issues and TV-induced tourism. *Tourism Management*, *26*(5), 763–776.

Connell, J. (2005b). What's the story in Balamory?: The impacts of a children's TV programme on small tourism enterprises on the Isle of Mull, Scotland. *Journal of Sustainable Tourism*, *13*(3), 228–255.

Croce, V. (2016). Can tourism confidence index improve tourism demand forecasts? *Journal of Tourism Futures*, *2*(1), 6–21.

Dragouni, M., Filis, G., Gavriilidis, K., & Santamaria, D. (2016). Sentiment, mood and outbound tourism demand. *Annals of Tourism Research*, *60*, 80–96.

Dumitrescu, E. I., & Hurlin, C. (2012). Testing for granger non-causality in heterogenous panels. *Economic Modelling*, *29*, 1450–1460.

Galariotis, E., & Giouvris, E. (2015). On the stock market liquidity and the business cycle: A multi country approach. *International Review of Financial Analysis*, *38*, 44–69.

Gounopoulos, D., Petmezas, D., & Santamaria, D. (2012). Forecasting tourist arrivals in Greece and the impact of macroeconomic shocks from the countries of tourists' origin. *Annals of Tourism Research*, *39*(2), 641–666.

Heo, J. (2002). The "Hanryu" phenomenon and the acceptability of Korean TV dramas in China. *Korean Journal of Broadcasting*, *16*(1), 496–529.

Hyundai Economic Research Institute. (2004). *'Hallyu' and its commercialization strategy*. Seoul.

Jones, C. (2016, November 14). ECB warns on protectionism and political risk and Trump victory. *Financial Times*. Retrieved February 26, 2017, from https://www.ft.com/content/8cc14c0a-aa5a-11e6-809d-c9f98a0cf216

Kah, J. A., & & Lee, S.-H. (2013). The value of Japanese Yen and Japanese Tourism in Korea International. *Journal of Digital Content Technology and its Applications*, *7* (12), 302–306.

Kang, H. K. (2009, May 6). Hallyu becoming global cultural icon. *The Korea Times*. Retrieved October 29, 2015, from http://koreatimes.co.kr/www/news/nation/2009/05/113_44463.html

KBS News. (2005). Economic effects of Yong Sama.

Kim, S. S., Agrusa, J., Lee, H., & Chon, K. (2007). Effects of Korean television dramas on the flow of Japanese tourists. *Tourism Management*, *28*, 1340–1353.

Kim, J., Lee, C. K., & Mjelde, J. W. (2016). Impact of economic policy on international tourism demand: The case of abenomics. *Current Issues in Tourism*, *7*(1), 1–18.

Kim, M. (2000, December 30). A year of money and accolades for Korean films. *Korea Herald*.

Kim, E. M., & Ryoo, J. (2007). South Korean culture goes global: K-Pop and the Korean wave. *Korean Social Science Journal*, *34*(1), 117–152.

Korean National Tourism Office. (2006). *Annual report of inbound and outbound tourism*. Seoul: Author.

Law, R. (2000). Back-propagation learning in improving the accuracy of neural network-based tourism demand forecasting. *Tourism Management*, *21*, 331–340.

Lee, C.-K. (2013). Major determinants of international tourism demand for South Korea: Inclusion of marketing variable. In D. R. Fesenmaier, J. T. O'Leary, & M. Uysal (Eds.), *Recent advances in tourism marketing research* (pp. 101–119). New York: Routledge.

Lee, S., & Bai, B. (2016). Influence of popular culture on special interest tourists' destination image. *Tourism Management*, *52*, 161–169.

Lee, W. Y. (2014, March 27). Hallyu buoys Korean fashion. *The Korea Times*. Retrieved October 29, 2015, from http://news.naver.com/main/read.nhn?mode=LSD&mid=sec&sid1=104&oid=044&aid=0000142819

Nair-Reichert, U., & Weinhold, D. (2001). Causality tests for cross-country panels: A look at FDI and economic growth in less developed countries. *Oxford Bulletin of Economics and Statistics*, *63*, 153–171.

Oh, C. O. (2005). The contribution of tourism development to economic growth in the Korean economy. *Tourism Management*, *26*, 39–44.

Otmazgin, N., & Lyan, I. (2013). Hallyu across the Desert: K-pop Fandom in Israel and Palestine. *Cross-Currents: East Asian History and Culture Review*, E-Journal No. 9.

Park, J. H. (2013, April 30). Korean traditional medicine set to lead medical tourism. *The Korea Times*. Retrieved October 29, 2015, from http://koreatimes.co.kr/www/news/culture/2013/04/135_134885.html

Park, J. H. (2014, March 4). Dramas, R&D fuel K-cosmetics. *The Korea Times*. Retrieved October 29, 2015, from http://koreatimes.co.kr/www/news/culture/2014/03/135_152713

Park, S. S. (2014, February 21). Anti-Hallyu voices growing in Japan. *The Korea Times*. Retrieved October 29, 2015, from http://koreatimes.co.kr/www/news/culture/2014/02/135_152045.html

Riley, R., Baker, D., & Van Doren, C. (1998). Movie induced tourism. *Annals of Tourism Research*, *25*(4), 919–935.

Smithers, Rebecca. (2016, October 11). Tourism booms in UK after post-referendum fall in pound. *The Guardian*. Retrieved from https://www.theguardian.com/business/2016/oct/11/tourism-booms-in-uk-after-post-referendum-fall-in-pound

Soap, sparkle and pop. (2014, August 9).*The Economist*. Retrieved October 29, from http://www.economist.com/news/books-and-arts/21611039-how-really-uncool-country-became-tastemaker-asia-soap-sparkle-and-pop

Song, H., Li, G., Witt, S. F., & Athanasopoulos, G. (2010). Forecasting tourist arrivals using time-varying parameter structural time series models. *International Journal of Forecasting*, *27*(3), 855–869.

Song, H., Li, G., Witt, S. F., & Fei, B. (2010). Tourism demand modelling and forecasting: How should demand be measured? *Tourism Economics*, *16*(1), 63–81.

Swamy, P. A. (1970). Efficient inference in a random coefficient regression model. *Econometrica*, *38*, 311–323.

Tooke, N., & Baker, M. (1996). Seeing is believing: The effect of film on visitor numbers in screened locations. *Tourism Management*, *17*(2), 87–94.

Uysal, M., & O'Leary, J. (1986). A canonical analysis of international tourism demand. *Annals of Tourism Research*, *13*, 651–655.

Visser, D. (2002, March 10). What Hip Asians want: A little bit of Seoul: from films to fashion, Korean pop culture becomes "Kim Chic" across continent. *Washington Post* (from the Associated Press). Retrieved from http://seattlepi.nwsource.com/national/56897_pac04.shtml

Weinhold, D. (1996). Tests de causalite sur donnees de panel: Une application a letude de la causalite entre l'investissement et la croissance. *Economie et Prevision*, *126*, 163–175.

Winkelmann, R. (2014). Unemployment and happiness. University of Zurich, Switzerland, and IZA, Germany. Retrieved October 29, 2015 from http://wol.iza.org/articles/unemployment-and-happiness-1.one-pager.pdf

Yap, G., & Allen, D. E. (2011). Investigating other leading indicators influencing Australian domestic tourism demand. *Mathematics and Computers in Simulation*, *81*(7), 1365–1374.

Yoon, J. Y. (2014, August 31). Hallyu fuels foreign investment in Korea. *The Korea Times*. Retrieved October 29, 2015, from https://www.koreatimes.co.kr/www/news/biz/2014/08/488_163871.html

References in Korean

대한무역투자진흥공사 (KOTRA)/한국문화산업교류재단 (KOFICE). (2016). 한류의 경제적 효과에 관한 연구, 연구보고서.

이상미. (2015). 한류문화가 중국인 관광객의 한식 인지도 및 구매의도에 미치는 영향, 한국콘텐츠학회논문지, Vol 15 (5), pp.514-522.

이형룡, 김유진, 김정매. (2013). 한류문화컨텐츠에 대한 이미지가 의료서비스 참여도에 미치는 영향 -일본, 중국 여성관광객을 대상으로 -, 고객만족 경영연구, 제 15권, 제 1호, p. 129-144.

한국수출입은행 (The Export-Import Bank of Korea). (2012). 한류수출 파급효과 분석 및 금융지원 방안, 중점연구 2012-1.

Local and international perspectives of the influence of creative experiences of Chinese traditional culture on revisit intentions

Chih-Hsing Liu

ABSTRACT
There are an increasing number of studies of Chinese traditional culture; however, until now, there has not been an empirical examination of the integrated perspectives on whether internationalized or localized viewpoints influence cultural tourism. To address this gap, this study develops an integrated and comparative model of tourist behaviours, where international or local tourists' creative experiences may influence their revisit intentions. Study 1 examines this proposition through 631 local tourists' perspectives and suggests that creative experiences may influence revisit intentions through authenticity. Further, the relationships between authenticity and revisit intentions may be strengthened by motivation. Study 2 investigates 381 foreign tourists' creative experiences to demonstrate that experiences may influence their revisit intentions through the destination's attractions. Further, the relationships between the destination attractions and revisit intentions may be strengthened by brand image. The mutual relations among the dimensions of creative experiences also have implications for tourists when making tourism decisions.

Introduction

Chinese culture has existed for thousands of years and has influenced people's daily lives. Recently, following the advancements in China, Chinese culture has not only been attracting foreign tourists but also influencing China's young generation to discover the true values of cultural influence (Jiang, Ryan, & Zhang, 2018). In tourism and hospitality research, cultural activities have been indicated to be an effective vehicle for national unity, influencing domestic tourists' behaviours related to national identity (Weaver, Kwek, & Wang, 2017), and cultural activities have also become an important part of international tourists' behaviours by providing memorable and creative experiences (Mo & Fu, 2017). Although the Chinese cultural influences are increasingly important and although the matter of *whether* culture tourism inspires domestic and international tourists' destination choices has received abundant research consideration, fewer is discovered about *how* creative cultural perceptions influence tourists' travel intentions under Chinese contexts and the *differences* among the effects (Hsu & Huang, 2016). Examining the *how* of creative culture perceptions of Chinese and Western cultures and the *differences* between them is a critical subject and opportunity to understand cultural diversity (Wong, McKercher, & Li, 2016), yet relatively fewer empirical studies have examined the role of creative experiences and the two cultures collective integration in trip intentions and Chinese value perceptions. This study tries to fill the tourism literature gaps by conducting a deeper investigation of Chinese cultural values within localized and internationalized perspectives and the *differences* in creative experience offerings and *how* they are formed.

Previous tourism literature regarding Chinese culture representation has suggested two important discoveries. First, for local tourists, cultural and creative tourism improve the image of traditional Chinese culture and generate multiplier effects for understanding valuable Chinese cultural resources and seeing those Chinese cultural resources as memorable assets (Hung, Lee, & Huang, 2016). Second, for foreign tourists, temples are representative of Chinese culture and rank as the most popular destinations among sightseeing destinations that fascinate the masses of worldwide visitors who want to experience local history, lifestyles, traditions and folk art (Chang & Liu, 2009). Santa-Cruz and López-Guzmán (2017) asserted that culture as a marketing tool helps local economic development and provides unique business opportunities and the international cultural brand advertising at fairs promote tourism. However, insufficient tourism studies have demonstrated *how* cultural experiences influence travel intentions in the setting of localized and internationalized comparisons. Culturally unique involvements are dissimilar with other events or festivals in terms of their inimitability with specific cultural distinctions, their undiscovered mysterious value, and their authentic experience in influencing the tourists' feelings and behaviours (Bendle et al., 2014). Thus the intentions of the current research were to add to the existing studies and explore the different processes, roles and mutual relationships among creative experiences in affecting revisit intentions and investigative this unsolved issue by assessing domestic and foreign tourists' viewpoints when visiting traditional Chinese cultural settings.

Figure 1 presents a theoretical conceptual model of the process of creative experiences affecting revisit intentions and the proposed hypotheses in the two distinctive field studies of localized and internationalized perspectives. For the first part of the study of localized perspectives, this study proposed that creative experiences may influence revisit intentions through authenticity. In the process of creative experiences, we argue that the escape function of creative experiences may inspire local tourists' peace of mind through the following: (1) spending time in a learning process, (2) experiencing a unique feeling and (3) interacting with a cultural and creative environment. More specifically, motivation plays an important role as a moderating effect because it acts as internal motivation through 'push' psychological forces and external motivation through 'pull' forces from the attraction of the destination (Yoon & Uysal, 2005). The second part of the study of the international perspective investigates how the creative experiences of culture tourism (Tan, Kung, & Luh, 2013) and the destination attraction for tourists (Richards, 2002) induce behavioural intentions towards cultural tourism. Specifically, given the combinations of multiple mediators, we assert that a unique creative experience may influence revisit intentions through the mechanisms of learning, interaction, peace of

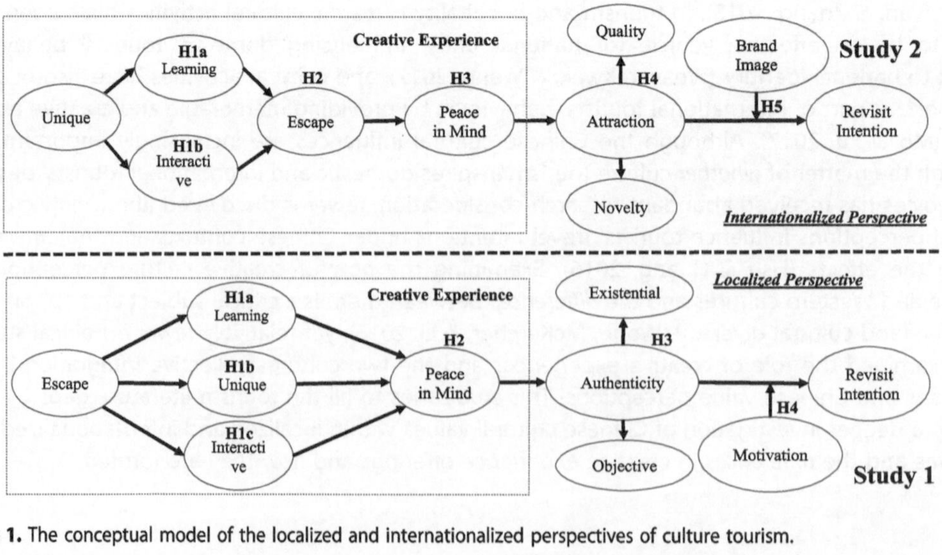

Figure 1. The conceptual model of the localized and internationalized perspectives of culture tourism.

mind and attraction. Furthermore, a brand image plays an important role as a moderating effect because it not only influences tourists' perceptions of historical sites and cultural destinations (Qu, Kim, & Im, 2011), but it also affects their evaluation of a destination's attraction (Hung et al., 2016).

This research contributes to the creative experiences, cultural tourism and revisit intentions literature in several ways. First, creative experiences refer to tourists' search for what they want from the process of co-creating tourism experiences with destinations (Tan et al., 2013). Thus industry practitioners and academic researchers need to know more about the critical factors that can increase culture influences in the generation of creative experiences. However, as noted, few studies examine these critical essentials factors. Therefore, this research conceptualizes the creative experience creation process and addresses these gaps in the tourism literature. Second, this study responds to the calls for understanding culture tourism by using comparative evaluations within an integrated framework with localized and internationalized perspectives (Mo & Fu, 2017). Specifically, this study considers the meditation–moderation processes to systematically examine various antecedents and consequences of tourists' creative experiences and how these factors interact in the context of intangible Chinese cultural destinations influencing revisit intentions. Third, culture scholars have proposed that culture has become one of the most noteworthy motivators of visitors when choosing destinations (Hsu & Huang, 2016; Santa-Cruz & López-Guzmán, 2017). By contrast, cultural scholars traditionally focused on single groups of tourists' behaviours, such as domestic tourists' creative experiences, memories and revisit intentions (Hung et al., 2016) or foreign tourists' images, motivations and revisit intentions. However, cross cultural and scholarly attention has begun to focus on tourists' heterogeneity (both domestic and foreign) (Cisneros-Martínez & Fernández-Morales, 2015; Whang, Yong, & Ko, 2016). Therefore, our novel approach is pertinent since it illuminates an unsolved issue that continues to puzzle cultural comparison researchers with a more in-depth investigation that uses empirical research to determine the relationship between the Chinese cultural perceptions of the domestic and foreign tourists in their travel decision-making process.

STUDY 1: Localized perspective of creative experiences affecting revisit intentions

Theoretical background and hypothesis development

Creative experiences and how they influence tourists' behaviours is a relatively new idea for practical managers and tourism scholars to discover the differences in traditional travel experiences (Hung et al., 2016). As the experience economy grows, tourism service or product providers need to create memorable events and experiences for tourists. Ali, Ryu, and Hussain (2016) argued that the consumption of experiences as co-creation values the generation process, which can provide remarkable memories through the atmosphere, surroundings and scenarios that the attractions provide. Many scholars have attempted to identify how culture fits into the various identified dimensions and mutual influences of tourists' creative experiences. For example, Ali et al. (2016) studied active cultural participation, such as the storytelling of local tales and 'batik painting', and identified five critical characteristics of the creative experience (e.g. escape, peace of mind, unique involvement, interactivity and learning). Moreover, in an extensive cultural study, Chen and Chen (2011) proposed that people tend to travel when they encounter troubles or failures, and the benefits of travel are that they provide escapes from daily life, improve intra- and interpersonal esteem and enhance peace of mind. We can apply the concept to cultural tourism, which is embedded in the traditional culture, since it provides a temporarily escape from the stresses of daily life and can increase peace of mind to re-establish a tourist's social status through the unique involvement, interactivity and learning through cultural activities (Smith, 2015). In addition, cultural tourism as a leisure activity satisfies domestic tourists' search for nostalgia, memorability and peace of mind through a unique atmosphere, learning and interactivity with culture heritage (Lee, 2015). Accordingly, the subsequent hypotheses were proposed.

Hypothesis 1a. Cultural learning mediates the relationship between escape and peace of mind.

Hypothesis 1b. Unique cultural involvement mediates the relationship between escape and peace of mind.

Hypothesis 1c. Cultural interactivity mediates the relationship between escape and peace of mind.

Peace of mind drives tourists' psychological safety and comfort when engaging in tourism (Chen & Chen, 2010, p. 30). Therefore, peace of mind can be defined as tourists' emotional responses to their favourite social–psychological behaviours. In the tourism context, social–psychological benefits result from tourists' participation in cultural or heritage tourism activities (McKercher, McKercher, & Du Cros, 2002). Wang (1999) noted that social–psychological benefits and authenticity are highly connected, especially in traditional cultures and places, which refer to an existential state of being, a psychological evaluation of the genuineness, and the reality or the uniqueness of culture activities. In other words, through interactivity, learning and unique involvement with traditional culture activities, tourists may receive social–psychological benefits and satisfy their peace of mind, which eventually results in an authentic experience. The result is consistent with Chhabra and Kim (2018), who observed that an authentic experience is generated from the values attached (e.g. interactivity, learning and uniqueness) to the objectivity (e.g. heritage or cultural activities), which are seen as important antecedents of tourists' psychological behaviours (Chhabra & Kim, 2018, p. 56). Thus hypothesis 2 is stated as follows:

Hypothesis 2a. Peace of mind mediates the relationship between cultural learning and authenticity experience.

Hypothesis 2b. Peace of mind mediates the relationship between unique cultural involvement and authenticity experience.

Hypothesis 2c. Peace of mind mediates the relationship between cultural interactivity and authenticity experience.

The issue of authenticity has been discussed for decades as a central tenant of cultural tourism studies (Kolar & Zabkar, 2010). In an early study, the concepts of authenticity defined as exploration of a specific social context and the search for authentic cultural activities that are 'unspoiled, pristine, genuine, untouched and traditional'. Recently, Moufahim and Lichrou (2019) asserted that authenticity is a combination of the concepts of existential and objective dimensions of authenticity. Existential authenticity refers to tourists' subjective experience, true self-concept and subjective feeling of reflecting their true self and behaviour (Kirillova, Lehto, & Cai, 2017, p. 14), and the type of objective authenticity that was originated and represented in the cultural events is concerned with confirming original historical truth and fact (Reisinger & Steiner, 2006). However, the modern tourist has become more sophisticated in their mindset and desire for authenticity, as noted by Ramkissoon and Uysal (2011), who further asserted that searching for traditional cultural does not only refer unspoiled, pristineness, genuineness and reliability of traditional culture but also to the interpretation of the true desire for it. Chung, Kim, Lee, and Kim (2018) argued that the true desire of the tourist mindset started to be transferred from the appeal for authenticity to the increased gratefulness of the tourism object and the revisit intentions, such as peace of mind, which is vital to the latter's success. In other words, Chung et al. (2018) asserted that seeking peace of mind represents another new tourism purpose, which is closer to all that is traditional, local, organic and authentic, and it helps in understanding tourists' revisit behaviours. Based on the reasoning from the previous literatures, the following hypothesis was proposed:

Hypothesis 3. The authenticity of an experience mediates the relationship between peace of mind and revisit intentions.

Motivation has been referred to as emotional and cognitive needs, including the internal psychological needs of feelings and instincts, and the external mental desires of knowledge or beliefs (Yoon & Uysal, 2005, p. 46). From an anthropological point of view, tourists are motivated to escape everyday routine life, relax, experience adventure and seek authentic experiences (Paulauskaite, Powell, Coca-

Stefaniak, & Morrison, 2017). In cultural tourism studies, maximizing the authenticity of travel experiences is crucial for a successful tourism organization (Chhabra & Kim, 2018). The evaluation of the cultural service and physical products and the psychological and mental interpretation of a destination's expressive attributes are necessary for tourists' behavioural explanations (Kolar & Zabkar, 2010; McKercher et al., 2002), which could be represented as travel motivations and destination revisit intentions. Ramkissoon and Uysal (2011) also asserted that travel motivations are a key influence in travellers' authentic experiences of cultural attractions and a further determinant of their behavioural intentions to revisit. Thus the following hypothesis was proposed:

Hypothesis 4. Motivations moderate the relationship between the authenticity of an experience and revisit intentions.

Sampling

The data were collected by a self-reported survey in Fujian Province's cultural and creative theme park, as it is an important cultural and creative industry development base in mainland China (Hua, Chen, Fang, & Wang, 2018). Ma, Li, and Chan (2018) asserted that Fujian Province has been recognized as one of the important Chinese cultural heritage conservation regions that is representative of traditional lifestyle patterns, cultural construction and historical features. The tested questionnaire was distributed by four well-trained research assistants. To increase the response rates and clarify any questionnaire issues, the research assistants stood away from participants when they were filling the questionnaire. In total, 660 surveys were distributed. After excluded multiple missing values and incomplete surveys, 631 valid surveys were used for further statistics analysis, which yielded a 95.61 response rate. The demographic information of participants is as follows. The Chinese tourists included 209 (33.12%) males and 422 (66.88%) females, with 279 respondents (44.22%) aged between 21 and 30 years. The other age categories were below 20 (88, 13.95%), 31–40 (174, 27.58%), 41–50 (67, 10.62%) and above 51 (23, 3.65%). Most tourists had a university degree (267, 42.31%), 165 (26.15%) had an MBA degree or above, and 199 (31.54%) had another type of degree or below a university degree.

Variables

All items were measured using a seven-point Likert-type scale. This response format was used for the variables of study 1, such as creative experiences, authenticity, motivations and revisit intentions, and the seven-point Likert scale ranged from 'Strongly disagree'(1) to 'Strongly agree'(7). First, five sub-dimensions that were adopted from Ali et al. (2016) were used to measure creative experiences. (1) *Escape* used three measurement items, and its Cronbach's α was .916. (2) *Learning* used four measurement items, and its Cronbach's α was .881. (3) *Unique involvement* used four measurement items and its Cronbach's α was .849. (4) *Interactive* used three measurement items, and its Cronbach's α was .897. (5) *Peace of mind* used four measurement items, and its Cronbach's α was .879. Second, following prior authenticity experience studies (e.g. Waitt, 2000; Zhou, Zhang, & Edelheim, 2013), this study used two sub-dimensions to measure the authenticity of tourist experiences. (1) *Existential* used three measurement items and its Cronbach's α was .874. (2) *Objective* used three measurement items, and its Cronbach's α was .861. Third, the tourists' motivations were rated using the three-item scale from Kolar and Zabkar (2010). Its Cronbach's α was .851. Fourth and finally, revisit intentions were measured using the three-item scale from Stylos, Vassiliadis C, Bellou, and Andronikidis (2016) and Bonn, Joseph-Mathews, Dai, Hayes, and Cave (2007), and its Cronbach's α was .940.

Confirmatory factor analyses (CFA)

In this study, we conducted CFA to verify domestic tourists' revisit intentions, authenticity, motivations and creative experiences. Our theorized second-order model of authentic experiences with the seven first-order variables (e.g. escape, learning, unique involvement, interactivity, peace of mind, motivations and revisit intention) was a better fit to the data ($\chi^2_{(403)} = 2.954$, CFI = .950, AGFI = .862, GFI = .888, IFI = .950 and RMSEA = .055) than an alternative model formed by setting the authenticity as the first-order model of *existential* and *objective* ($\chi^2_{(399)} = 3.602$, CFI = .934, AGFI = .838, GFI = .869, IFI = .934 and RMSEA = .064). The CFA results thus reveal that the second-order model is more appropriate for the theorizing and hypothesis testing.

Results

The descriptive statistics of study 1, such as the means, standard deviations (S.D.) and correlations, are shown in Table 1. Further, the high correlation among variables was observed, which raised the concern of multicollinearity. As Li and Liu (2018) suggested, if the variance inflation factor (VIF) is below 10, multicollinearity may not exist. The VIF values in study 1 ranged from 1.80 to 3.88, which met the acceptable level and suggested the absence of substantial multicollinearity in this study.

Furthermore, the descriptive Statistics of the measuring items and values of composite reliabilities (CRs), averages of the variances extracted (AVEs) are shown in Table 2. When estimating the variables' validity and reliability, Yoon and Uysal (2005) suggested that the CR should be above the value of .60 and that the AVE should be above .5. Table 2 shows that the CRs were between .821 and .941 and that the AVEs were between .537 and .841, which supported the variables' convergent validity in study 1.

Hypotheses testing

This study used structural equation modelling (SEM) to test the proposed hypotheses. Figure 2 presents the summarized standardized path estimate results and shows that the hypothesized model achieved a good fit ($\chi^2 = 1312.182$, df = 313, $p < .001$, CFI = .929, AGFI = .833, GFI = .862, IFI = .929 and RMSEA = .071). Hypothesis 1a to Hypothesis 1c proposed that a creative escape experience is positively related to peace of mind through culture learning ($\beta = .109$, $p < .05$), uniqueness ($\beta = .425$, $p < .001$) and interactivity ($\beta = .198$, $p < .01$). Accordingly, Hypothesis 1a, Hypothesis 1b and Hypothesis 1c were supported.

Hypothesis 2 proposed that peace of mind mediates the relationship between (a) *cultural learning*, (b) *unique cultural involvement*, (c) *cultural interactivity* and authenticity. Consistent with this prediction, interactivity ($\beta = .144$, $p < .001$), learning ($\beta = .526$, $p < .001$) and uniqueness ($\beta = .290$, $p < .001$) were positively related to peace of mind, while peace of mind was related to authenticity ($\beta = .734$, $p < .001$), which confirmed the indirect effect of peace of mind (e.g. *cultural learning* to authenticity $\beta = .106$, $p < .05$; *unique cultural involvement* to authenticity $\beta = .386$, $p < .001$ and *cultural*

Table 1. Descriptive statistics of the means, standard deviations (S.D.) and correlations – *Localized Perspective.*

Variables	Mean	S.D.	1.	2.	3.	4.	5.	6.	7.	8.	9.	VIF
1. Revisit intention	4.843	1.605	(.940)									
2. Motivation	5.307	1.366	.382	(.851)								1.80
Authenticity												
3. Existential	5.281	1.244	.483	.640	(.874)							3.88
4. Objective	5.433	1.199	.476	.556	.810	(.861)						3.20
Creative experience												
5. Peace in mind	5.179	1.201	.480	.469	.607	.585	(.879)					2.69
6. Escape	5.236	1.295	.530	.520	.694	.650	.625	(.916)				2.67
7. Learning	4.631	1.345	.483	.512	.587	.575	.604	.659	(.881)			2.46
8. Interactive	4.724	1.308	.529	.407	.523	.543	.654	.583	.628	(.897)		2.44
9. Unique	4.690	1.144	.492	.478	.572	.533	.677	.665	.683	.648	(.849)	2.23

*N = 631.

**The diagonal data is the values of Cronbach's *α* for each variable.

Table 2. Descriptive statistics of the measuring items – *Localized Perspective.*

Measure items	Mean	S.D.	Factor loading	AVE	CR
1. Revisit intention				**.841**	**.941**
I would revisit this facility in the future	4.931	1.698	.870		
If given the opportunity, I would return to this facility	4.811	1.713	.940		
I am loyal to this cultural destination	4.787	1.685	.939		
2. Motivation				**.672**	**.859**
I would like to discover new places and things about Chinese culture	5.358	1.537	.858		
I would like to increase my knowledge about Chinese culture	5.177	1.534	.720		
I am interested in visiting cultural and historical attractions/events	4.633	1.802	.873		
Authenticity					
3. Existential				**.700**	**.875**
I can be immersed in the local atmosphere with local cultural characteristics	5.144	1.429	.843		
I can feel an atmosphere of a long history of traditional culture	5.267	1.399	.871		
I enjoy the culture of this destination as displayed by the architecture, couplets, plaques, steles, cliff-side carvings, and other characteristics	5.432	1.346	.794		
4. Objective				**.670**	**.859**
I feel a sense of the long history of the site from its appearance	5.473	1.295	.825		
Locals retain traditional handicrafts manufacturing skills that have not been over-commercialized and packaged	5.304	1.360	.803		
The architecture retains the traditional ethnic customs and artistic beauty	5.522	1.408	.827		
Creative experience					
5. Peace in mind				**.669**	**.889**
The location and activities were physically comfortable	4.973	1.466	.726		
I felt relaxed during the activities	5.267	1.382	.911		
I felt safe during the activities	5.302	1.361	.902		
My privacy was assured during the activities	5.175	1.396	.711		
6. Escape				**.787**	**.917**
The activities made me forget my daily routine	4.984	1.461	.872		
I felt that I was playing a different role during the activities whenever help was needed	4.556	1.436	.915		
I felt important and respected throughout the activities	4.633	1.409	.874		
7. Learning				**.725**	**.913**
The activities increased my skills	4.326	1.513	.791		
I learned a lot through these activities	4.572	1.485	.908		
These activities stimulated my curiosity to learn new things	4.809	1.495	.860		
It was a real learning experience	4.817	1.563	.843		
8. Interactive				**.752**	**.901**
The staff/instructors were friendly and engaging	4.226	1.743	.809		
The staff/instructors were knowledgeable	4.155	1.716	.910		
The staff/instructors had the ability to control the situation	4.664	1.647	.879		
9. Unique				**.537**	**.821**
I was involved in something that I really liked doing	5.057	1.407	.611		
I did something unique and memorable	4.798	1.507	.717		
I had a 'once in a lifetime' experience	4.294	1.655	.795		
I did something new and different	4.540	1.584	.793		

Note. Standard Deviations (S.D.), Composite Reliabilities (CR) and Averages of the Variances Extracted (AVE).

interactivity to authenticity $\beta = .213$, $p < .001$). Thus Hypothesis 2 a, b and c was supported. The final Hypothesis 3 asserted that authenticity mediates the relationship between peace of mind and revisit intentions. The indirect effects shown in Figure 2 indicated that peace of mind may influence revisit intentions through authenticity ($\beta = .415$, $p < .001$).

Hypothesis 4, which proposed that motivations moderate the relationship between authenticity and revisit intentions, was supported because the interaction term between motivations and authenticity on revisit intentions was significant ($\beta = .418$, $p < .001$). This study followed Dong, Seo, and Bartol (2014) and plotted the relationships between authenticity and revisit intentions at 1 *SD* above (high motivation) and 1 *SD* below (low motivation) the mean of motivations. The plot in Figure 3 shows that when tourists had higher (rather than lower) motivations, their authenticity was more positively associated with the revisit intentions of local tourists. The simple slope test confirmed the findings and strongly supported Hypothesis 4.

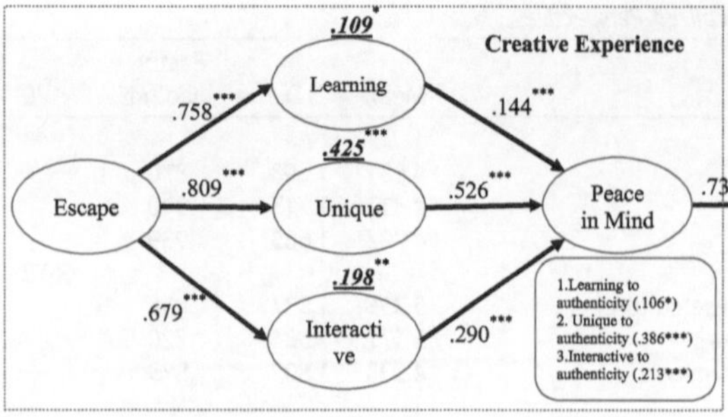

Figure 2 The results of the internationalized perspective of culture tourism. Note. The underlined numbers are standardized coefficient that represent the indirect effects among variables.

To confirm the results of this study, an alternate first-order model of existential and objective experiences was used instead the second-order model of authenticity that is shown in Figure 2. Figure 4 depicts the alternate models. However, the results of Figure 4 show the worst model fit ($\chi^2 = 1647{,}071$, df = 313, $p < .001$, CFI = .905, AGFI = .784, GFI = .821, IFI = .906 and RMSEA = .082) compared with Figure 2 ($\chi^2 = 1312.182$, df = 313, $p < .001$, CFI = .929, AGFI = .833, GFI = .862, IFI = .929 and RMSEA = .071), thus confirming our findings and suggesting that the second-order model best represents the impact of the authenticity of cultural tourism on revisit intentions.

STUDY 2: Internationalized perspective of creative experiences and revisit intentions

Theoretical background and hypothesis development

Following the fast growth of the Asian tourism industry, the Chinese culture reflects the increasing attraction of international tourist arrivals (Kim & Prideaux, 2005). A review of the cross cultural literature identified various critical factors that may influence foreign tourists' travel intentions. These critical factors such as 'seeking uniqueness', 'escape from daily life', 'cultural learning', 'interacting with different people' and feeling 'peace of mind' in the final stage are commonly cited in the cultural literature (Huang & Liu, 2018). Kim and Prideaux (2005) asserted that foreign tourists showed a higher preference for 'unique experiences', which is more of a foundational attribute than other motivational factors when engaging in cultural tourism. When summarizing the views of foreign tourists' perspectives about Chinese culture, Huang and Liu (2018) stated that unlike domestic tourists, foreign

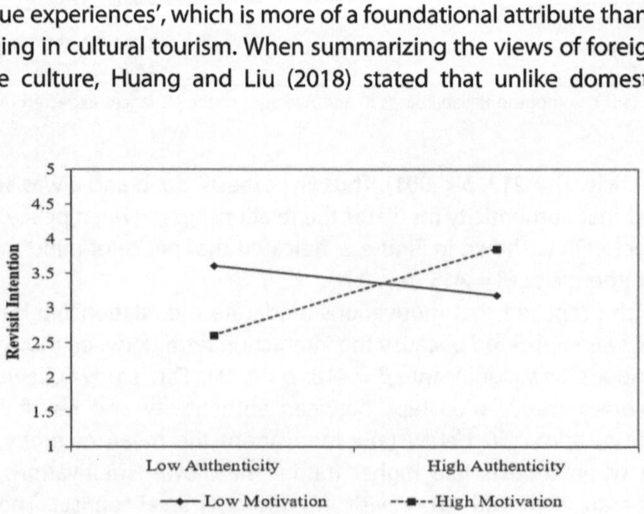

Figure 3. Interaction of motivations and authenticity on revisit intentions.

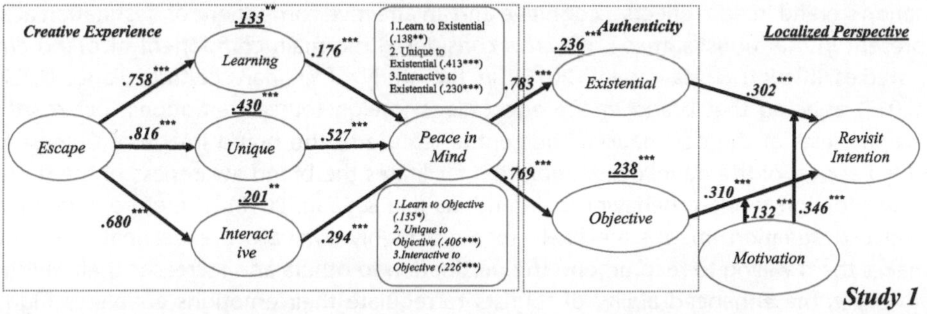

Figure 4 The results of the localized perspective of culture tourism. Note. The underlined numbers are standardized coefficient that represent the indirect effects among variables.

tourists' primary motivations for culture tourism normally focus on enjoying and trying unique experiences and enjoying different things. In other words, the 'seeking uniqueness' motivation considers the fundamental purpose of participation in culture tourism. Further, culture tourism creates opportunities for direct interactions and learning with local residents or other tourists to exchange knowledge about the culture's history, origins and development (Bonn et al., 2007; Santa-Cruz & López-Guzmán, 2017). Additionally, seeking unique experiences in different cultural experiences enhances the interactivity and learning of cultural knowledge and, subsequently, provides the means to escape life's troubles for foreign tourists (Huang & Liu, 2018).

Hypothesis 1a. Learning mediates the relationship between unique and escape experiences.

Hypothesis 1b. Interactivity mediates the relationship between unique and escape experiences.

In considering the economic costs, peace of mind is the main concern for foreign tourists regarding safety and comfort when experiencing and traveling in China (Xu & Chan, 2010). Chinese culture makes China a destination attraction because of its unique and charming historical and cultural features that other countries do not possess (Yang, Lin, & Han, 2010). Liu (2016) asserted that the *quality* of the culture's appearance and *novelty* or uniqueness of the culture appear to determine the destination's attraction for foreign tourists. Goossens (2000) indicated that quality appearance and novelty seeking are 'pull factors' and fundamental features of destination attraction that encourage tourists to visit. Furthermore, Chinese culture, in foreign tourists' minds, is an unfamiliar feature that they know little about; it represents a different cultural evaluation process from Western culture; thus although they may have read travel information on social media, they still have a great deal of confusion and interest about the culture (Becken, Jin, Zhang, & Gao, 2017). Furthermore, because they may misunderstand the current developing situation of the Chinese environment, foreign tourists might be concerned about travel safety before they experience Chinese culture and see the changes in the country (Jin & Sparks, 2017). However, after experiencing different social and cultural contexts in China, the perception changes and becomes a likely venue for the understanding of mutual well-being and the enhancement of their perceived destination satisfaction (Fu, Cai, & Lehto, 2017). In other words, through the interactivity and learning with the local culture, foreign tourists better understand Chinese culture, which encourages their potential cultural enjoyment, helps them escape their daily affairs (Whang et al., 2016) and helps them escape from the daily lives and regulation (Morgan & Lugosi, 2010). This will have a positive impact on their future travel choices and feelings of peace of mind (Chen & Chen, 2010).

Hypothesis 2. The escape experience mediates the relationship between learning, interactivity and peace of mind.

Hypothesis 3. Peace of mind mediates the relationship between escape and destination attraction.

Hypothesis 4. Destination attraction mediates the relationship between peace of mind and revisit intentions.

A destination's brand image reflects a cognitive and an affective component of destination attraction that is present in the tourist's image, which is considered the main component of brand equity in tourism studies (Kladou & Mavragani, 2015). In the study of culinary culture, Jiang, Li, Liu, and Chang (2017) asserted that brand image encourages foreign tourists' emotions towards the self-expressive interests of the destination. The central feature of the brand image of Chinese culture is the added appeal of the travel destination that facilitates the brand awareness of the destination and changes foreign tourists' behavioural intentions (Liu & Fang, 2018). Therefore, foreign tourists with positive destination images are likely to more highly evaluate the destination's attraction, which makes them willing to recommend the destination to others and increases their revisit intentions. Moreover, the enhanced ability of tourists to regulate their emotions enables a high brand image that will increase the positive consequences derived from the destination's attraction and their revisit intentions.

Hypothesis 5. Brand image moderates the relationship between destination attraction and revisit intentions.

Sampling

We tested our hypotheses at famous temples in the southern, northern and central regions of Taiwan because the traditional temple structure and appearance represent Chinese culture and provide different cultural experiences for tourists (Shuo, Ryan, & Liu, 2009). Jiang et al. (2018) asserted that religion has existed in China for thousands of years and has become an irreplaceable part of Chinese culture. Therefore, selecting famous temples for the questionnaire distribution location not only reflects the appropriateness of the sample selection with respect to Chinese cultural representation (Li, Ryan, & Cave, 2016) but also fits our research purpose of discussing how culture differences and creative experiences can influence foreign visitors' behaviours, especially when the Taiwanese government invests substantial resources in maintaining and developing traditional Chinese cultural sites (Huang & Liu, 2018). In the data collection planning, research assistants visited each temple to observe the appropriate time for data collection. Then, we visited the temple management department to obtain approval for data collection. We provided a guarantee that the data would only be used for academic purposes and that the study would not involve the disclosure of any personal information. Further, we also guaranteed that the data collection would not interference with the religious activities when foreign tourists visit the temple.

In the data collection phase, six well-trained research assistants were hired to clarify the unclear research items when foreign tourists were confused or had a question about our survey. Another purpose of the hired research assistants was to check the data effectiveness and stand close to participants when they were completing the survey to increase the response rate. A total of 450 questionnaires were issued. After removing the multiple missing values or useless surveys, 381 useable valid responses were completed, which yielded an 84.67% response rate. The demographic information of participants is as follows. The foreign tourists included 168 (44.09%) males and 213 (55.91%) females. They were categorized into age ranges of below 20 (23, 6.03%), 21–30 (160, 41.99%), 31–40 (136, 35.70%), 41–50 (51, 13.39%) and above 51 (11, 2.89%). The nationalities of tourist were Hong Kong/Macau (71, 18.64%), Japan (127, 33.32%), Korea (23, 6.04%), Europe (112, 29.40%) and other countries (48, 12.60%). Most foreign tourists had a university degree (228, 59.84%), 50 (13.12%) had an MBA degree or above and 104 (27.04%) had another degree or below a university degree.

Variables

The measuring items for foreign tourists that were used in this study were all measured using a seven-point Likert-type scale. The response format for the variables of study 2, such as creative experiences, attraction, brand image and revisit intentions, were assessed using the seven-point Likert scale ranging from 'Strongly disagree' (1) to 'Strongly agree' (7). The measures used in this study were selected from international journals; thus the items were selected in English. After several rounds of discussion with our research teams, we selected the suitable items and translated them into Chinese. To ensure that the original meaning was maintained, the 'back-translation' method was used by three experts who had abundant research experience in cultural and creative studies (Liu, 2018). Furthermore, to ensure that the information about the measures was clear and easy to read, we asked two non-experts to revise the first part of the manuscript to enhance the readability and clarify the sentences. Additionally, the participants were intended to be mainly foreign tourists; according to a previous study, when investigating foreign tourists' tourism experiences in Taiwan, different versions of the questionnaire should be prepared (Jiang et al., 2017). Therefore, the questionnaire was prepared in three languages: simple Chinese, Japanese and English. Moreover, to confirm the content of the questionnaire, we also sent the revised survey to a professional editing company to help us correct the grammar, sentencing and phrasing to ensure its quality. The values of the descriptive statistics and reliability of internal consistency of measuring items are shown in Table 2.

First, the five sub-dimensions that were used to measure creative experiences were adopted from Ali et al. (2016). (1) *Escape* used three measurement items, and its Cronbach's α was .845. (2) *Learning* used four measurement items, and its Cronbach's α was .740. (3) *Unique involvement* used four measurement items, and its Cronbach's α was .824. (4) *Interactive* used three measurement items, and its Cronbach's α was .826. (5) *Peace of mind* used four measurement items, and its Cronbach's α was .838. Second, following the prior destination attraction studies (e.g. Liu, 2016), this study used two subdimensions to measure the destination attraction for tourists' expectation. (1) *Quality* used four measurement items, and its Cronbach's α was .847. (2) *Novelty* used five measurement items, and its Cronbach's α was .904. Third, tourists rated the brand image using the four-item scale from Boo, Busser, and Baloglu (2009). The Cronbach's α was .863. Fourth and finally, revisit intentions were measured using the three-item scale from Stylos et al. (2016) and Bonn et al. (2007), and its Cronbach's α was .722.

Confirmatory factor analyses (CFA)

In this study, we conducted CFA to verify foreign tourists' revisit intentions, destination attractions, brand images and creative experiences. Our theorized second-order model of destination attraction with seven first-order variables (i.e. escape, learning, uniqueness, interactivity, peace of mind, brand image and revisit intentions) was a better fit to the data ($\chi^2_{(550)} = 2.306$, CFI = .916, AGFI = .810, GFI = .841, IFI = .917 and RMSEA = .059) than the alternative model formed by setting the destination attraction as the first-order model of *quality* and *novelty* ($\chi^2_{(564)} = 2.954$, CFI = .913, AGFI = .808, GFI = .837, IFI = .914 and RMSEA = .059). The CFA results thus reveal that the second-order model is more appropriate for the theorizing and hypothesis testing regarding foreign tourists.

Results

Following the format of the descriptive statistics of study 1, study 2 also shows the means, standard deviations (S.D.), composite reliabilities (CRs), averages of the variances extracted (AVEs) and correlations, which are shown in Table 3. Further, a high correlation among variables was observed, which raised the concern of multicollinearity. This study followed the Li and Liu (2018) procedure to test the multicollinearity by using the variance inflation factor (VIF). The VIFs in study 2 ranged from 2.54 to 3.91, which met the acceptable level and showed that multicollinearity was not a concern in this study.

Furthermore, the descriptive statistics of the measuring items and values of composite reliabilities (CRs), averages of the variances extracted (AVEs) are shown in Table 4. When estimating the variables' validity and reliability, Yoon and Uysal (2005) suggested that the CR should be above the value of .60 and that the AVE should be above .5. Table 4 shows that the CRs were between

Table 3. Descriptive statistics of the means, standard deviations (S.D.) and correlations – *Internationalized Perspective.*

Variables	Mean	S.D.	1.	2.	3.	4.	5.	6.	7.	8.	9.	VIF
1. Revisit intention	4.795	1.042	(.722)									
2. Brand image	4.711	1.065	.638	(.863)								2.89
Attraction												
3. Quality	4.896	1.036	.593	.751	(.847)							3.91
4. Novelty	4.955	1.031	.541	.707	.822	(.904)						3.82
Creative experience												
5. Peace in mind	4.732	1.049	.581	.678	.650	.691	(.838)					3.13
6. Escape	4.558	.999	.638	.664	.632	.653	.773	(.845)				3.49
7. Learning	4.791	.962	.644	.591	.601	.644	.653	.683	(.740)			2.80
8. Interactive	4.781	1.091	.655	.643	.629	.671	.637	.671	.681	(.826)		2.54
9. Unique	4.703	.972	.625	.612	.589	.577	.661	.736	.728	.642	(.824)	2.89

*$N = 381$.
**The diagonal data is the values of Cronbach's α for each variable.

.750 and .884 and that the AVEs were between .507 and .632, which supported the variables' convergent validity in study 2.

Hypotheses testing

The software of AMOS 18 was used to performing the structural equation modelling (SEM) to test the proposed hypotheses. Figure 5 presents the summarized standardized path estimate results and shows that the hypothesized model achieved a good fit ($\chi^2 = 1189.282$, df = 455, $p < .001$, CFI = .901, AGFI = .813, GFI = .839, IFI = .902 and RMSEA = .065). First, Hypothesis 1a and Hypothesis 1b proposed that uniqueness is positively related to escape experiences through learning ($\beta = .653$, $p < .001$) and interactivity ($\beta = .235$, $p < .001$). Accordingly, Hypothesis 1a and Hypothesis 1b were supported.

Second, Hypothesis 2 proposed that escape experiences mediate the relationship between interactivity, learning and peace of mind. Consistent with this prediction, interactivity ($\beta = .235$, $p < .001$) and learning ($\beta = .653$, $p < .001$) were positively related to escape experiences, while escape experiences were related to peace of mind ($\beta = .941$, $p < .001$), which confirmed the indirect effect of peace of mind (e.g. learning to peace of mind $\beta = .630$, $p < .001$ and interactivity to peace of mind $\beta = .271$, $p < .001$). Thus Hypothesis 2 was supported. Third, Hypothesis 3 proposed that escape experiences are positively related to destination attraction through peace of mind ($\beta = .820$, $p < .001$). Accordingly, Hypothesis 3 was supported. Finally, Hypothesis 4 asserted that destination attraction mediates the relationship between peace of mind and revisit intentions. The indirect effects shown in Figure 5 indicated that peace of mind may influence revisit intentions through destination attraction ($\beta = .635$, $p < .001$).

Hypothesis 5, which proposed that brand image moderated the relationship between destination attraction and revisit intentions, was supported because the interaction term between brand image and destination attraction on revisit intentions was significant ($\beta = .742$, $p < .001$). We followed Dong et al. (2014) examination of the moderating effects and plotted the relationships between

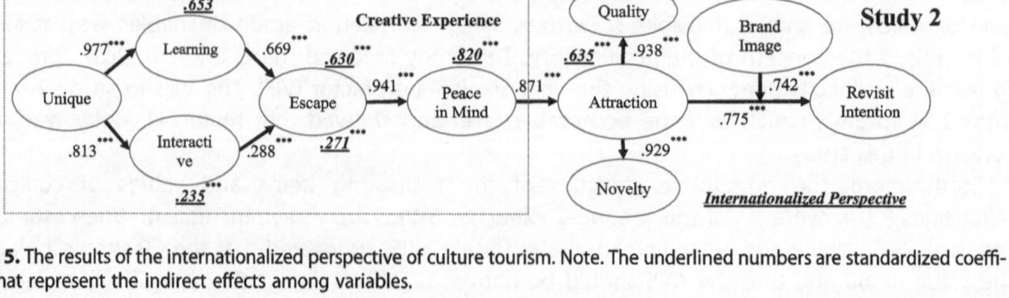

Figure 5. The results of the internationalized perspective of culture tourism. Note. The underlined numbers are standardized coefficient that represent the indirect effects among variables.

Table 4. Descriptive Statistics of the measuring items – *Internationalized Perspective.*

Measure items	Mean	S.D.	Factor loading	AVE	CR
1. Revisit intention				**.507**	**.750**
I would revisit this facility in the future	4.338	1.385	.770		
If given the opportunity, I would return to this facility	4.425	1.202	.798		
I am loyal to this cultural destination	4.832	1.234	.539		
2. Brand image				**.611**	**.863**
This destination fits my personality	4.763	1.259	.816		
My friends would think highly of me if I visited this destination	4.658	1.276	.808		
The image of this destination is consistent with my own self-image	4.740	1.253	.730		
Visiting this destination reflects who I am	4.682	1.269	.770		
Attraction					
3. Quality				**.632**	**.872**
The presentation of the cultural and historical events is visually attractive	4.837	1.262	.729		
Chinese culture offers a variety of cultural and historical attractions/events	4.881	1.207	.826		
Chinese culture is well explained through the cultural and historical attractions/ events	4.979	1.275	.801		
Chinese culture serves the interest of the cultural and historical attractions/events	4.889	1.255	.819		
4. Novelty				**.604**	**.884**
Chinese culture is unique compared with other countries	4.842	1.221	.698		
I was curious about cultural and historical attractions/events	4.989	1.279	.808		
Visiting cultural and historical attractions/events can lead to experience new and different things	4.989	1.250	.819		
Cultural and historical attractions/events are interesting and exciting	4.942	1.257	.791		
I enjoy the Chinese culture because of the cultural and historical attractions/events	4.866	1.258	.765		
Creative experience					
5. Peace in mind				**.551**	**.830**
The location and activities were physically comfortable	4.606	1.253	.712		
I felt relaxed during the activities	4.784	1.233	.793		
I felt safe during the activities	4.787	1.343	.740		
My privacy was assured during the activities	4.750	1.282	.721		
6. Escape				**.568**	**.798**
The activities made me forget my daily routine	4.338	1.385	.772		
I felt that I was playing a different role during the activities whenever help was needed	4.425	1.202	.730		
I felt important and respected throughout the activities	4.832	1.234	.758		
7. Learning				**.512**	**.807**
The activities increased my skills	4.561	1.299	.679		
I learned a lot through these activities	4.837	1.146	.719		
These activities stimulated my curiosity to learn new things	4.763	1.195	.678		
It was a real learning experience	4.774	1.214	.780		
8. Interactive				**.621**	**.831**
The staff/instructors were friendly and engaging	4.779	1.264	.740		
The staff/instructors were knowledgeable	4.711	1.260	.832		
The staff/instructors had the ability to control the situation	4.853	1.275	.790		
9. Unique				**.521**	**.813**
I was involved in something that I really liked doing	4.679	1.225	.698		
I did something unique and memorable	4.624	1.153	.738		
I had a 'once in a lifetime' experience	4.771	1.290	.780		
I did something new and different	4.779	1.327	.667		

Note. standard deviations (S.D.), composite reliabilities (CR) and averages of the variances extracted (AVE).

authenticity and revisit intentions at 1 *SD* above (high motivation) and 1 *SD* below (low motivation) the mean of brand equity. The plot in Figure 6 shows that when tourists had higher (rather than lower) brand images, their evaluations of the destination attraction were more positively associated with the revisit intentions of foreign tourists. The simple slope test confirmed the findings and strongly supported Hypothesis 5.

To confirm the results of this study, alternate first-order models of the quality and novelty of the destination were conducted instead of the second-order model of destination attraction, which is shown in Figure 5. Figure 7 depicts the alternate models. However, the results of Figure 7 show the worst model fit ($\chi^2 = 1271.526$, df = 455, $p < .001$, CFI = .890, AGFI = .795, GFI = .823, IFI = .891

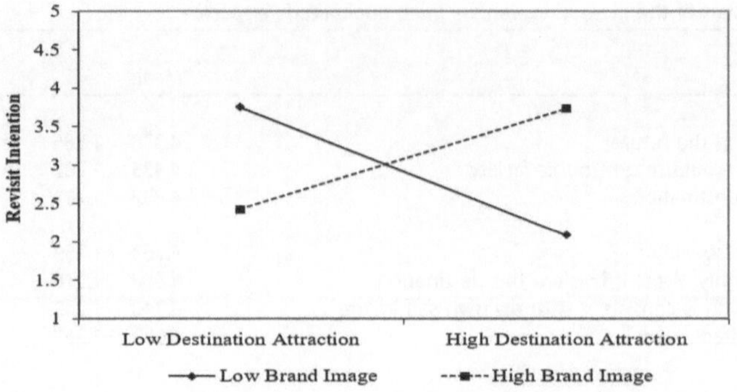

Figure 6. Interaction of brand image and destination attraction on revisit intentions.

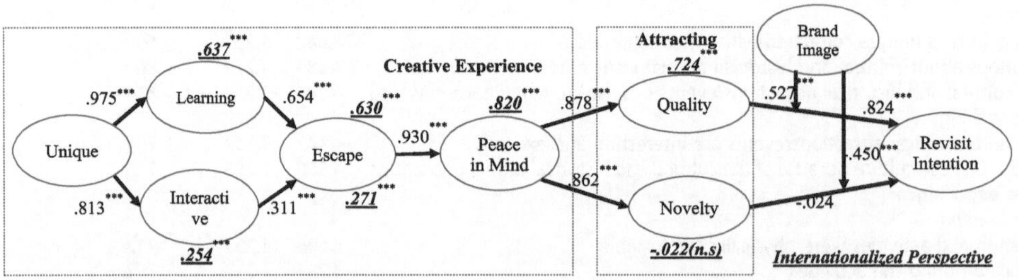

Figure 7. The results of the internationalized perspective of the cultural tourism-alternative model (first-order model). Note. The underlined numbers are standardized coefficient that represent the indirect effects among variables.

and RMSEA = .062) compared with Figure 5 (χ^2=1189.282, df = 455, p < .001, CFI = .901, AGFI = .813, GFI = .839, IFI = .902 and RMSEA = .065). This confirmed our findings and suggested that the second-order model best represents the impact of culture tourism and destination attraction on to the revisit intentions of foreign tourists.

Discussion

Two studies were conducted to better understand when and how culture perception and creative experiences influence local and international behavioural intentions. By discovering the local tourists' behavioural intentions, the results show that creative experiences influence local tourists' authenticity (Kirillova et al., 2017): increased culture creative experiences, such as escape, learning, uniqueness and interactivity, for local tourists result in the strongest relation to peace of mind and the resulting increased experience of authenticity. In addition, in accordance with the idea that motivation is the foundational attribute that influences travel intentions (Albayrak & Caber, 2018), we found that culture motivations exert a stronger effect on the relationship between authenticity and revisit intentions. Additionally, in Study 2, the proposed integrated model asserted that foreign tourists' revisit intentions may be influenced by creative experiences through the destination attraction. Further, the mediation process also asserted that there may be mutual relationships among creative experiences, and the foreign tourists' peace of mind may influence the role of uniqueness through the experiences of learning, interactivity and opportunities for escape.

Theoretical implications

This study makes several contributions to the cultural tourism literature as follows. First, it proves that creative experiences not only create memorable events (Hung et al., 2016) but also present more opportunities to be involved in local cultural actives and to feel like a part of the environment (Ali et al., 2016). However, despite the important roles of creative experiences, until now, few studies have demonstrated how different cultural perceptions influence behavioural intentions, especially while assessing the localized and internationalized perspectives. This study is the first to assess the localized and internationalized perspective under Chinese culture contexts and thus provides opportunities for future tourism studies. Additionally, with the attraction of Chinese culture (Weaver et al., 2017; Whang et al., 2016), illuminating the cultural travel behaviours provides more meaningful cultural tourism (Lin, 2017). As such, this study provides novel insights and a comparative cultural setting that advances the understanding of cultural tourism decision-making processes. In addition, it provides a set of complex, fundamental emotional analysis procedures that demonstrate how these critical behavioural mechanisms inspire local and foreign tourists' souls and minds simultaneously.

Second, this study extends the existing tourism studies of the multiple sequences and interrelated mechanisms of creative experiences that have not been assessed in cultural tourism (Huang & Liu, 2018), especially with the cultural learning behavioural setting of the comparative sample of local and foreign tourists. Recently, internationalization has been an increasing focus of academics and researchers, which not only brought increased tourism opportunities but also has become a future trend of tourism industry development that often only considers the local perspective and thus may include incomplete analysis (Liu & Fang, 2018). Particularly, the mediation–moderation model has been introduced in many tourism studies (Huang & Liu, 2017); however, until now, it has not been used to examine the comparative studies of culture differences. The current studies offer new perceptions for the cultural tourism literature from local and international perspectives in the setting of the mediation-moderation modelling. On one hand, the results disclose the critical moderating role of motivations with the localized perspective and the role of brand image with the internationalized perspective that connect the relationships between creative experiences and revisit intentions. On the other hand, the proposed theoretical mediation model explicitly predicts the different relationships of creative experiences (Huang & Liu, 2018) and suggests that there may be different directional predictions among creative experiences with local and foreign tourists perceptions about Chinese culture; these singularities have not been discovered by the previous tourism literature.

Finally, the exposed outcomes of this study provided an answer that has not been offered in the existing literature on local and foreign tourists' travel intentions under the cultural tourism setting by recognizing creative experiences as a significant context for the influence of the destination's attractiveness, authenticity and behavioural intentions (Huang & Liu, 2018; Kim, 2014). Specifically, brand images and motivations, which are operationalized as catalysts of the revisit intentions of tourists' destination choices, may provide answers to the unresolved question in cultural tourism studies regarding the appropriate critical characteristics of behavioural intentions. This research also provides a compass for the cultural and creative tourism studies by observing the differences between the localized and internationalized perspective of cultural experiences by detecting and considering the plentiful perspectives that were not assessed in previous studies (Huang & Liu, 2018).

Managerial implications

The results provide various managerial implications for cultural tourism development and management when promoting the distinctiveness of Chinese culture for local and foreign tourists. First, for domestic tourists, in addition to creative experiences, it may be beneficial to offer multiple recreational opportunities and provide existential and objective authenticity while preserving local traditions (Girish & Chen, 2017). This is because the authenticity can help managers to understand domestic tourists' motivations and behaviours, as well as provide strategic and tactical implications for destination management (Chhabra & Kim, 2018). Further, for foreign tourists, the brand image of Chinese culture is vital because it assists foreign tourists in advancing their interests for different cultural experiences (Liu & Fang, 2018). Managers must be aware that the brand image influences and emphasizes specific Chinese traditional cultural characteristics that may help to attract certain types of foreign tourists when the brand image fits with their individual imaginations and requirements (Matzler, Strobl, Stokburger-Sauer, Bobovnicky, & Bauer, 2016). Second, this study considers young tourists that also have high interests in traditional Chinese culture. Thus it might be helpful for tourism organizations to allocate limited resources to educate young people to improve their perception of Chinese culture and the awareness of the importance of cultural protection and sustainability (Gardiner & Kwek, 2017). Funk and Bruun (2007) asserted that cultural education evokes tourists' perceptions of cultural protection information, increases the awareness of the importance of traditional cultural sustainability, and influences the tourists' direct and indirect evaluations of cultural values. Therefore, to attract the younger generations, tourism managers can generate educational cultural environments, such as demonstrating the cultural evaluation process, training employees to be good story tellers, using new technology to reveal how culture has been extended in traditional ways with video demonstrations that enhances tourists' creative experiences and increasing their revisit intentions.

Third and finally, the results with the localized and internationalized perspectives indicate that promotional strategies or policies should be different when targeting tourists who are interested in Chinese culture but have divergent cultural backgrounds. Additionally, tourism managers should follow Funk and Bruun's (2007) cultural clusters to categorize the worldwide and local tourism marketing strategies when attracting different types of cultural tourists. For example, local tourists with a similar cultural background should be targeted by utilizing a motivational encouragement approach that does not highlight the cultural opportunities and knowledge of the cultural tourism, but rather focuses on traditional cultural extensions and protections (Wu & Pearce, 2014). Additionally, for tourists of nationalities within different cultural background clusters, tourism managers should empathize employing a brand image improvement strategy that increases their visit intentions, their strong culture learning interests and their knowledge of the destination (Huang & Liu, 2018).

Limitations and suggestion for future study

Despite the contributions of this study, there were also several limitations that should be addressed by future tourism studies. First, the survey was conducted using domestic and foreign tourists under Chinese cultural contexts. Future studies should, consequently, check the robustness of these findings by conducting more research on multiple cultural activities, such as cultural festivals (Savinovic, Kim, & Long, 2012; Sofield & Sivan, 2003) or heritage tourism (Light, 2017) with global and local perspectives. Different cultural activities might shed further light upon the explanations regarding why the results of this investigation have global and local perspectives that are distinct from those hypothesized. Second, the survey that was used in this study only focused on Chinese cultures. Future research might include different cultural and country settings, which might yield the discovery of greater variance in the creative experience dimensions. Further research could also use extra cultural experience measures in order to interpret the possibility of cultural differences from various countries, which could help recognize further effects. Third, this study is implemented in the context of a survey with multi-dimensional measures of domestic and foreign tourists. While the results of our measure present high reliability and validity, the application of multi-dimensional measures has been used in many tourism studies (Girish & Chen, 2017; Huang & Liu, 2017; Liu & Fang, 2018), which suggested that future studies should take and extend these main and subdimensions and test the mediation–moderation effect of cultural tourism on creative experiences and tourist behavioural intentions. Despite these limitations, this research delivers a first insight into how creative experiences impact the relationships between cultural values, motivations, brand images and visit intentions in a Chinese cultural tourism context with localized and internationalized perspective, which could be valuable for cultural tourism development in the future.

Disclosure statement

No potential conflict of interest was reported by the author.

References

Albayrak, T., & Caber, M. (2018). Examining the relationship between tourist motivation and satisfaction by two competing methods. *Tourism Management, 69*, 201–213.

Ali, F., Ryu, K., & Hussain, K. (2016). Influence of experiences on memories, satisfaction and behavioral intentions: A study of creative tourism. *Journal of Travel & Tourism Marketing, 33*(1), 85–100.

Becken, S., Jin, X., Zhang, C., & Gao, J. (2017). Urban air pollution in China: Destination image and risk perceptions. *Journal of Sustainable Tourism, 25*(1), 130–147.

Behr, D. (2017). Assessing the use of back translation: The shortcomings of back translation as a quality testing method. *International Journal of Social Research Methodology, 20*(6), 573–584.

Bonn, M. A., Joseph-Mathews, S. M., Dai, M., Hayes, S., & Cave, J. (2007). Heritage/cultural attraction atmospherics: Creating the right environment for the heritage/cultural visitor. *Journal of Travel Research, 45*(3), 345–354.

Boo, S., Busser, J., & Baloglu, S. (2009). A model of customer-based brand equity and its application to multiple destinations. *Tourism Management, 30*(2), 219–231.

Chang, L. Y., & Liu, W. (2009). Temple fairs in Taiwan: Environmental strategies and competitive advantage for cultural tourism. *Tourism Management, 30*(6), 900–904.

Chen, L. J., & Chen, J. S. (2011). The motivations and expectations of international volunteer tourists: A case study of 'Chinese Village traditions'. *Tourism Management, 32*(2), 435–442.

Chhabra, D., & Kim, E. G. (2018). Brand authenticity of heritage festivals. *Annals of Tourism Research, 68*, 55–57.

Chung, J. Y., Kim, J. S., Lee, C. K., & Kim, M. J. (2018). Slow-food-seeking behaviour, authentic experience, and perceived slow value of a slow-life festival. *Current Issues in Tourism, 21*(2), 123–127.

Cisneros-Martínez, J. D., & Fernández-Morales, A. (2015). Cultural tourism as tourist segment for reducing seasonality in a coastal area: The case study of Andalusia. *Current Issues in Tourism, 18*(8), 765–784.

Dong, Y., Seo, M. G., & Bartol, K. M. (2014). No pain, no gain: An affect-based model of developmental job experience and the buffering effects of emotional intelligence. *Academy of Management Journal, 57*(4), 1056–1077.

Fu, X., Cai, L., & Lehto, X. (2017). Framing Chinese tourist motivations through the lenses of Confucianism. *Journal of Travel & Tourism Marketing, 34*(2), 149–170.

Funk, D. C., & Bruun, T. J. (2007). The role of socio-psychological and culture-education motives in marketing international sport tourism: A cross-cultural perspective. *Tourism Management, 28*(3), 806–819.

Gardiner, S., & Kwek, A. (2017). Chinese participation in adventure tourism: A study of generation Y international students' perceptions. *Journal of Travel Research, 56*(4), 496–506.

Girish, V. G., & Chen, C. F. (2017). Authenticity, experience, and loyalty in the festival context: Evidence from the San Fermin festival, Spain. *Current Issues in Tourism, 20*(15), 1551–1556.

Goossens, C. (2000). Tourism information and pleasure motivation. *Annals of Tourism Research, 27*(2), 301–321.

Hsu, C. H., & Huang, S. S. (2016). Reconfiguring Chinese cultural values and their tourism implications. *Tourism Management, 54*, 230–242.

Hua, L., Chen, C., Fang, H., & Wang, X. (2018). 3D documentation on Chinese Hakka Tulou and Internet-based virtual experience for cultural tourism: A case study of Yongding County, Fujian. *Journal of Cultural Heritage, 29*, 173–179.

Huang, C. E., & Liu, C. H. S. (2018). The creative experience and its impact on brand image and travel benefits: The moderating role of culture learning. *Tourism Management Perspectives, 28*, 144–155.

Huang, Y. C., & Liu, C. H. S. (2017). Moderating and mediating roles of environmental concern and ecotourism experience for revisit intention. *International Journal of Contemporary Hospitality Management, 29*(7), 1854–1872.

Hung, W. L., Lee, Y. J., & Huang, P. H. (2016). Creative experiences, memorability and revisit intention in creative tourism. *Current Issues in Tourism, 19*(8), 763–770.

Jiang, W. H., Li, Y. Q., Liu, C. H., & Chang, Y. P. (2017). Validating a multidimensional perspective of brand equity on motivation, expectation, and behavioural intention: A practical examination of culinary tourism. *Asia Pacific Journal of Tourism Research, 22*(5), 524–539.

Jiang, T., Ryan, C., & Zhang, C. (2018). The spiritual or secular tourist? The experience of Zen meditation in Chinese temples. *Tourism Management, 65*, 187–199.

Jin, X. C., & Sparks, B. (2017). Barriers to offering special interest tour products to the Chinese outbound group market. *Tourism Management, 59*, 205–215.

Kim, J. H. (2014). The antecedents of memorable tourism experiences: The development of a scale to measure the destination attributes associated with memorable experiences. *Tourism Management, 44*, 34–45.

Kim, S. S., & Prideaux, B. (2005). Marketing implications arising from a comparative study of international pleasure tourist motivations and other travel-related characteristics of visitors to Korea. *Tourism Management, 26*(3), 347–357.

Kirillova, K., Lehto, X. Y., & Cai, L. (2017). Existential authenticity and anxiety as outcomes: The tourist in the experience economy. *International Journal of Tourism Research, 19*(1), 13–26.

Kladou, S., & Mavragani, E. (2015). Assessing destination image: An online marketing approach and the case of TripAdvisor. *Journal of Destination Marketing & Management, 4*(3), 187–193.

Kolar, T., & Zabkar, V. (2010). A consumer-based model of authenticity: An oxymoron or the foundation of cultural heritage marketing? *Tourism Management, 31*(5), 652–664.

Lee, Y. J. (2015). Creating memorable experiences in a reuse heritage site. *Annals of Tourism Research, 55*, 155–170.

Li, M., Cai, L. A., Lehto, X. Y., & Huang, J. (2010). A missing link in understanding revisit intention—The role of motivation and image. *Journal of Travel & Tourism Marketing, 27*(4), 335–348.

Li, Y. Q., & Liu, C. H. S. (2018). The role of problem identification and intellectual capital in the management of hotels' competitive advantage-an integrated framework. *International Journal of Hospitality Management*.

Light, D. (2017). Progress in dark tourism and thanatourism research: An uneasy relationship with heritage tourism. *Tourism Management, 61*, 275–301.

Liu, C. H. S. (2016). The relationships among brand equity, culinary attraction, and foreign tourist satisfaction. *Journal of Travel & Tourism Marketing, 33*(8), 1143–1161.

Liu, C. H. S. (2018). Examining social capital, organizational learning and knowledge transfer in cultural and creative industries of practice. *Tourism Management, 64*, 258–270.

Liu, C. H., & Fang, Y. P. (2018). Conceptualizing, validating, and managing brand equity for tourist satisfaction. *Journal of Hospitality & Tourism Research, 42*(6), 960–978.

Ma, H., Li, S., & Chan, C. S. (2018). Analytic Hierarchy process (AHP)-based assessment of the value of non-World heritage Tulou: A case study of Pinghe County, Fujian Province. *Tourism Management Perspectives, 26*, 67–77.

Matzler, K., Strobl, A., Stokburger-Sauer, N., Bobovnicky, A., & Bauer, F. (2016). Brand personality and culture: The role of cultural differences on the impact of brand personality perceptions on tourists' visit intentions. *Tourism Management, 52*, 507–520.

Mo, J., & Fu, Y. (2017). A comparisonal study on satisfaction of Chinese and foreign visitors about performance arts of intangible cultural heritage—taking impression of Liu Sanjie as example. *Journal of Service Science and Management, 10*(04), 376–387.

Moufahim, M., & Lichrou, M. (2019). Pilgrimage, consumption and rituals: Spiritual authenticity in a Shia Muslim pilgrimage. *Tourism Management, 70*, 322–332.

Paulauskaite, D., Powell, R., Coca-Stefaniak, J. A., & Morrison, A. M. (2017). Living like a local: Authentic tourism experiences and the sharing economy. *International Journal of Tourism Research, 19*(6), 619–628.

Qu, H., Kim, L. H., & Im, H. H. (2011). A model of destination branding: Integrating the concepts of the branding and destination image. *Tourism Management, 32*(3), 465–476.

Reisinger, Y., & Steiner, C. (2006). Reconceptualising interpretation: The role of tour guides in authentic tourism. *Current Issues in Tourism, 9*(6), 481–498.

Santa-Cruz, F. G., & López-Guzmán, T. (2017). Culture, tourism and World heritage sites. *Tourism Management Perspectives, 24*, 111–116.

Savinovic, A., Kim, S., & Long, P. (2012). Audience members' motivation, satisfaction, and intention to re-visit an ethnic minority cultural festival. *Journal of Travel & Tourism Marketing, 29*(7), 682–694.

Shuo, Y. S. S., Ryan, C., & Liu, G. M. (2009). Taoism, temples and tourists: The case of Mazu pilgrimage tourism. *Tourism Management, 30*(4), 581–588.

Smith, S. (2015). A sense of place: Place, culture and tourism. *Tourism Recreation Research, 40*(2), 220–233.

Sofield, T. H., & Sivan, A. (2003). From cultural festival to international sport-the Hong Kong dragon boat races. *Journal of Sport Tourism, 8*(1), 9–20.

Stylos, N., Vassiliadis C, A., Bellou, V., & Andronikidis, A. (2016). Destination images, holistic images and personal normative beliefs: Predictors of intention to revisit a destination. *Tourism Management, 53*, 40–60.

Tan, S. K., Kung, S. F., & Luh, D. B. (2013). A model of 'creative experience' in creative tourism. *Annals of Tourism Research, 41*, 153–174.

Waitt, G. (2000). Consuming heritage perceived historical authenticity. *Annals of Tourism Research, 27*(4), 835–862.

Wang, N. (1999). Rethinking authenticity in tourism experience. *Annals of Tourism Research, 26*(2), 349–370.

Weaver, D. B., Kwek, A., & Wang, Y. (2017). Cultural connectedness and visitor segmentation in diaspora Chinese tourism. *Tourism Management, 63*, 302–314.

Whang, H., Yong, S., & Ko, E. (2016). Pop culture, destination images, and visit intentions: Theory and research on travel motivations of Chinese and Russian tourists. *Journal of Business Research, 69*(2), 631–641.

Wong, I. A., McKercher, B., & Li, X. (2016). East meets West: Tourist interest in hybrid culture at postcolonial destinations. *Journal of Travel Research, 55*(5), 628–642.

Wu, M. Y., & Pearce, P. L. (2014). Chinese recreational vehicle users in Australia: A netnographic study of tourist motivation. *Tourism Management, 43*, 22–35.

Xu, J., & Chan, A. (2010). Service experience and package tours. *Asia Pacific Journal of Tourism Research, 15*(2), 177–194.

Yang, C. H., Lin, H. L., & Han, C. C. (2010). Analysis of international tourist arrivals in China: The role of World heritage sites. *Tourism Management, 31*(6), 827–837.

Yoon, Y., & Uysal, M. (2005). An examination of the effects of motivation and satisfaction on destination loyalty: A structural model. *Tourism Management, 26*(1), 45–56.

Zhou, Q., Zhang, J., & Edelheim, J. R. (2013). Rethinking traditional Chinese culture: A consumer-based model regarding theauthenticity of Chinese calligraphic landscape. *Tourism Management, 36*(1), 99–112.

A study of segmentation of cannabis-oriented tourists from China based on motivation

Jun Wen, Fang Meng, Tianyu Ying and Yaniv Belhassen

ABSTRACT

In the wake of the early liberal policies of the Dutch government, coffee shops in Amsterdam have become famous for their outright sale and free use of cannabis-related products. Based on a sample of 654 Chinese tourists who consumed cannabis during their vacation in the city, this study proposes three clusters of cannabis tourists: cannabis enthusiasts, diversionists/recreationists, and the curious. Significant differences in sociodemographic characteristics and behavioural intentions were detected among these clusters. In light of the commercialization of cannabis in other destinations across the world, the research proceeded to discuss theoretical and practical implications for marketers and managers in the tourism industry.

1. Introduction

Cannabis consumption in the context of tourism has been increasingly taking place with the legalization and commercialization of cannabis in destinations such as Colorado, Washington, Spain, and Uruguay (Webster, 2017). This shift toward cannabis consumption in tourism destinations makes cannabis-oriented tourists an important segment that deserves a rigorous examination for marketing and managerial purposes.

As the normalization of the use of cannabis and other recreational drugs in the West makes cannabis consumption a common activity in leisure time to many individuals in Western societies, drug use in some countries like China is illegal and strictly forbidden. In history, China had prevalent drug use, originated from the illegal trafficking of opium from British India to China in 1760s and led to severe social problems (Fang, Wang, Shi, Liu, & Lu, 2006; Lowinger, 1977). Drug use/addiction is considered as one of the causes and features of the 'Century of Humiliation' in China's history, which lasts from the eve of the First Opium War in 1840 to the foundation of People's Republic of China in 1949 (Scott, 2008). After 1949, the Chinese government carried out strict laws, strong drug control, and anti-drug acts to prohibit drug abuse (Lu, Fang, & Wang, 2008). Drug use of any kind, including cannabis and marijuana, could lead to police detention for up to 15 days and mandatory treatment in rehabilitation centres. The general public are also widely educated to stay away from any kind of drugs.

Nowadays, with the fast expand of the Chinese outbound tourism market, Chinese tourists have much more exposure and access to drugs when they travel overseas, particularly to destinations where drugs are legally and commercially available. Considering the differences in cultural and legal environment between the home country and destinations, it is particularly interesting to examine Chinese drug tourist segments in overseas destinations as a deviant or marginal behaviour from their normal daily life. In addition, the existing literature on drug tourism are predominantly situated in a Western discourse, which demonstrates a clear research gap and lack of understanding of characteristics and profiles of drug tourists from a global perspective, particularly in the fast-growing Asian market such as China.

Amsterdam, which is the geographical focus of this study, is regarded as the worldwide capital of 'cannabis culture' and attracts millions of tourists every year (Belhassen, Santos, & Uriely, 2007; Kleiman, 2015; Korf, 2002). The city hosts nearly 200 such coffee shops out of the 700 coffee shops available all over the Netherlands (MacCoun, 2011). It is estimated that nearly 1.5 million tourists visit these coffee shops every year and consume drugs, and this demand for drugs is expected to increase at the rate of 23.4% over the next few years (Coffee shops, 2016; MacCoun, 2011).

Recent statistics have shown significant growth of inbound Chinese tourists to Netherlands, a well-known drug tourism destination worldwide. The number of Chinese tourists in the Netherlands increased by 152% over the past 10 years (NBTC, 2015). Approximately 250,000 Chinese tourists arrived in the Netherlands during 2016, 15% higher than the previous year. Chinese tourists accounted for the sharpest increase in inbound tourism in Amsterdam for the year 2015, and the number of overnight stays in hotels in the Netherlands by Chinese tourists soared by 33% to a total of 436,000 (CBS, 2016). Furthermore, China is the number one tourism source market in the world with a spending of $261.1 billion in 2016, and has ranked No. 1 in tourism expenditure worldwide since 2012 (UNWTO, 2017). With the increase of Chinese outbound tourists, countries all over the world have targeted to this lucrative market and drug tourism unavoidably becomes one of the attractions in Netherlands (BBC, 2012).

Therefore, it would be reasonable to assume, as already shown elsewhere (Wen, Meng, Ying, Qi, & Lockyer, 2018), that many of these tourists are attracted to Amsterdam due to its association with cannabis and the accompanying opportunity to consume it. In recognition of the need for specific market research for Chinese outbound tourism markets, and as part of a larger research project related to the drug use of Chinese tourists in Amsterdam, this study aims to identify and cluster this market of Chinese tourists who come to Amsterdam specifically to consume cannabis.

2. Literature review

2.1. Drugs and tourism

Drug tourism is a type of special interest tourism that has garnered scholarly attention throughout the past two decades. Drug tourism has myriad definitions; it is often considered a non-institutionalised form of tourism (Belhassen et al., 2007). According to Valdez and Sifaneck (1997), drug tourism is 'the phenomenon by which persons become attracted to a particular location because of the accessibility of licit or illicit drugs and related services' (p. 880). Uriely and Belhassen (2005) pointed out that definitions of drug tourism should account for the multifaceted nature of tourism, noting that drug tourism involves travellers who deem 'their drug taking merely as a by-product of their travel experience and tourists who become aware of the accessibility of drugs only during their stay in a particular destination' (p. 244). Wen et al. (2018) investigated drug tourists' motivations based on a definition framing drug consumption as a major reason for travel (i.e. Chinese tourists labelled drug-oriented tourists). The same definition applies in the present study.

Drug tourism is not new; the phenomenon has been explored from various psychological and social frameworks. However, most studies have concerned the West aside from work by Wen et al. (2018), which examined from the Chinese market perspective. As most scholars have depicted drug tourism as a deviant tourist behaviour or a Western lifestyle prevalent in developed countries (Bandyopadhyay, 2013; Uriely & Belhassen, 2006), drug use in a tourism context tends to be classified as marginal tourism (Uriely & Belhassen, 2005). Bandyopadhyay (2013) called for a 'paradigm shift' in studies of sensitive tourism activities (e.g. sex tourism) to unveil Asian tourists' escapades in the West.

Although Wen et al. (2018) investigated the motivations underlying Chinese cannabis-oriented tourists' recreational consumption of cannabis in licensed cafés in Amsterdam, tourist segments related to drug tourism remain unexplored in both Asian and Western contexts. Therefore, this study intends to explore drug tourist segments based on their motivations by using a Chinese sample in an overseas destination to fill this research gap.

2.2. Motivations of cannabis consumption among Chinese tourists

Tourist motivation is regarded as the fundamental psychological reason why people wish to travel to a particular destination and/or participate in tourism-based activities (Crompton, 1979). In other words, motivation is a both compelling and impelling factor behind tourist behaviours, and the study of travel motivation is to seek the 'Why?' of one's travel behaviour (Hsu, Cai, & Wong, 2007; Baloglu & Uysal, 1996).

The analysis of the currently available drug tourism literature indicates a few relevant tourist motivations. Specifically, Belhassen et al. (2007) suggested four umbrella motivations for consuming cannabis while on vacation: *extermination, diversion, search for authenticity of cannabis*, and *purchasing*. Based on this conceptual framework, Wen et al. (2018) developed and validated a scale of drug tourism motivation based on a sample of Chinese tourists who travelled to Amsterdam to consume commercial cannabis. The study employed a mixed-methods approach and collected two rounds of quantitative data for scale development and empirical testing. A six-factor motivation scale was developed, comprising *spiritual and emotional healing, social prestige, relaxation and escape, cannabis authenticity, commercial availability*, and *cannabis experimentation*. To develop cannabis tourist clusters for Chinese tourists, the same six-dimension scale is used in this study to segment drug tourists according to their sociodemographic information.

It is particularly important to note that studies conducted among Chinese tourists have found the market to be largely influenced by ego enhancement attributes such as prestige (Hua & Yoo, 2011; Li, Xu, & Weaver, 2009; Lu, 2011). In other words, the tourism literature has suggested that Chinese tourists attribute high value to the social recognition and prestige associated with impressing their friends and family by visiting well-known destinations (Hui, Wan, & Chi, 2008) and sharing those experiences with others. For example, according to Zhang and Lam (1999), prestige and enhancement of human relationships are the primary consensus domains for mainland Chinese travellers. Furthermore, based on their investigations, Li et al. (2009) found prestige to be the second major influential factor among Chinese tourists travelling to the United States. Consistent with this finding, Li, Wen, and Leung (2011) identified prestige as one of four key motivators for Chinese tourism to Hong Kong.

2.3. Tourist segmentation

Market segmentation has been widely used in tourism studies to categorize diversified tourists and thus identify niche markets for tourism products and services (Mohsin & Ryan, 2004; Park & Yoon, 2009; Rid, Ezeuduji, & Pröbstl-Haider, 2014). The purpose of segmentation is to tailor marketing practices (i.e. developing, promoting, and delivering purpose-designed products) to the needs of specific target groups (Park & Yoon, 2009). In other words, segmentation in marketing practices primarily consists of demographic, geographic, behavioural, and psychographic methods (Kotler, Bowen, & Makens, 2014). In the tourism and hospitality literature, several segmentation variables have been considered in different approaches. Sociodemographic variables, including nationality and age, are commonly used to segment tourist markets (Mohsin & Ryan, 2004). Behavioural variables such as expenditure and sought benefits, as well as psychographic variables such as attitudes, motivation, and lifestyle, are also widely employed in segmentation (Pesonen, 2012; Shani, Wang, Hutchinson, & Lai, 2010). Compared to the sociodemographic segmentation approach, behavioural and psychographic criteria have more effective predictive power on tourism behaviour, with motivation a significant factor in understanding tourists' needs, preferences, expectations, and/or desired outcomes

(Dryglas & Salamaga, 2017; Li et al., 2009; Park & Yoon, 2009; Rid et al., 2014). Motivation-based segmentation has recently been applied in a variety of tourism market contexts, such as rural tourism (Park & Yoon, 2009; Rid et al., 2014), cultural/festival tourism (Báez & Devesa, 2014; Chiang, Wang, Lee, & Chen, 2015), marine tourism, natural tourism (Fung & Jim, 2015), and health/well-being tourism (Dryglas & Salamaga, 2017). However, due to limited empirical research on the role and influence of the commercialization of cannabis in tourism destinations, studies on the motivations and segmentation of cannabis tourists is relatively scarce.

3. Methods

3.1. Research design, study population, and study procedure

The research design of this study consisted of three major components. First, a comprehensive motivation measurement scale for cannabis consumption in tourism was developed and validated. This scale was then used as the framework for segmenting a sample of 654 Chinese tourists who consumed cannabis during their vacation in Amsterdam, comprising the study population, into three clusters. In the third phase of this study, tourists were examined for differences in sociodemographic characteristics and behavioural intentions across segments.

The present study adopts Churchill's (1979) four-step procedure to develop an instrument for measuring cannabis tourism motivation, which include (1) initial item generation, (2) item purification, (3) measurement scale dimensionality determination, and (4) measurement scale reliability assessment and construct validation. A two-part survey questionnaire was designed to measure the motivations of Chinese tourists travelling to Amsterdam. The first section measured motivations and behavioural intentions, and the second section investigated respondents' sociodemographic characteristics. The motivation measurement items in the survey were derived from two sources. First, 12 measurement items were adopted from relevant tourism and leisure motivation studies, with some targeted revisions made by the authors to ensure content validity. Second, due to the exploratory nature of drug tourism motivation research, a qualitative method was used to develop the motivation measurement items. With the permission of 15 coffee shop owners/managers in Amsterdam, the leading author and five university students conducted 10 in-depth interviews with Chinese tourists who smoked cannabis in Amsterdam coffee shops. The generated motivation items were reviewed by another 15 Chinese tourists who consumed cannabis during their vacation and a panel of experts to ensure face validity. The panel consisted of five academics with research expertise in tourism and five tourism professionals who had work experience in China and overseas.

The motivation measurement items were pre-tested in Amsterdam among Chinese tourists. As a result, 22 motivation items were used in the survey, measured on a 7-point Likert scale (1 = 'entirely disagree' to 7 = 'entirely agree'). In addition, four items measuring tourists' travel intentions were adapted from Zeithaml, Berry, and Parasuraman (1996). The second part of the questionnaire included 12 sociodemographic questions related to tourists' age, educational background, annual income, marital status, travel type, information sources, gender, and previous experience with drug use. Four screening questions were included in the final survey questionnaire to select Chinese tourists who were mainly driven by cannabis-related consumption and experiences in Amsterdam: (1) 'Cannabis smoking is one of my major leisure activities during this trip'; (2) 'Cannabis smoking is one important part of this travelling experience'; (3) 'I am mainly driven to take this trip by cannabis smoking'; and (4) 'I travel to Amsterdam due to the accessibility of commercial cannabis in coffee shops.' Only those respondents who answered all four questions affirmatively were asked to complete the survey for this study.

The survey questionnaire was developed in English, translated into Chinese, and reviewed by bilingual researchers to check language accuracy. Then, multiple Chinese tourism experts, including academics, travel operators, and industry leaders, were invited to assess the items' content validity and provide comments on the clarity and readability of the questionnaire. The survey instruments were edited accordingly for formal data collection.

3.2. Data collection

This survey was conducted among Chinese tourists who visited coffee shops (drug retail outlets) in Amsterdam. Data collection took place between February 2014 and October 2016. Chinese tourists were randomly approached and intercepted while consuming cannabis in coffee shops or when exiting coffee shops. Those who confirmed they had consumed drugs were invited to participate in this study. This strategy allowed the respondents to reflect instantly on their drug-related experiences while the experiences were still fresh in their minds. After a brief explanation and assurance of confidentiality, those who agreed to complete the survey questionnaire were asked to move to a more private place to fill out the survey to guarantee respondents' privacy and anonymity. Any participant who felt uncomfortable completing the questionnaire had the right to withdraw from the study at any point. Completed questionnaires were screened for inappropriate respondents (who failed meet the criteria of the four filter questions) and missing data. Out of the 700 questionnaires distributed, 654 were deemed usable responses and included for further analysis.

3.3. Data analysis

A total of 654 questionnaires were collected and analysed through four stages in IBM SPSS version 24. First, a profile of the tourists was developed using descriptive statistics, which helped to contextualize respondents who participated in the survey. Second, dimensions of tourists' motivations were examined. Third, K-means cluster analysis was employed to classify the sample based on drug tourism motivation dimensions. Finally, analysis of variance (ANOVA), Scheffé multiple-range tests, and multiple discriminant analysis with chi-square tests were employed to examine the differences between clusters with respect to respondents' sociodemographic characteristics and behavioural intentions.

4. Results

4.1. Sociodemographic profile of Chinese drug tourists

The results indicate that our respondents were diverse in their demographic characteristics and behavioural intentions. Table 1 presents the sociodemographic profile of the respondents. Although Chinese cannabis smokers are predominately male (Chen, Zhong, Collins, Li, & Peto, 1997), nearly half of respondents in the current study were women (43.7%). Many respondents were educated through college and above (81.8%). Most tourists were under 35 years old (93.8%) and not married (61.6%). The largest age group of respondents was 26–35 years old (54.8%). Over 73.2% of the tourists had an undergraduate degree. The majority (85.5%) earned an annual income of USD 30,600 (200,000 Chinese Yuan) or less. Slightly over half of the tourists (51.7%) visited Amsterdam along with a small group of friends, and less than 20% travelled alone. Over 70% of respondents obtained cannabis-related information from Internet sources such as search engines (31.0%) and online forums (40.0%). Most (80%) respondents reported having never tried cannabis prior to their trip.

4.2. Drug tourist motivation scale

Motivation items were analysed using exploratory factor analysis and confirmatory factor analysis (CFA) on two split samples based on two-stage data collection (i.e. before or after July 2015). Six motivations were identified and validated for the measurement scale: *spiritual/emotional healing, social prestige, relaxation and escape, cannabis authenticity, commercial cannabis availability*, and *cannabis experimentation* (Wen et al., 2018). Detailed CFA results are presented in Table 2.

Table 1. Respondent profiles ($N = 654$).

Demographic characteristics	Frequency	Percentage (%)
Age		
18–25	255	39.0
26–35	358	54.8
36 and above	41	6.2
Gender		
Male	350	53.5
Female	304	46.5
Educational level		
High school (no GED)	21	3.2
High school (GED)/Vocational school	98	15.0
College/University	479	73.2
Postgraduate	56	8.6
Annual income		
¥100,000 or under (US$ 15,300 or under)	278	42.5
¥100,001–200,000 (US$ 15,300–30,600)	281	43.0
¥200,001–300,000 (US$ 30,600–45,900)	76	11.6
¥300,001–500,000 (US$ 45,900–76,500)	19	2.9
Marital status		
Married	251	38.4
Single	403	61.6
Travel type[a]		
Alone	117	17.9
Small group with friends	338	51.7
Small group with strangers	44	6.7
Large group with friends	116	17.7
Large group with strangers	39	6.0
Information sources		
Internet search	203	31.0
Relevant Internet forum	260	40.0
Friends' word of mouth	116	17.7
Advertisement	39	6.0
Others	36	5.3
Previous cannabis experience		
Yes	133	20.0
No	521	80.0

[a]Small group: < 10 tourists in one group; large group: ≥ 10 tourists in one group.

4.3. Motivation-based drug tourist segmentation

A series of non-hierarchical K-means cluster analyses were conducted based on the six drug tourism motivations identified by Wen et al. (2018). After comparing different solutions with the number of clusters ranging from two to six, a three-cluster solution (Table 3) was considered most appropriate in terms of respondent differentiation and meaningful cluster interpretation. Each cluster was labelled following a thorough interpretation of the final cluster scores and based on the relevant tourism literature on cannabis use (Belhassen et al., 2007).

Cluster 1 had 171 cases, accounting for 26.1% of respondents. Respondents in this category had relatively high scores across all motivation factors, which largely related to cannabis experience overall, and were therefore labelled *cannabis enthusiasts*. Cluster 2 had 286 cases, representing about 43.7% of the sample. These respondents were labelled *diversionists/recreationists* as they were pleasure-oriented and had relatively higher scores on the motivation dimensions of cannabis experimentation and relaxation and escape compared to other motivational attributes. Cluster 3 had 197 cases, accounting for 30.1% of respondents. Respondents belonging to this segment were labelled *the curious* for having the highest score on the motivation of cannabis authenticity, indicating their curiosity regarding cannabis culture in Amsterdam; they rated other motivation attributes below the mid-point of 4, much lower than the other two clusters.

A series of multivariate analyses were then conducted to determine the validity of the three generated clusters. The results of one-way ANOVA tests indicated statistically significant differences

Table 2. Performance of final measurement scale from CFA.

	CR	AVE	Factor Loading	S.E.	C.R.	*SD*
Spiritual/emotional healing	0.898	0.641				
To find something greater than myself			0.757	–	–	1.198
To have more imagination			0.781	0.081	14.206	1.334
To obtain the power to see inside myself			0.761	0.080	13.783	1.311
To get a wider philosophical and spiritual understanding			0.939	0.082	17.208	1.370
To have a different mindset			0.748	0.087	13.527	1.427
Social prestige	0.902	0.699				
To show my socio-economic status			0.777	–	–	1.600
To show my experience to others			0.913	0.069	17.481	1.641
To experience what others did not			0.890	0.068	17.064	1.611
I think smoking cannabis is a fashion.			0.752	0.067	13.925	1.547
Relaxation and escape	0.888	0.668				
Smoking cannabis makes me feel high			0.739	–	–	1.413
To relieve daily boredom and busyness			0.867	0.077	15.413	1.431
To get away from a stressful social environment			0.938	0.079	16.328	1.432
To temporarily escape from family			0.703	0.077	12.301	1.414
Cannabis authenticity	0.869	0.741				
To learn about the local cannabis culture			0.871	0.067	16.906	1.465
To learn about the local cannabis industry			0.914	0.066	17.418	1.415
To learn about the local cannabis users			0.794			1.424
Commercially available cannabis	0.837	0.634				
Commercially available cannabis itself makes me feel safe to use cannabis.			0.730	0.057	12.299	1.122
Commercially available cannabis itself makes me unworried about legal risk.			0.898	0.082	13.425	1.434
Commercially available cannabis itself makes me unworried about social risk.			0.750	–	–	1.554
Cannabis experimentation	0.863	0.759				
To fulfil the need for an diversion of ordinary life			0.844	–	–	1.392
To temporarily experience a crazy lifestyle			0.898	0.118	9.192	1.422

Note: CR = composite reliability; AVE = average variance extracted; C.R. = critical ratio; SD = standard deviation; all items = $p < 0.001$.

among the three clusters at $p < 0.001$. In addition, Scheffé multiple-range tests were employed to examine differences between clusters with respect to each of the six motivational attributes. The results of Scheffé tests revealed statistically significant differences between clusters, thus representing an appropriate categorization. Further, a multiple discriminant analysis was employed to validate the results of the three clusters (three criteria in this case). The results of both discriminant functions were subjected to a chi-square test to determine their significance; the two functions were found to be statistically significant. Function 1 (Wilk's lambda = 0.19, chi-square = 1069.71***, *df* = 12), with an eigenvalue of 3.06, explained 91.6% of the variance. Function 2 (Wilk's lambda = 0.78, chi-square =

Table 3. Means of motivation factors among the three clusters.[a,b]

	Cluster				
	Cluster 1: *Cannabis enthusiasts* (*n* = 171)	Cluster 2: *Diversionist/ Recreationists* (*n* = 286)	Cluster 3: *The curious* (*n* = 197)	*M*	*F*-value & significance
Spiritual/emotional healing	5.44H	4.30M	3.57L	4.44	207.98***
Social prestige	5.07H	3.31L	3.01L	3.80	230.74***
Relaxation and escape	5.79H	5.01H	3.86L	4.89	237.52***
Cannabis authenticity	5.32H	4.10M	4.07M	4.50	87.12***
Commercial cannabis availability	5.35H	4.55M	3.72L	4.54	130.12***
Cannabis experimentation	5.57H	5.11H	3.66L	4.78	224.94***

[a]Mean values measured on 7-point Likert scale (1: strongly disagree, 4: neutral, 7: strongly agree).
[b]H, M, and L indicate high, medium, or low levels of combined means for each segment.
***$p < 0.001$.

Table 4. Summary of discriminant analysis results.

Function	Eigenvalue	Percent of variance explained by function	Canonical correlation	Wilks' lambda	Chi-square	*df*	Significance
1	3.06	91.6	0.87	0.19	1069.71	12	0.001***
2	0.28	8.4	0.47	0.78	160.49	5	0.001***
Discriminant loading		Function 1	Function 2				
Spiritual/emotional healing		0.29	−0.27				
Social prestige		0.47	0.62				
Relaxation and escape		0.48	−0.2				
Cannabis authenticity		0.26	0.51				
Commercial cannabis availability		0.38	−0.18				
Cannabis experimentation		0.52	−0.53				

Note: 93.0% of original grouped cases correctly classified; 92.2% of cross-validated grouped cases correctly classified; ***$p < 0.001$.

160.49***, $df = 5$), with an eigenvalue of 0.28, explained nearly 8.4% of the variance. These results enhanced our confidence in the reliability and validity of the three clusters, as illustrated in Table 4. To determine the significance of each of the five motivational factors, a Wilk's lambda test and a univariate *F* test were conducted. Figure 1 presents a visual representation of the three tourist clusters based on two discriminant functions.

4.4. Demographic profiling of clusters

The three clusters were profiled based on sociodemographic variables (age, gender, education, annual income, marital status, travel type, information source, and drug use history) and a behavioural variable (intention to revisit Amsterdam). The differences among these tourist segments in terms of sociodemographic variables and future behavioural intentions were assessed using one-way ANOVA and chi-square tests (Table 5). Each cluster was cross-tabulated with drug tourists' sociodemographic characteristics and future behavioural intentions. The chi-square tests indicated that all

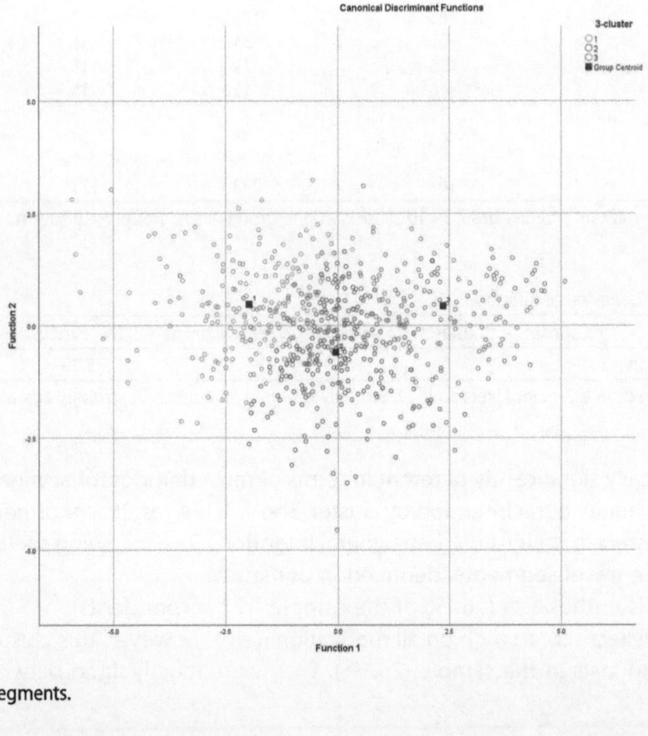

Figure 1. Drug tourist segments.

Table 5. Sociodemographic profile of Chinese tourist segments ($N = 654$).

Characteristics	Cannabis enthusiasts (n = 171)	Diversionists/ Recreationists (n = 286)	The curious (n = 197)	Statistics*
Age				
18–25	43	119	93	$\chi^2 = 22.40$, df = 4, $p < 0.001$
26–35	111	153	94	
36–45	17	14	10	
Gender				
Male	91	146	113	n.s.
Female	80	140	84	
Education				
High school (no GED)	7	6	8	$\chi^2 = 27.06$, df = 6, $p < 0.001$
High school (GED)/Vocational school	17	40	41	
College/University	120	226	133	
Postgraduate	27	14	15	
*Annual income**				
¥100,000 or under (US$ 15,300 or under)	46	129	103	$\chi^2 = 29.35$, df = 6, $p < 0.001$
¥100,001–200,000 (US$ 15,300–30,600)	88	125	68	
¥200,001–300,000 (US$ 30,600–45,900)	30	25	21	
¥300,001–500,000 (US$ 45,900–76,500)	7	7	5	
Marital status				
Married	44	111	96	$\chi^2 = 27.82$, df = 2, $p < 0.01$
Single	127	175	101	
*Travel type**				
Alone	22	62	33	n.s.
Small group with friends	92	138	108	
Small group with strangers	16	14	14	
Large group with friends	31	54	31	
Large group with strangers	10	18	11	
Information source				
Internet search	49	95	59	n.s.
Relevant Internet forum	85	103	72	
Friends' word of mouth	26	54	36	
Advertisement	5	19	15	
Others	6	15	15	
Drug-use history				
Yes	64	43	26	$\chi^2 = 40.02$, df = 2, $p < 0.001$
No	107	243	171	

Notes: n.s. indicates 'not significant'; *Small group: <10 tourists in one group; large group: ≥ 10 tourists in one group.

Table 6. ANOVA of future behavioural intention.

Variables	Cannabis enthusiasts	Diversionists/Recreationists	The curious	F ratio	Significance
Future behavioural intention	5.11	4.01	3.29	131.86	0.001***

Note: Mean values measured on a 7-point Likert scale (1: strongly disagree, 4: neutral, 7: strongly agree), ***$p < .001$.

clusters were statistically significantly different in terms of most demographic variables. Furthermore, Table 6 shows future behavioural intention by cluster. The ANOVA results confirmed significant differences among the clusters in their future behavioural intentions. The following sections present a brief summary of the three travel segments identified in our study.

Cluster 1: Cannabis enthusiasts (26.1% of the sample, 171 respondents)

This group of travellers scored high on all motivation items; however, this cluster represented the smallest number of tourists in the sample (26.1%). They were mostly aged between 26 and 35 and held an educational degree from a college or university. Many tourists in this segment were single (74.3%), travelled with a small group of friends, and were highly active on Internet travel forums. This group exhibited the highest intentions to revisit Amsterdam.

Cluster 2: Diversionists/Recreationists (43.7% of the sample, 284 respondents)

This cluster accounted for the largest number of tourists. They travelled and consumed cannabis mostly for the sake of experiencing/experimenting with the local cannabis culture in Amsterdam as well as relaxation, pleasure, and to escape from stressful social environments. However, this group scored lowest on the social prestige dimension compared to other motivations. In terms of this cluster's demographic characteristics, most had college or university education (79%) and low income (45.1% earned <$15,300 per year). Similar to Cluster 1, this group preferred to travel with a small group rather than alone. They mostly collected travel information from Internet search engines and online forums. In addition, unlike Cluster 1, tourists in Cluster 2 were less often single (38.8%).

Cluster 3: The curious (30.2% of the sample, 197 respondents)

Compared to other motivational attributes, this group scored highest on the cannabis authenticity dimension but represented the smallest number of tourists in the sample (30.2%); they were predominantly male (57.4%) and single (51.3%). This group was also the youngest (47.2% were under 25 years old), earned the lowest income (52.3% earned <$15,300 per year), and demonstrated the lowest intentions to revisit Amsterdam compared to the other two clusters.

5. Discussion and conclusions

In light of the legalization and commercialization of cannabis in many countries, a better understanding of the motivations of tourists attracted to such destinations is sought. Whereas the literature on cannabis consumption has been mostly situated in Western discourse, a growing number of Chinese tourists appear attracted to Amsterdam (Wen et al., 2018). The results of this study offer an empirically based segmentation of Chinese cannabis tourists in Amsterdam.

Three distinct groups were identified based on drug tourists' motivation factors, namely *cannabis enthusiasts*, *diversionists/recreationists*, and *the curious*. The first cluster, *cannabis enthusiasts*, was strongly motivated by all six motivation attributes in the study, indicating strong and multifaceted desires to be involved in drug tourism overseas. The second cluster, *diversionists/recreationists*, was highly motivated by two dimensions – *cannabis experimentation* and *relaxation and escape* – and moderately by *spiritual/emotional healing*, *cannabis authenticity*, and *commercial cannabis availability*. Surprisingly, tourists in this cluster showed low motivation with regard to social prestige, inconsistent with previous studies in the Chinese context (e.g. Hua & Yoo, 2011; Hui et al., 2008; Li et al., 2009; Li et al., 2011; Lu, 2011; Wen et al., 2018). This finding further indicates that most respondents were unenthusiastic about the social prestige attribute (i.e. 'showing-off motive') when involved in deviant tourism activities that are considered illegal or immoral in Chinese society, such as sex tourism, drug tourism, or gambling. Another explanation is that all drugs are strictly prohibited in Chinese society, and the general public has been educated to keep away from any drugs. The third cluster, *the curious*, was more likely to be moderately motivated by cannabis authenticity; social prestige was also rated lowest among all motivation factors.

The identification of Cluster 2, *diversionists/recreationists*, and Cluster 3, *the curious*, may be of key interest because social prestige in these two segments was identified as the weakest motivational factor, which has rarely been demonstrated in general Chinese outbound tourist motivation studies. Even more surprisingly, Clusters 2 and 3 were also the largest in this research sample, accounting for about 74% of total respondents. Therefore, drug tourism operators should be noted that this particular tourist market is demotivated by social prestige. Given the sensitivity of drug use in China, it would be a very private issue to experience drugs. Very limited Chinese drug tourists would choose to share their drug-taking experience with others; social prestige does not apply in the drug-taking overseas. In fact, even though many Chinese tourists are curious about commercial drugs, under the pressure of Chinese social norms and traditional attitudes toward cannabis, most of them would choose to stay away from such activities or keep a low profile about their drug-tourism expeirence. These findings may help other similar overseas tourism operators to better understand Chinese tourists' motivations and behaviours in the context of sensitive activities such as drug tourism. This study sets a benchmark for a more comprehensive understanding of how social prestige motivates differentially based on various types of tourism activities in different tourism destinations.

While tourists of Cluster One are strongly motivated by all six motivation dimensions, Cluster Three, the second largest group in the sample, clearly shows that they are motivated by cannabis authenticity and gaining knowledge on the local cannabis culture. These two clusters demonstrate that Chinese drug tourists desire to 'experience all' and seek authenticity out of their normal daily life and society during the overseas travel.

The market segments and socio-demographic profiles described above can be used to develop marketing strategies and develop niche target markets as part of a diversification strategy. Considering the differences between the West and China in terms of law enforcement, social norm and attitude toward cannabis use, under the new phenomenon of normalization of cannabis and recreational drugs in more and more destinations, this study is one of the first attempts to explore the under-researched Chinese market on this sensitive topic of drug-tourism. The research's findings hold critical implications for both Chinese overseas market and cannabis tourism destinations operators and stakeholders.

In addtion, relevant studies have long been undertaken using qualitative approach to examine tourists' experiences of using drugs during trips (e.g. Belhassen et al., 2007; Uriely & Belhassen, 2005, 2006). This study is one of the very limited attempts to examine the drug tourists' motivation using empirical, quantitative research method through questionnaire. This study took great efforts in selecting participants whose major motivation is drug use experience among Chinese outbound tourists. Therefore, the motivation measurement specifically apply to the drug tourism context, and the market segmentation is based on quantitative data from the Chinese tourism market. From the managerial perspective, given the empirical support for the existence of contrasting Chinese drug tourist segments, this research also suggests that there is a need for tourism operators and local tourism board in drug tourism destinations to better understand and tailor strategies to manage tourists who are motivated by the local commercial cannabis.

This finding is not only valuable for drug tourism, but also for other kinds of tourism such as sex tourism, and LGBT tourism in the Chinese context. Such tourism activities are attracting more Chinese outbound tourists to engage at overseas destinations. However, their motivations and segments are under-explored due to the influences of Chinese cultures and stereotypes on deviant activities such as drug, commercial sex, and LGBT. This research will help relevant tourism operators and marketers to develop marketing strategies to better attract and serve more potential special interest tourists from the Chinese market.

Although this study provides insights into the drug consumption behaviours of Chinese outbound tourists, a few limitations need to be addressed as directions for future research. As this investigation focused on Chinese tourists in Amsterdam, Netherlands, findings may not be generalizable to the drug consumption behaviours of tourists from other cultural backgrounds and to other destinations associated with cannabis consumption. Future studies are therefore encouraged to further verify the segmentation framework of drug tourists in different cultural contexts and to different destinations associated with drug consumption, such as Jamaica, Egypt, and India. A cross-cultural comparison could generate a more insightful understanding of drug tourism as a sociocultural phenomenon. Moreover, future research should continue to explore the influences of cultural and social aspects on Chinese tourists' overseas drug consumption experiences. These influences could include motivation and various other relevant issues such as perceived risks and constraints, perceptions and attitudes toward psychoactive substances, and risk-taking behaviours associated with drug consumption. This study also enacted stringent criteria to capture Chinese cannabis-oriented tourists mainly motivated by the prospect of consuming drugs when travelling to Amsterdam. Other types of

Chinese tourists who did not meet the criteria but nevertheless used cannabis while travelling were not included in this sample; thus, more moderately drug-related tourist segmentations were not accounted for. Subsequent studies could include tourists whose drug consumption was a lesser aspect of their travel experience to derive more details about these segments. Local tourism practitioners in similar overseas destinations can also benefit from such empirical discovery.

Acknowledgement

The authors wish to thank Professor Chris Ryan from University of Waikato for his insightful comments to help improve the quality of this paper.

Disclosure statement

No potential conflict of interest was reported by the authors.

References

Báez, A., & Devesa, M. (2014). Segmenting and profiling attendees of a film festival. *International Journal of Event and Festival Management*, *5*(2), 96–115.

Baloglu, S., & Uysal, M. (1996). Market segments of push and pull motivations: A canonical correlation approach. *International Journal of Contemporary Hospitality Management*, *8*(3), 32–38.

Bandyopadhyay, R. (2013). A paradigm shift in sex tourism research. *Tourism Management*, *6*, 1–2.

BBC. (2012). *Amsterdam tourist cannabis ban rejected by mayor*. Retrieved from http://www.bbc.com/news/world-europe-20165371

Belhassen, Y., Santos, C., & Uriely, N. (2007). Cannabis usage in tourism: A sociological perspective. *Leisure Studies*, *26*(3), 303–319.

Berry, L., & Parasuraman, A. (1996). The behavioral consequences of service quality. *Journal of Marketing*, *60*, 31–46.

CBS. (2016, April 8). The Netherlands more popular with Chinese tourists. *Statistics Netherlands*. Retrieved from https://www.cbs.nl/en-gb/news/2016/14/the-netherlands-more-popular-with-chinese-tourists

Chen, Z., Zhong, X., Collins, R., Li, W., & Peto, R. (1997). Early health effects of the emerging tobacco epidemic in China: A 16-year prospective study. *The Journal of the American Medical Association*, *278*(18), 1500–1504.

Chiang, C., Wang, M.-Y., Lee, C., & Chen, Y. (2015). Assessing travel motivations of cultural tourists: A factor-cluster segmentation analysis. *Journal of Information and Optimization Sciences*, *36*(3), 269–282.

Churchill, G. (1979). A paradigm for developing better measures of marketing constructs. *Journal of Marketing Research*, *XVI*(February), 64–73.

Coffee shops. (2016). Retrieved from http://amsterdam.org/en/coffeeshops.php

Crompton, J. (1979). Motivations for pleasure vacation. *Annals of Tourism Research*, *6*(4), 408–424.

Dryglas, D., & Salamaga, M. (2017). Applying destination attribute segmentation to health tourists: A case study of Polish spa resorts. *Journal of Travel & Tourism Marketing*, *34*(4), 503–514.

Fang, Y., Wang, Y., Shi, J., Liu, Z., & Lu, L. (2006). Recent trends in drug abuse in China. *Acta Pharmacologica Sinica*, *27*, 140–144.

Fung, C., & Jim, C. (2015). Segmentation by motivation of Hong Kong Global Geopark visitors in relation to sustainable nature-based tourism. *International Journal of Sustainable Development and World Ecology*, *22*(1), 76–88.

Hsu, C., Cai, L., & Wong, K. (2007). A model of senior tourism motivations-Anecdotes from Beijing and Shanghai. *Tourism Management*, *28*(5), 1262–1273.

Hua, Y., & Yoo, J. (2011). Travel motivations of mainland Chinese travelers to the United States. *Journal of China Tourism Research*, *7*(4), 355–376.

Hui, T., Wan, D., & Chi, M. (2008). Chinese tourists visiting Singapore: Pull and push factors. *Advances in Hospitality and Leisure*, *4*(1), 229–244.

Kleiman, M. (2015). Legal commercial cannabis sales in Colorado and Washington: What can we learn? *Brookings Institution*. Retrieved from https://www.brookings.edu/wp-content/uploads/2016/07/Kleiman-Wash-and-Co-final.pdf

Korf, D. (2002). Dutch coffee shops and trends in cannabis use. *Addictive Behaviors*, *27*(6), 851–866.

Kotler, P., Bowen, J., & Makens, J. (2014). *Marketing for hospitality and tourism*. Upper Saddle River, NJ: Pearson Education.

Li, M., Wen, T., & Leung, A. (2011). An exploratory study of the travel motivation of Chinese female outbound tourists. *Journal of China Tourism Research*, *7*(4), 411–424.

Li, X., Xu, Y., & Weaver, P. (2009). Motivation segmentation of Chinese tourists visiting the US. *Tourism Analysis*, *14*(4), 515–520.

Lowinger, P. (1977). The solution to narcotic addiction in the People's Republic of China. *The American Journal of Drug and Alcohol Abuse*, *4*(2), 165–178.

Lu, Z. (2011). The study of Chinese tourists' motivations to Canada. *Journal of China Tourism Research*, *7*(4), 345–354.

Lu, L., Fang, Y., & Wang, X. (2008). Drug abuse in China: Past, present and future. *Cellular and Molecular Neurobiology, 28*(4), 479–490.

MacCoun, R. (2011). What can we learn from the Dutch cannabis coffeeshop system? *Addiction, 106*(11), 1899–1910.

Mohsin, A., & Ryan, C. (2004). Determinants of destination choice: The role of socio-demographic variables. *Tourism Recreation Research, 29*(3), 27–33.

NBTC. (2015). *2014 inbound tourism survey*. Retrieved from http://www.nbtc.nl/en/home/research-statistics/inbound-tourism-en.htm

Park, D., & Yoon, Y. (2009). Segmentation by motivation in rural tourism: A Korean case study. *Tourism Management, 30*(1), 99–108.

Pesonen, J. (2012). Social media channel segmentation of tourists. *e-Review of Tourism Research, 10*(2), 67–71.

Rid, W., Ezeuduji, I., & Pröbstl-Haider, U. (2014). Segmentation by motivation for rural tourism activities in the Gambia. *Tourism Management, 40*, 102–116.

Scott, D. (2008). *China and the international system, 1840–1949: Power, presence, and perceptions in a century of humiliation.* Albany, NY: State University of New York Press.

Shani, A., Wang, Y., Hutchinson, J., & Lai, F. (2010). Applying expenditure-based segmentation on special-interest tourists: The case of golf travelers. *Journal of Travel Research, 49*(3), 337–350.

UNWTO. (2017). Chinese tourists spent 12% more in travelling abroad in 2016. *UNWTO Press Release* (PR No.: PR 17046, 12 April, 2017). Retrieved from http://media.unwto.org/press-release/2017-04-12/chinese-tourists-spent-12-more-travelling-abroad-2016

Uriely, N., & Belhassen, Y. (2005). Drugs and tourists' experiences. *Journal of Travel Research, 43*(3), 238–246.

Uriely, N., & Belhassen, Y. (2006). Drugs and risk-taking in tourism. *Annals of Tourism Research, 33*(2), 339–359.

Valdez, A., & Sifaneck, S. (1997). Drug tourists and drug policy on the US-Mexican border: An ethnographic investigation of the acquisition of prescription drugs. *Journal of Drug Issues, 27*(4), 879–898.

Webster, D. (2017). How to go green for your travel destinations in 2018. *Spokesman Review*. Retrieved from http://www.spokesman.com/stories/2017/dec/29/how-go-green-your-travel-destinations-2018

Wen, J., Meng, F., Ying, T., Qi, H., & Lockyer, T. (2018). Drug tourism motivation of Chinese outbound tourists: Scale development and validation. *Tourism Management, 64*, 233–244.

Zhang, H., & Lam, T. (1999). An analysis of mainland Chinese visitors' motivations to visit Hong Kong. *Tourism Management, 20*(5), 587–594.

Appendix

Cannabis Tourist Motivation Questionnaire

1 = Entirely disagree; 2 = Mostly disagree; 3 = Somewhat disagree; 4 = Neutral;
5 = Somewhat agree; 6 = Mostly agree; 7 = Entirely agree

I consumed and experienced commercial cannabis in Amsterdam because:	Entirely disagree	2	3	4	5	6	Entirely agree
1. Commercial cannabis itself makes me feel safe to use cannabis.	1	2	3	4	5	6	7
2. Commercial cannabis itself makes me unworried about any legal risk.	1	2	3	4	5	6	7
3. Commercial cannabis itself makes me unworried about any social risk.	1	2	3	4	5	6	7
4. To learn about the local cannabis culture	1	2	3	4	5	6	7
5. To learn about the local cannabis industry	1	2	3	4	5	6	7
6. To learn about the local cannabis users	1	2	3	4	5	6	7
7. To have a look at real cannabis	1	2	3	4	5	6	7
8. Smoking cannabis makes me feel high.	1	2	3	4	5	6	7
9. To get away from a stressful social environment	1	2	3	4	5	6	7
10. To relieve daily boredom and busyness	1	2	3	4	5	6	7
11. To temporarily escape from family	1	2	3	4	5	6	7
12. To temporarily experience a crazy lifestyle	1	2	3	4	5	6	7
13. To fulfil the need for an inversion of ordinary life	1	2	3	4	5	6	7
14. To find something greater than myself	1	2	3	4	5	6	7
15. To have more imagination	1	2	3	4	5	6	7
16. To obtain the power to see inside myself	1	2	3	4	5	6	7
17. To get a wider philosophical and spiritual understanding	1	2	3	4	5	6	7
18. To have a different mindset	1	2	3	4	5	6	7
19. To show my socioeconomic status	1	2	3	4	5	6	7
20. To show my experience to others	1	2	3	4	5	6	7
21. To experience what others did not visit	1	2	3	4	5	6	7
22. I think smoking cannabis is a fashion.	1	2	3	4	5	6	7

1 = Entirely disagree; 2 = Mostly disagree; 3 = Somewhat disagree; 4 = Neutral;
5 = Somewhat agree; 6 = Mostly agree; 7 = Entirely agree

In the next 12 months:	Entirely disagree	2	3	4	5	6	Entirely agree
23. I plan to travel again to experience overseas cannabis tourism.	1	2	3	4	5	6	7
24. I would travel again to experience overseas cannabis tourism.	1	2	3	4	5	6	7
25. I am eager to travel again to experience overseas cannabis tourism.	1	2	3	4	5	6	7
26. I will recommend experiencing overseas cannabis tourism to my friends.	1	2	3	4	5	6	7

Demographic Information:

27. Age:
☐ 18-25 ☐ 26-35 ☐ 36 and above

28. Educational background:
☐Secondary ☐School/vocational school ☐College/university ☐Postgraduate

29. Annual income in RMB (1 RMB ≈ 0.16 USD):
☐ 100,000 or under ☐ 100,001 – 200,000 ☐ 200,001 – 300,000 ☐ 300,001 – 500,000

30. Marital status:
☐ Single ☐ Married

31. Travel types in Amsterdam for the purpose of commercial cannabis:
☐ Alone ☐ Small group with friends ☐ Small group with strangers
☐ Big group with friends ☐ Big group with strangers

Note: Small group: <10 tourists in one group Big group: ≥ 10 tourists in one group

32. Information sources on travelling to Amsterdam on the purpose of commercial cannabis:
☐Internet search ☐Relevant Internet forum ☐Friends' word of mouth
☐Advertisement ☐Others

33. Gender:
☐ Male ☐ Female

34. Did you smoke cannabis before:
☐ Yes ☐ No

Filter Questions:

35. Cannabis smoking is one of my major leisure activities during this trip.

☐ Yes ☐ No

36. Cannabis smoking is one important part of this travelling experience.

☐ Yes ☐ No

37. I am mainly driven by cannabis smoking to take this trip.

☐ Yes ☐ No

38. I regard myself as a cannabis tourist.

☐ Yes ☐ No

中国大麻游客在荷兰阿姆斯特丹(Amsterdam)旅游合法体验吸食大麻动机的问卷调查

1=非常不赞同； 2=很不赞同； 3=不赞同； 4=不知道； 5=赞同； 6=很赞同； 7=非常赞同

在 Amsterdam 购买体验大麻是因为：	非常 不赞同	2	3	4	5	6	非常 赞同
1. 商业大麻使我感到吸食大麻是安全的	1	2	3	4	5	6	7
2. 商业大麻使我不担心相关的法律风险	1	2	3	4	5	6	7
3. 商业大麻使我不担心相关的社会风险	1	2	3	4	5	6	7
4. 为了了解当地的大麻文化	1	2	3	4	5	6	7
5. 为了了解当地的大麻产业	1	2	3	4	5	6	7
6. 为了了解当地的大麻吸食者	1	2	3	4	5	6	7
7. 为了亲眼看看大麻	1	2	3	4	5	6	7
8. 吸食大麻让我感到兴奋	1	2	3	4	5	6	7
9. 为了暂时逃离身边紧张的社会环境	1	2	3	4	5	6	7
10. 为了暂时逃离日常的琐事和忙碌	1	2	3	4	5	6	7
11. 为了暂时逃离来自家庭的束缚	1	2	3	4	5	6	7
12. 为了暂时体验一种疯狂的生活方式	1	2	3	4	5	6	7
13. 为了颠覆常规生活	1	2	3	4	5	6	7
14. 为了在精神世界里寻找比自身更有价值的东西	1	2	3	4	5	6	7
15. 为了使自己更有想象力	1	2	3	4	5	6	7
16. 为了了解自己的内心深处	1	2	3	4	5	6	7
17. 为了获取更广阔的精神认知	1	2	3	4	5	6	7
18. 为了获取不同的思维模式	1	2	3	4	5	6	7
19. 体现自己优越的社会与经济地位	1	2	3	4	5	6	7
20. 体现自己比别人经验丰富	1	2	3	4	5	6	7
21. 体现自己比别人见多识广	1	2	3	4	5	6	7
22. 我认为吸食大麻是一种时尚	1	2	3	4	5	6	7

1=非常不赞同； 2=很不赞同； 3=不赞同； 4=不知道； 5=赞同； 6=很赞同； 7=非常赞同

在未来的一年内：	非常 不赞同	2	3	4	5	6	非常 赞同
23. 我计划再去境外体验大麻旅游	1	2	3	4	5	6	7
24. 我会再去境外体验大麻旅游	1	2	3	4	5	6	7
25. 我很渴望再去境外体验大麻旅游	1	2	3	4	5	6	7
26. 我会推荐朋友境外体验大麻旅游	1	2	3	4	5	6	7

人口统计学信息

27. 您的年龄

☐ 18-25 ☐ 26-35 ☐ 36 和以上

28. 您的教育程度

☐ 初中及以下 ☐ 高中或中专 ☐ 大专或本科 ☐研究生及以上

29. 请大概描述您的年收入

☐ 低于 10 万元人民币 ☐ 10—20 万 ☐ 20—30 万 ☐ 30—50 万

30. 您的婚姻状况

☐ 单身 ☐
已婚

31. 您在前往阿姆斯特丹旅行并吸食体验大麻时比较倾向于何种出游方式？*

☐ 独自一人前往 ☐ 和好友组小团 ☐ 和陌生人组小团

☐ 和好友一起参加大团 ☐ 参加陌生人大团

32. 您关于选择到阿姆斯特丹旅游吸食体验大麻的信息主要通过哪些渠道获得的？

☐ 网上搜索 ☐ 相关网上论坛 ☐ 朋友口耳相传 ☐ 广告推介

☐ 其他

33. 性别

☐ 男 ☐ 女

34. 您之前有过任何吸食体验大麻的经历？

☐ 有 ☐ 没有

35. 我认为吸食体验大麻是我此行主要的休闲活动之一

☐ 是 ☐ 否

36. 我认为吸食体验大麻是我此行重要的旅游体验之一

☐ 是 ☐ 否

37. 我认为吸食体验大麻是我此行的主要因素

☐ 是 ☐ 否

38. 我认为自己是个大麻（为了体验吸食大麻为目的）旅游者

☐ 是 ☐ 否

备注： *小团：< 10 人一个团；大团： ≥ 10 人一个团

A journey of hope: an institutional perspective of Japanese outbound reproductive tourism

I-Chieh Michelle Yang

ABSTRACT

As one of the most medically advanced nations in the world, it is observed that more Japanese couples are travelling abroad for Assisted Reproductive Technologies (ART). At present, at least one in six Japanese couples suffer from infertility and requires the use of ART. However, the usage of ART has raised several ethical questions. Notably, Japanese society objects to the use of donor eggs and surrogacy. Thus, reproductive tourism that offers prohibited procedures in another country becomes a viable alternative. Drawing upon institutional theory, specifically regulative, normative and cultural-cognitive pillars, this paper demonstrates that the emerging phenomenon of outbound reproductive tourism from Japan can be seen as an outcome of the interplay of institutional forces. A conceptual model of reproductive tourism is also proposed to advance the existing nebulous understanding of the phenomenon.

Introduction

Recently, Yomiuri Shimbun (2017), the largest media agency in Japan reported a sharp increase in Japanese couples travelling to Taiwan for In-Vitro Fertilization (IVF) treatments. According to their telephone survey with 72 fertility clinics in Taiwan, it was found that between 2014 and 2016, 177 couples travelled to Taiwan for IVF and 110 babies were successfully conceived and born. In 2014, only 17 couples visited Taiwan to obtain care, but within three years, the number has increased to 88, indicating a fivefold increment in outbound reproductive tourism. The report also specified that the availability of donor eggs and lower costs were the key motivations to travel out of Japan. This finding resonates with Hibino and colleagues' previous assertion that more Japanese couples would travel for fertility care for donor eggs and cost saving (Hibino, Shimazono, Kambayashi, Hitomi, & Nakamura, 2013; Hibino & Shimazono, 2014). This observed trend is certainly not novel, as has been previously discussed more infertile Japanese couples are found to be engaging in fertility care abroad (Hibino & Shimazono, 2014; Ikemoto, 2009; Ito, 2015; Shimazono & Hibino, 2013). Such act of travelling to another country for fertility or reproductive treatments is generally referred to as 'reproductive tourism' in existing literature. At present, the Japan Society of Obstetrics and Gynecology (JSOG) prohibits the use of donor eggs and embryos and surrogacy in fertility treatments. Thus, despite having the highest number of fertility clinics and the highest number of IVF cycles performed in the world (Collins, 2002), scholars (e.g. Ikemoto, 2009, 2018; Martin, 2015; Utsunomiya, 2015) noted that Japan would continue to be a prominent source of reproductive tourists, specifically those who seek egg and embryo donation and surrogacy.

This article has been republished with minor changes. These changes do not impact the academic content of the article.

Hibino and Shimazono (2014) examined the attitude towards reproductive tourism amongst infertile Japanese couples and found that as high as 76% of the respondents would consider reproductive tourism, while 22% of them have already received treatment abroad. This study asserted that couples who have engaged in reproductive tourism were motivated to circumvent Japan's stringent restriction on egg and sperm donation (Hibino & Shimazono, 2014). It is found in a government study that between 2009 and 2012, the clinics that participated in the survey had handled 117 donor egg births (Takeshita et al., 2003; as cited in Utsunomiya, 2015, p. 128). Given that the usage of donated eggs is prohibited under JSOG's guideline, it is reasonable to construe that the 117 cases were in collaboration with fertility centres located outside Japan. In a recent interview, Professor Hibino avows that the understanding of Japanese outbound reproductive tourism remains an under-researched topic, with most of the understanding reported in the media (Brasor, 2017). Furthermore, there is, at present, no reliable statistics on the number of Japanese who have travelled abroad. The phenomenon of outbound Japanese reproductive tourism remains empirically uninvestigated. More important, in Speier's (2016) book – *Fertility Holidays: IVF Tourism and the Reproduction of* Whiteness, the phenomenon is conceptualized as a 'reproduction of whiteness' to describe the popularity of reproductive tourism in North America and Europe (p. 8). Hudson and colleagues (Hudson & Culley, 2015; Hudson et al., 2011) observed that much of existing empirical works on reproductive tourism focus predominantly on the European, Middle Eastern and North American contexts. There is a dearth of attention paid to Asia, a region where reproductive tourism is an emerging phenomenon, thus presenting a critical literature gap to be addressed in this paper.

Furthermore, while a nascent body of work has briefly illustrated that the prohibition of third-party material such as donor egg and embryo as well as the high costs have propelled outbound travel, understanding why reproductive tourists travel based on national context remains nebulous. This paper puts forth the notion that the law-evasion/reproductive tourism nexus is too reductionist in nature. There is a need to examine the institutional forces that shape fertility practices in Japan to obtain a more concise understanding of why Japanese citizens choose to travel abroad to seek treatment. For these couples, breaking away from the accepted norm in Japan to seek overseas treatment that is deemed immoral and unacceptable in Japan may entail ponderations that are beyond just restrictions and costs. Anchoring on these apparent literature gaps, this paper seeks to explore the driving institutional forces that precipitate the growth of outbound Japanese reproductive tourism. Drawing on institution theory, this paper's primary aim is to understand why, despite being one of the global leaders in reproductive medicine (Irizawa & Gotoh, 2016), Japanese citizens resort to reproductive tourism. The paper is structured as such – introduction, literature review, institutional theory, discussion and conclusion.

Literature review

Reproductive tourism

Reproductive tourism essentially refers to the act of travelling to another country to receive assisted reproductive treatments. Connell (2011) avers that reproductive tourism involves access to reproductive technology in a foreign destination and sometimes enables cheaper, more efficient and comprehensive services for couples to evade restrictive regulations, long waiting lists and high costs. As a part of the wider trend of medical tourism (Ikemoto, 2009), the phenomenon has amassed substantial scholarly attention in recent decades, presumably due to the global rise in infertility and emergence of medical tourism. With globalization and the rapid advancement of information technology, medical tourism as a form of travel has burgeoned in recent decades. Several commentaries have illustrated that individuals engage in medical tourism to seek different medical procedures, based on a wide range of motivations. However, even though the current findings of travel motivations for medical tourism appears to be similar across various contexts, Gerrits (2018) exposited that specific characteristics of the tourist's country may also influence travel motivation, for example, in

Sunni Islam countries such as the UAE, all forms of third-party assisted ART are prohibited, prompting infertile couples to travel to other Islamic countries such as Iran, where the dominant Shi'a Islam policy has more liberal policies towards ART (Moghimehfar & Nasr-Esfahani, 2011). However, while reproductive tourism may seem to be a mere component of medical tourism, as Ikemoto (2009, 2018) notes, there are important distinctions between reproductive tourism and other forms of medical tourism. For instance, cosmetic tourism, which refers to the act of travelling to another country to obtain cosmetic enhancements on one's body, is a prevalent form of medical tourism due to its low-risk and elective nature. According to Viladrich and Baron-Faust (2014), the medical tourist's body is the only medium and endpoint of cosmetic tourism. Conversely, reproductive tourism involves the reproduction of human life, which raises different yet important bioethical questions as to the child's status and future welfare. Furthermore, this form of medical tourism involves not just the infertile couples but also other stakeholders such as third-party gamete donors or surrogates (Ikemoto, 2018). As such, reproductive tourism is a complicated variation of medical tourism. More importantly, the act of travelling to another country to receive fertility care, such as IVF has been discussed and empirically examined across various disciplines, with a preponderant of works on the travel motivation of reproductive tourists. The existing body of literature has highlighted a multitude of travel motivations of reproductive tourists, encapsulating, *inter alia*, circumvention of legal restrictions and cost saving.

Circumvention of legal restrictions

Pennings (2002) defines reproductive tourism as 'the travelling by candidate service recipients from one institution, jurisdiction or country where treatment is not available to another institution, jurisdiction or country where they can obtain the kind of medically assisted reproduction they desire' (p. 337). A plethora of commentaries has collectively insinuated that the key impetus of reproductive tourism is law (Inhorn, Shrivastav, & Patrizio, 2012; Storrow, 2011). Accordingly, Whittaker (2010) notes, reproductive tourism is largely characterized by travelling to countries where regulations allow fertility treatments that are not available in the patient's home country. The varying legal perspectives toward ART across nations has facilitated the growth of reproductive tourism, as some countries with a rigorous legal framework of ART became departure nations and others with minimal or no legislation of ART became reproductive tourism destinations.

Most of the legal bans on ART revolve around the use of third-party gamete (egg or sperm). In countries such as Italy, Turkey and the Middle East where religious doctrines play an influential role in public policy, third-party involved reproductive assistance such as sperm and egg donation, embryo donation and surrogacy are prohibited (Zanini, 2011). To exemplify, the legal framework of ART in Italy has been described to be 'one of the strictest' in the world. Reproductive medicine is stringently regulated under the *Law 40* which was implemented in 2004 due to Catholic lobbying. This overarching law forbids a wide continuum of ART procedures, such as limiting ART to heterosexual married couples; prohibiting the use of donor eggs, sperm, uteri and surrogacy as well PGD. What is noteworthy is that prior to the arrival of *Law 40*, Italy was known to be a destination country in Europe with the highest number of IVF pregnancies amongst postmenopausal women due to the negligible legislations, however, as the law became legally-binding in 2004, Italy's previous position as a destination country is now reversed (Bartolucci, 2008; Ikemoto, 2009; Pennings, 2002). Zanini (2011) propounds that the enactment of *Law 40* precipitated a quadruplet increase in outbound reproductive tourism, with 70% of the Italian couples in her study citing law evasion as the main travel motivation. The Italian respondents in her study appear to feel betrayed and abandoned by their government and the church, as they felt forced to leave their country to seek for fertility treatment elsewhere.

On the other hand, some commentators note that even in countries where ART is fully or partially covered under national healthcare plans, citizens are sometimes still denied access to ART procedures due to certain demographic restrictions. Factors such as sexual orientation, marital status and age are some of the common reasons that precipitate reproductive tourism abroad (Ferraretti, Pennings,

Gianaroli, Natali, & Magli, 2010; Pennings, 2004, 2009; Smith-Cavros, 2010). For example, even though France is one of the six countries in the world where ART is fully covered by public funding (Blyth & Farrand, 2005), the provision is not extended to lesbian couples, single women and women of a certain age (Pennings, 2002). Along the same vein, other countries also impose age restrictions on infertile patients. In the UK, where IVF is free for most citizens except for women who are over 40 years of age, as a result, thousands of women have reportedly travelled abroad. As Smith-Cavros (2010) argues, over 40 is precisely when a women's fertility declines, making them the demographic that requires governmental support to achieve pregnancy. The imposition of restrictions based on an individual's demographic factors has rendered it inevitable for them to access reproductive care in another country (Utsunomiya, 2015).

Cost saving

Given that ART is a relatively new and sophisticated method in reproductive medicine, the costs involved can be considerably high. Deonandan (2015) categorized medical tourists into price-sensitive and quality-sensitive polar and averred that reproductive tourists are often more price-sensitive given the expensive costs of ART. To illustrate, Wu, Odisho, Washington, Katz, and Smith (2014) state that the average price of an IVF cycle is the highest in the United States. On average, the total charge for an IVF cycle amount to be $9,500, accounting for almost 20% of the median American family income (Collins, 2002). Additionally, donor eggs may cost upward of $30,000 in some American clinics (Smith-Cavros, 2010). Given that subsidies are not universal in all states, ART is largely viewed as an elective procedure and most patients have to pay for fertility care from their own pocket. As a result, only 1% of infertile Americans conceive through ART in the country (Spar, 2005). The high cost in America has also instigated outbound reproductive tourism. Lee (2005) illustrates how collaboration between an American clinic and a Romanian clinic facilitates the use of donated egg which is fertilized in Bucharest and subsequently shipped back to the United States. In so doing, the American couples who engage in the service can save approximately $18,000 and have a greater probability of successful conception. Whittaker and Speier (2010) similarly found that Americans also travel far to Thailand to obtain cheaper reproductive care, where the price of one IVF cycle costs approximately 80,000 Thai Baht, one-third of the price in the United States (p. 369).

As illustrated, the affordability of ART can potentially jeopardize one's hope to become a parent (Davidson, 2010). As the cost of reproductive treatments can vary amongst countries, some couples would resort to travelling to a destination where the cost of ART is affordable to pursue their desire to become parents (Ferraretti et al., 2010; Ikemoto, 2009; Smith-Cavros, 2010). While countries such as Belgium, France and Sweden have formally recognized infertility as a medical condition and provide substantial subsidies, most countries still view ART as an elective and discretionary procedure and provide insufficient financial assistance to couples (Blyth & Farrand, 2005). Thus, Pennings (2002, 2004) underscores that reproductive tourism can be seen as a solution to income inequity amongst infertile couples and allow less-affluent individuals to achieve parenthood. Nevertheless, despite the broad scholarly acknowledgement that reproductive tourism is an emerging component of medical tourism, the extant body of knowledge is focused on policy and regulatory aspects of the phenomenon. There is less work on understanding how national ethos and bioethical factors have influenced policies and regulations on ART. The differences in regulation across nations have resulted in some countries becoming destinations for reproductive tourists while others become source countries. Thus, it is important to understand context-specific travel motivations.

Assisted reproduction in Japan

As a 'fertility powerhouse' (Ito, 2015), Japan has the highest number of registered fertility treatment clinics in the world (more than 600) (Collins, 2002; Maeda et al., 2014), and the demand for ART continues to increase annually (Mayeda, 2006; Saito et al., 2017). It is however interesting to note that the number of IVF cycles performed in Japan is the highest in the world; yet, its success rate is also reportedly the lowest. Such discrepancy is surprising due to Japan's renowned reputation as a world leader in reproductive medicine (Irizawa & Gotoh, 2016; Utsunomiya, 2015). One plausible reason could be the prohibition of the use of donor eggs and embryos in Japan, which limits the type of treatments available in the country, thereby lowering the success rates of conception.

Rise in infertility

Recently, a rise in infertility has been observed amongst couples of reproductive age in Japan. Cases of infertility are on a persistent rise from 1 in 10 Japanese couples in 2009 to 1 in six couples in 2015 (Hayashitani & Suzui, 2009; Ito, 2015). Correspondingly, in the latest National Fertility Survey in 2015 (NIPSSR, 2015), the inability to conceive accounts for 23.5% of the reasons why couples did not produce the intended number of children they want, a conspicuous rise from 16.3% in the 2005 survey. It is also vital to note that infertility is a stigmatized topic in Japan that is rarely discussed, even though there are more scholarly attempts to document the rise of infertility in recent years, the actual statistics and reasons of infertility remains undetermined.

One of the most cited reasons for the rise in infertility would be *bankonka* – the postponement of marriage (Kubo, 2009; Shimazono & Hibino, 2013). As is the case with most developed countries, the trend of late marriage and postponement of childbearing are prevalent in Japan. Whereas in other low-fertility countries such as France and Germany, extra-marital childbirth is common, but the same does not apply in the Japanese context. Mainly, Japanese society remains conservative about marriage and childbirth, in spite of its economic progress and modernization. In Japan, marriage is viewed as a 'package' that is tied with childbearing (Shirahase, 2000; Tsuya, 2005). Thus, as more Japanese couples marry and try for conception from the early to late 30s, their reproductive life span is reduced (Kubo, 2009; Kumagai, 2015). In an attitude survey with 2007 infertile patients, Shimazono and Hibino (2013) found that advanced age was the major cause of infertility in Japan. Age-associated conditions also lead to other complications such as a decline in ovarian function. Nevertheless, the rise in cases of infertility has increased the demand for ART services, with more ART cycles reportedly conducted in Japan. JSOG reported in 2008 that the number of treatment cycles performed in the country has increased significantly by 15% between 2007 and 2008. Similarly, more babies are also born through the use of ART, with the proportion of babies conceived through assisted methods accounting for 2% of live births in 2008 and subsequently increased to 5% in 2015 (Ooki, 2011; The Japan Times, 2017).

Third-party assisted reproduction in Japan

In reproductive medicine, third-party assisted reproduction can refer to donor conception or surrogacy. On the one hand, donor conception refers to the attempt to conceive a baby using either sperm or eggs from the third party, or both, to create embryos. The first successful third-party assisted conception via Artificial Insemination of Donor Sperm (AID) through animal husbandry technique was achieved in 1948 at Keio University (Shirai, 2010; Tsuge, 2005). The practice of AID was met with insignificant resistance and was not openly opposed in Japan – both JSOG and the government did not express their position on AID or establish any guidelines to regulate the practice (Shinagawa, 1996). As a result, physicians were allowed to conduct AID so long as they adhered to their respective institution's code of conduct (Sotoya, 2012). AID was the *de facto* reproductive method in Japan until it was replaced by IVF in 1983. Nevertheless, when IVF was first introduced, JSOG specified that '*the mother must be capable of sustaining the removal, implantation and gestation of the mature ovum*' (Mori, 1992, p. 294), implying that donor sperm is allowed but upholds a stance against the use of donor egg and donor embryo. However, when a pair of twin boys was born after being conceived through IVF via donor egg in 1998, it instigated a nationwide debate, as the twins were conceived using sperm from their father and donated ovum from the mother's sister. This epoch-making event resulted in Dr. Netsu's, the physician who performed this controversial IVF procedure, expulsion from JSOG (Morioka & Motegi, 2005).

Surrogacy, on the other hand, is a procedure which allows women who suffer from uterine infertility or medical conditions that rendered her unable to carry the foetus in order to achieve motherhood by using a surrogate to carry and deliver the child on her behalf (Kisu, Banno, Mihara, Iida, & Yoshimura, 2011; Suzuki et al., 2006). Surrogacy is similarly prohibited by JSOG. In 2001, Dr. Netsu (the same physician who conducted first IVF using donated egg), announced that he had successfully conducted the first local surrogacy in Japan (Gunning, 2003; Kisu et al., 2011). This event instantly exacerbated Japan's existing quandaries on the ethicality of third-party assisted reproduction and engendered violent reaction from medical associations such as JSOG, the ART committee and the special committee who unanimously condemned the procedure (Mayeda, 2006; Tsuge, 2005). In summation, it is apparent that Japan's prohibition of donor egg, donor embryo and surrogacy have rendered it difficult for infertile patients to access the necessary care. Shimazono and Hibino (2013) assert that the demand for donor oocytes is 'left largely unmet' (p. 340) and compelled these couples to become reproductive tourists. The aforementioned review reveals that Japan has stringent regulations for ART. The reasons for the conservative policies of such a medically advanced nation towards ART requires a deeper understanding of the country's various institutional factors, specifically its normative and cultural dimensions.

Institutional theory

Fundamentally, institutional theory seeks to understand how institutional forces or interactions can affect individual interests and result in a change of behaviours (Powell & Colyvas, 2008). As one of the most prominent scholars in institutional theory, Scott (1995) defined institutions as 'social institutions that have attained a high degree of resilience' (p. 33). Institutions are embedded in political, social or economic realms that structure incentives in human exchange (North, 1990). Scott (1995) further proposed the tripartite pillars of institutional theory – the regulative pillar, the normative pillar and the cultural-cognitive pillar, and asserted that the interaction of these components of an institution provides an understanding of human motivations and behaviours in a society. First, the regulative pillar reflects the formal laws and rules of a society that would influence behaviour through force and expedience. Second, the normative pillar is embedded in the values and norms of a society, where members are expected to perform certain roles. It reflects the fundamental assumption of human nature and behaviours (Ramsey, Leonel, Gomes, & Monteiro, 2011). Finally, the cultural-cognitive pillar involves the creation of shared belief and knowledge of a social reality. It is through this pillar that members construct meanings of their respective identities (Scott, 2003).

As a prominent theory used to study human motivations and behaviours, this paper seeks to explore the trend of outbound Japanese reproductive tourism through the lens of institutional theory. The justification for using this theory is twofold, first, as Scott (2008) highlights, institutional theory seeks to understand phenomena in a context-specific framework within a given scope. The importance of context is highlighted in the regulatory and cultural differences between countries that can produce varying motivations for couples to engage in reproductive tourism. Each country has its distinct approach and regulatory framework towards the use of third-party assisted reproduction; therefore, some countries with stringent regulations tend to become a source country for reproductive tourists whereas countries with relatively relaxed regulations often become the destination for reproductive tourism. As Ferraretti et al. (2010) and Hudson and Culley (2015) enumerate, reproductive tourism can be perceived as the outcome of the interplay of different institutional factors (e.g.: ethical, religious and legal dimensions). The second reason for employing institutional theory would be the novelty of it in reproductive tourism scholarship. At present, studies of reproductive tourism tend to lack theoretical underpinnings, thus, this paper is the first attempt to employ institutional theory to examine the global phenomenon.

Regulative pillar

The regulative pillar essentially views institutions with regulative functions, with the capacity to establish formal rules to influence individual behaviours. Several authors have referred to such institutions as formal institutions or macro-level institutions. Indicators of the regulative pillar include rules, laws and sanctions, which are often employed by the state to constrain and regulate behaviours. Primarily, institutions are believed to have the ability and authority to constrain and regulate individual behaviours. In a broad sense, institutions exist in society to reduce uncertainty for individuals and define boundaries of legitimacy (Peng, Sun, Pinkham, & Chen, 2009). The regulative process entails the establishment of rules and constructs formal systems to monitor individual adherence and conformity. In the context of this study, the phenomenon of infertile Japanese couples travelling abroad for fertility treatments can be examined as an outcome of influences from the regulative institutions.

Following the reporting of Dr. Netsu's controversial procedures, the extensive reaction from the society prompted responses from eminent Japanese medical associations and policy-makers. Numerous meetings were held to develop guidelines to oversee the use of ART (Gunning, 2003). Both JSOG and Japan Society of Fertility and Sterility (JSFS) reviewed the existing guideline on ART and subsequently amended their guideline to include the prohibition of donor egg and embryo as well as surrogacy. The stance to limit third-party assisted conception by eminent medical associations reflects the emphasis of traditional family structure amongst Japanese citizens. The belief that parents need to have a genetic or blood relationship with the child is entrenched in Japanese culture. It is also imperative to note that as a medically advanced nation with a growing demand for ART, there is no regulatory or legal framework promulgated to oversee the practice of ART in Japan. To date, fertility experts in Japan have exercised voluntary self-regulation and discipline in abiding with JSOG's guideline, which is by the far the most impactful set of guidelines in Japan's obstetrics milieu. It should be emphasized, however, that the guideline has no actual legal enforcement, therefore, despite going against JSGO's principle, Dr. Netsu was not legally prosecuted (Mayeda, 2006). As can be seen in Dr Netsu's case, JSOG's guideline is followed by the majority of Japanese physicians who are afraid of the punishment from the society and ostracism from fellow members.

Another regulative factor that might have propelled outbound reproductive tourism is the revision of ART subsidies. Maeda et al. (2014) examined age-specific cost for ART in Japan found that the average cost of ART (excluding costs relating to diagnosis, consultation, physical examination) incurred for one live birth itself is JPY1,974,000 (approximately USD18,000) and the cost increases with the woman's age. The costs associated with fertility treatments in Japan appears to be one of the highest in the world. Recognizing this potential problem, the Japanese government launched the first subsidy programme for infertile couples in 2004. This governmental support initially appeared to ease the financial burden on couples who require ART, however, with the revision of the subsidy in 2016, fewer couples are eligible for the subsidy. Under the newly revised subsidy, an age cap is imposed, restricting women who are 43 years or older to receive the assistance and the maximum number of IVF cycles to be subsidized is also reduced from 10 to 6. Japanese women who are 39 years old or younger can receive a subsidy for up to 6 IVF cycles, whereas patients aged between 40 and 42 years are eligible to a maximum of 3 IVF cycles. Couples who require the use of ART now face greater problems to access fertility care. Maeda et al. (2014) avowed that, due to the prohibition in the use of donor gametes or embryos in Japan and increasing cost of ART, Japanese couples may increasingly be interested in travelling abroad for fertility treatment.

Normative pillar

The normative pillar posits that normative rules introduce a 'prescriptive, evaluative and obligatory dimension into social life' (Scott, 1995, p. 38). The normative system essentially consists of both values and norms, which reflects a society's assumption about human nature and human behaviour (Ramsey et al., 2011). According to Scott (2003), the normative component of a society reflects national ethos and the expectations of how human should behave. At one end of the spectrum, value refers to the ideal standards of behaviour in a society. Hall and Jenkins (1995) postulate that 'values are the overarching criteria people use to make decisions and an ideology is a systematic representation of those values' (p. 45). In the case of human reproduction, it appears that the Japanese population still emphasize traditional family values and to have genetic ties with one's child. As exemplified in both scandals involving Dr. Yahiro Netsu, the Japanese society was largely critical of the procedures involving third-party reproduction. For the first time since the AID was performed in the country, the appropriateness and ethicality of third-party reproduction dominated public discourse, with overwhelming concerns pointing to the extent to which artificial medical manipulation should be allowed to intervene human reproduction.

A predominant concern amongst opponents of ART pertains to the possible erosion of Japan's traditional family values and structure. Importantly, the Japanese society attaches great significance to genetic and blood ties and the advent of ART has raised concerns on complicating familial relationships and bloodlines (Feldman, 1985; Mayeda, 2006). To illustrate, Shirai (2010) asserts that the ingrained social dogma which dictates that the couples who raise a child (social parents) should also be the biological and birth parents has caused a general discomfort in using third-party reproduction amongst infertile couples. Mainly, third-party conception implies that a child can have multiple parents, from his or her social parents, biological parents to birth mother, the use of donor gamete or surrogacy may potentially destruct families. The child born through such an arrangement might also suffer from identity confusion and complicate family relationships (Sotoya, 2012; Tsuge, 2005). Thus, when the use of ART can potentially complicate family or parent–child relationship, individuals who resort to ART for conception are viewed negatively. This view is shared by King (2009), who further explicates that the continual development of ART challenges the Japanese notion of family and parenting and propelled nationwide contestations.

In addition to these national ethos, the norms of any given society also define the ways in which things should be done and normative rules are typically internalized by individuals. In a news article published in The Washington Post, Japanese women would rather visit *Suitengu Shrine* – a known shrine in Tokyo where couples would pray for a solution to their fertility woes, than to resort to artificial reproductive procedures (Jordan & Sullivan, 1998). Quintessential of Japanese norms, while infertile couples in other medically advanced nations would seek ART, their Japanese counterparts would opt for prayers or visits to fertility shrines. Apropos of Yamagata et al.'s opinion polls that were funded by the Ministry of Health, Wealth and Labor (MHWL) in 1999 and 2003, social approval of donor IVF is considerably low in Japan. Specifically, amongst infertile individuals, 75.9% responded that 'I wouldn't use it even if my partner wanted to' in 1999 and the disapproval increased to 79.3% in 2003. 'Family relationships would be unnatural' (47.4%) and 'I don't think it's a natural way to conceive' (25.5%) were cited as the top reasons to oppose donor conception (Yamagata, Hoshi, Hirata, & Takeda, 2003). More recent empirical evidence was provided by Shimazono and Hibino (2013), whose attitude survey amongst infertile couples shows that the majority of them (84.6%) object to the use of third-party gamete. This illustrates that even for individuals who might require the use of donor insemination, they would choose to not use it.

In terms of surrogacy, the same opinion poll by Yamagata et al. (2003) found that the percentage of respondents who would not opt for surrogacy even if their spouse wishes to account for 75%. The reluctance to use surrogacy amongst infertile couples is supported by Saito and Matsuo (2009) and Shimazono and Hibino (2013) who found that surrogacy is still not widely accepted even for Japanese individuals who face fertility issues,. It is also noteworthy that, even in modern day Japan, the notion of '*onaka wo itameta ko*' (the baby who made my belly hurt), still prevails amongst Japanese women. In her work that examines the infertility experience of Japanese women, Shirai (2010) found that there is still an ingrained pursuit of having a child that has a genetic or biological relationship, more importantly, as the respondents in her study have shared, they felt strongly about having a child to whom she herself gave birth to (Shirai, 2010, p. 22). It is therefore evident that the general attitude towards third-party assisted reproduction is negative as the nature of the procedures has may alter the notion of parenthood and made it difficult for Japanese women to accept (Jordan & Sullivan, 1998). Nevertheless, these surveys demonstrate that the perception towards third-party assisted reproduction remains negative amongst Japanese. According to Scott (1995), the basis of compliance for normative pillar is a social obligation. The normative system of a society defines the role of members, specifically in collectivist societies such as Japan, social values and norms are typically acknowledged as the way of life and internalized by members of the society. The preference for natural conception amongst infertile couples in Japan reflects an entrenched belief to maintain genetic ties and to avoid engaging in behaviours that contradict with Japanese normative system.

Cultural-cognitive pillar

The normative framework focuses on the influence of norms and social expectations of member's roles whereas the cultural-cognitive pillar looks at institutions as one that shapes social identities. As Scott (1995) explains, the regulative and normative institutions in the society provide objective conditions for individuals to make sense of their world; however, their subjective interpretation of these conditions will alter their behaviours accordingly. Therefore, even though formal constraints such as laws and normative systems provide Japanese individuals with some guidance, they also consider the cultural-cognitive elements embedded in everyday life to construct their respective social reality. Ideally, as members of a collectivist society, a Japanese individual may seek to behave in conventional ways (as described in the section of the normative pillar) that will not cause themselves to be seen as deviant or different. The crux of the cultural-cognitive pillar lies in the construction of social identities, specifically, how we perceive ourselves and how we can best fulfil our individual interests in a given situation. Parenthetically speaking, the cultural-cognitive component reflects the shared social knowledge and cognitive categories (stereotype of individual role) (Ramsey et al., 2011, p. 23).

In Japan, singlehood is often associated with negative connotations. Terms such as *'parasite singles'* have been coined to describe single individuals as 'parasites' who are not contributing to the country's declining marriage and childbirth rates. By and large, women who remain single are often described as selfish and irresponsible in the public discourse (Akizuki & Kai, 2008). Edwards (1989) contends that Japanese tend to view themselves as being incompetent or immature when they are single, and it is through marriage that one attains adulthood. As a result, an individual's identity of self is largely shaped by their marital status. The Japanese society is characterized as a *'kaikon shakai'* (a marriage-oriented society), thus, insofar as the phenomenon of late marriage prevails in contemporary Japan, the majority of Japanese individuals still intend to get married and start families at some point in their life (Tokuhiro, 2010). Through in-depth interviews, Tokuhiro (2010) attested that young Japanese women continue to believe and internalize the belief that marriage is an important personal milestone. Furthermore, the attainment of maturity and adulthood through marriage is also associated with new social responsibilities as a couple – to have babies. In present-day Japan, marriage is viewed as a 'package' that comes with childbearing (Shirahase, 2000; Tsuya, 2005), thus childless couples who are married for a long time are under considerable pressure to conceive.

Tsuya (2005) further averred that despite the strong influence of Westernization in the Japanese society, deeply-rooted distinction of gender roles and adulthood remain pertinent and prevalent. In present-day Japan, married women are often expected to return to the domestic sphere to produce babies. Career-oriented women who seek to advance their careers face mounting pressure to choose either work or family (Atoh, 2008; Retherford, Ogawa, & Sakamoto, 1996). As more women view work as a source of self-fulfillment or actualization, the opportunity cost involved in leaving work for marriage and family building is seemingly high (Boling, 2008). While policymakers have initiated reforms to introduce maternity leave to increase the fertility rate, only 21.4% of new mothers used the paid leave in 2010 (NIPSSR, 2015). Some scholars have explained that the low usage of maternity leave in

Japan is due to the Japanese belief that a mother should dedicate their full attention to her young child (Raymo & Iwasawa, 2008). The belief of '*sansaiji shinwa*' (three-year-old myth) propounds that a mother's presence and care during the first three years of a child's life is vital for his or her development and personal wellbeing (Oishi, 2008). Such entrenched belief of a mother's role has put pressure on Japanese women to leave the workplace (Tokuhiro, 2010). In her study to examine how Japanese women face a tradeoff between work and motherhood, Harvey (2016) noted that her respondents who were childless reacted negatively when they were asked if they have 'decided' to have children. Specifically, many of them did not see their childless situation as a matter of personal choice, but an outcome of their social and work environments. This study purports the strong desire of Japanese women to become mothers regardless of the importance of their work. She further highlights that the social identities of Japanese women face contradiction with the socio-economic realities of Japan. As elucidated previously, postponement of marriage and childbearing would reduce the effective reproductive life span of a woman, thus, as Japanese women spend longer time for career advancement, they are more likely to face difficulties in natural conception (Kubo, 2009).

Discussion

Having elucidated the social and legal environments on the practice of third-party assisted reproduction, marriage and childbirth vis-à-vis the three institutional pillars – regulative, normative and cultural-cognitive in Japan, this section turns to discuss the plausible interactions of these institutional forces in propelling outbound reproductive tourism. To better capture the meaning and nature of a phenomenon, Scott (2008) propounded that it is important to identify the dimensions of the three pillars and more importantly, how these dimensions interact with each other. Accruing from the aforementioned expositions, a conceptual model is developed to account for the interaction of institutional factors that propel outbound reproductive tourism in Japan (Figure 1).

Firstly, alluding to the cultural-cognitive pillar, a Japanese individual's social identity is largely shaped by the ingrained shared beliefs and conceptions. As a *kaikon shakai* (marriage-oriented

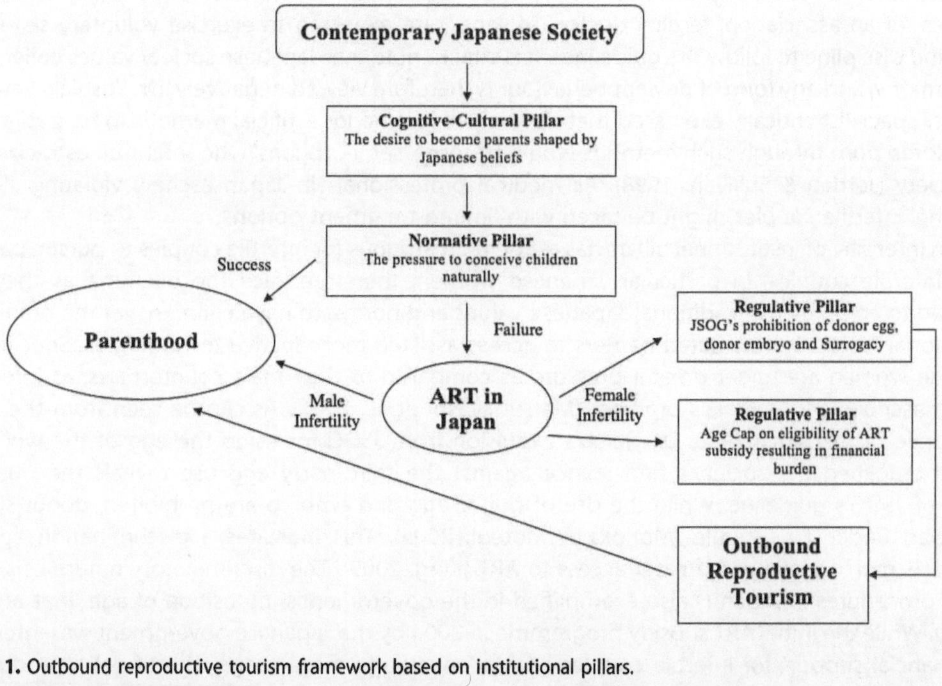

Figure 1. Outbound reproductive tourism framework based on institutional pillars.

society), Japanese individuals seem to have this entrenched belief that marriage and childbirth are personal goals that should be attained at some point of their life. Childless women commonly experience two reactions from their social environment – sympathy and criticism. On the one hand, some women are often sympathized for their childless situation, as infertile women are regarded as abnormal and lonely. On the other hand, some others receive criticism such as being selfish and are put under considerable pressure (Akizuki & Kai, 2008; Matsubayashi, Hosaka, Izumi, Suzuki, & Makino, 2001). However, as members of the Japanese normative system, Japanese women are also expected to conceive naturally. Given that the ability to conceive might be limited by a couple's physical conditions, in some instances, ART might be required to achieve pregnancy. The use of ART, particularly with donor gametes or surrogacy, however, violates the normative system of their Japanese society which believes in having genetic ties with children and natural conception. Akizuki and Kai (2008) demonstrated that infertile Japanese women often face scrutiny over their childless situation and are concomitantly discouraged from using ART. Several respondents were told by their elders, most often their mother or mother-in-law to not opt for IVF because it is 'unnatural and damaging' (p. 2740). According to Scott (2003), the normative and cultural-cognitive institutional pillars achieve individual compliance through social obligations and isomorphism. A Japanese person would thus try to behave in a conventional manner and would resort to mimetic behaviour (to follow the behaviour of other members of the society) (Scott, 2003) – to get married and to have children naturally. As exemplified in the opinion polls pertaining to third-party assisted reproduction, it is conspicuous that the willingness to adopt such procedures amongst infertile Japanese couples remain low, as they would attempt for conception in accordance with Japanese normative system.

Another plausible interplay of institutional pillar points to the influence of normative systems in shaping JSOG's stern guideline. As Takeshita et al. (2003) enumerate, legislation in Japan largely hinges on the social background of its citizens. JSOG's prohibition of donor eggs and embryo and surrogacy is influenced by the normative system of the country. While at present, the Japanese government has yet institutionalized legislation towards reproductive medicine, JSOG's guideline alone serves as a prominent framework for practitioners. Evident from Dr. Netsu's expulsion from JSOG, it is clear that the guideline, despite having no legal enforcement, is abided by the majority of fertility physicians who are afraid of punishment from society and being ostracized from fellow members. Such abidance of JSOG's guidelines can also be explained via the normative pillar, where, as members of an association, fertility doctors in Japan are expected to exercise voluntary self-regulation and discipline to follow the guidelines. It is vital to note that Japanese society values collectivity and harmony, and any form of deviant behaviour is therefore viewed negatively. Dr. Yoshiaki Sawada, a fertility specialist critically expressed that he rarely recommends artificial methods to his patients as the children born through such methods would become 'semi-orphans' who might be ostracized by the society (Jordan & Sullivan, 1998). As medical professionals in Japan eschew violating JSOG's guideline, infertile couples might be faced with limited treatment options.

Such interplay of institutional pillars has rendered it arduous for infertile couples to pursue parenthood. Infertile couples, in particular Japanese women, therefore, face the dilemma as they are expected to abide by the traditional Japanese values and norms, to have children, yet the regulative institutional factors have erected barriers to access assisted reproductive technologies. Specifically, Japanese women are under greater pressure as compared to their male counterparts, as infertility is often ascribed as a female's problem (Matsubayashi et al., 2001). As can be seen from the wide acceptance of AID since 1948, Dr. Netsu's expulsion from JSOG for using the egg of the woman's sister exemplified the society's firm stance against the third-party egg use reveals the dualistic nature of JSOG's guideline. While the use of donor egg and embryo are prohibited, donor sperm is allowed under JSOG's rule (Morioka & Motegi, 2005). This indicates a discrimination against women as men are allowed greater access to ART (King, 2009). The discriminatory nature of access to ART procedures in Japan is also exemplified in the government's imposition of age limit on ART subsidy. While the initial ART subsidy programme in 2004 by the Japanese government was intended as a financial support for infertile couples and to encourage more usage of artificial reproductive methods, the revision in 2016 imposed an age limit on female patients and reduced the number of deductible IVF cycles to a mere 6 for women under 40 years old and 3 for women aged between 40 and 42. The revision of the subsidy seemingly gives Japanese women greater pressure to race against their age in order to receive the governmental support. In addition, it also renders more Japanese women ineligible for the programme, and might give rise to outbound reproductive tourism in search of cheaper alternatives.

Furthermore, as King (2009) avers, the traditional burden to carry the family name is often shouldered by women. Childless women in Japan often describe feeling 'deeply tormented' and 'worthless' due to their circumstance (p. 198). These illustrations also highlight that the current Japanese medical environment provides infertile women with little or no aid to achieve motherhood (King, 2009, p. 201). As the number of infertile women increases, the pressure from the Japanese government to increase birth rate as well as the pressure from society to produce biological children have led them to an impasse – to give up parenthood or to travel abroad to access care that befits their conditions. Pennings (2004) views reproductive tourism as a 'safety valve' that allow infertile couples to reduce the moral conflict they face (between their national value and their desire to conceive), thereby contributing to the 'peaceful coexistence' of different religious and moral perspectives (p. 127). As illustrated in this paper, reproductive tourists do not typically travel for a single reason, but rather, it is a complex interaction of legal, cultural and ethical factors that propelled their quest for parenthood in a foreign country (Inhorn et al., 2012).

Hibino et al. (2013) congruently opined that the practice of reproductive medicine in Japan such as the lack of legal framework and inability to use donated gametes or embryos for IVF has 'forced' (p. 478) infertile Japanese patients to travel overseas to obtain fertility care. Since the late 2000s, many agencies have recognized the potential growth in this form of tourism and established their respective consultation firms and networks to facilitate outbound reproductive tourism. Whereas the United States used to be the primary destination choice amongst Japanese couples, of late, alternative destinations in nearby Asian countries such as India and Thailand have emerged to provide greater treatment options. Perhaps one of the biggest appeals of visiting fellow Asian countries would be the shorter travel distance and lower cost (Parasakou & George, 2017; Shimazono & Hibino, 2013). Japanese agents would source for fertility clinics in the host destination and market them as a cheaper option with a comparable standard of treatment to the United States (Hibino, 2016).

Conclusion, implications and future research

In essence, the emerging phenomenon of outbound reproductive tourism from Japan can be seen as an outcome of the interplay of institutional forces. As illustrated in this paper, Japan's own standards and perception of reproduction are influenced by its cultural ethos, which espouse natural conception and blood ties. In terms of theoretical implication, this conceptual paper extends the burgeoning academic interest in reproductive tourism and the broader phenomenon of medical tourism by drawing on the premise of institutional theory. As a prominent theory in the sociology discipline, the analysis of travel motivations through the institutional lens provides a deeper understanding of the Japanese society in general, and how societal forces influence human motivation. It is vital to note that the existing body of literature tends to examine travel motivation of medical tourists on the individual level. Matters pertinent to human reproductions are embodied in the society and cultural dimensions of a country. Specifically, for a collectivist society such as Japan where childbearing and fertility have a symbolic value to the individual and their families, employing institutional theory extends current understanding of behaviours in reproductive tourism.

As demonstrated in the paper, individuals' agencies are embedded in the wider social framework, thus, it is imperative to understand how institutional forces in society affect behaviour – it reveals that the interplay of regulative forces (JSOG's guideline), normative forces (the emphasis on natural conception and family genetic ties), and cultural-cognitive forces (what it means to be an adult, what it means to be a Japanese woman and mother) have propelled this new form of travel. King (2009) underscored that a family's fertility decisions involve several social, cultural and legal implications. Apart from overcoming the social expectation to conceive naturally and to accept one's infertile reality, couples also face difficulties to access the required treatment in Japan, such as finding a doctor who is willing to perform such procedures.

Furthermore, as the world's most aged nation, Japan is presently undergoing major demographic shifts, with high level of life expectancy and low level of fertility rate. Decades of population ageing exacerbated by the low birthrate has put Japan's long-term survival in question, specifically in terms of its ability to provide an adequate supply for labour and talent for future economic growth. For decades, policymakers have inaugurated several policies to provide social support for Japanese women. However, fertility rate remains low and the prominent figures in Japan often ascribe the country's fertility woes to Japanese women's selfish pursuit of career advancement. Indisputably, the Japanese society has achieved remarkable social developments by improving gender equality in education and employment domains, however, there appears to be a profound contradiction between the country's socio-economic reality and the traditional stereotype of women's roles and responsibilities. Issues pertinent to infertility and the use of ART continue to be a gendered topic, with conspicuous discrimination towards infertile women.

This paper contributes to the extant body of knowledge in several ways. Firstly, it addresses a gap in the current literature that tends to accentuate the empirical understanding of medical tourism, with limited reference to theory. This paper addresses the lack of theoretical underpinnings in the wider medical tourism scholarship by drawing on the premise of institutional theory. Notably, through institutional theory, the key cultural and normative factors that characterize Japan's existing ART regulations are identified. It also shows the applicability of institutional theory in examining reproductive tourism. Secondly, with its focus on Japan, this study addresses a gap in the existing literature that is predominantly on the European, Middle Eastern and North American contexts (Hudson & Culley, 2015), with minimal attention on Asia, a region where medical tourism and specifically reproductive tourism is a growing. Hopefully, this paper would encourage further academic attention on reproductive tourism in Asian countries. More importantly, the main contribution of the paper is highlighting the importance of reproductive tourism as a niche that has a great impact on societies, both in the departure and destination countries. Taking direction from scholars who propose the need for a paradigm shift towards understanding the sociological and anthropological aspects of tourism, the paper shows how identities, motivations and behaviours are embedded in societal institutions, hence the importance of including this aspect into tourism research.

This study provides critical implications for policy-makers in Japan and destination countries. As underscored previously, protracted decline in fertility rate is a national woe for Japanese policy-makers who, on the one hand, hope to increase fertility, yet on the other hand, is also wary about permitting ART procedures that contradict the Japanese culture. The growing number of Japanese couples travelling abroad for such procedures deserve greater governmental attention. Importantly, issues related to post-treatment care in Japan, the communication between fertility physicians in the destination and in Japan need to be formally institutionalized to provide better care for expectant mothers. Conspicuously, too, there appears to be a discrepancy between the regulative framework and rising demand for ART in Japan. In order to support couples, it is vital for Japanese policy-makers to revise its existing regulations and provide greater financial assistance to its citizens. Since Japanese couples prefer to visit destinations such as Thailand, India and Malaysia due to their relatively unrestrictive regulations, there is an imperative need to establish stringent guidelines and standards to protect the welfare of foreign reproductive tourists. It is also vital for these countries to review the extant phenomenon in their countries to ensure that the provision of health services to foreigners would not comprise the accessibility of care for local people.

Nevertheless, there are many questions to be addressed empirically to further the understanding of outbound reproductive tourism. As the country continues to suffer from low fertility rates and limited availability of fertility treatments, it is plausible that the demand for outbound reproductive tourism will continue to grow. Future studies could explore deeper into understanding the complex travel decision-making process of Japanese couples, such as the choice of destination and risk assessment. More importantly, the outbound reproductive tourism journey can be explored in three stages – pre-travel, during travel and post-travel to better capture the experience of such tourist. In addition, more research effort can be devoted to explore the supply side of reproductive tourism – specifically, how countries and fertility clinics attract potential reproductive tourists.

Acknowledgements

I would like to thank the reviewers of this paper for their helpful comments to help improve this paper. I am also indebted to my mentor, Professor Christina Lee, Federation University of Australia for her constructive feedback and constant encouragement. Without her, this paper would not be possible.

Disclosure statement

No potential conflict of interest was reported by the author.

References

Akizuki, Y., & Kai, I. (2008). Infertile Japanese women's Positive and negative social interactions within their social networks. *Human Reproduction, 23*(12), 2737–2743.

Atoh, M. (2008). Japan's population growth during the past 100 years. In F. Coulmas, H. Conrad, A. Schad-Seifert, & G. Vogt (Eds.), *The demographic challenge: A handbook about Japan* (pp. 5–21). Leiden: Brill Academic Publishers.

Bartolucci, R. (2008). Cross-border reproductive care: Italy, a case example. *Human Reproduction, 23*(Suppl 1), i88.

Blyth, E., & Farrand, A. (2005). Reproductive tourism – A price worth paying for reproductive Autonomy? *Critical Social Policy, 25*, 91–114.

Boling, P. (2008). Demography, culture, and policy: Understanding Japan's low fertility. *Population and Development Review, 34*(2), 307–326.

Brasor, P. (2017). Japanese tradition denies surrogacy. *The Japan Times*. Retrieved from https://www.japantimes.co.jp/news/2017/05/06/national/media-national/japanese-tradition-denies-surrogacy/#.Wts1H9Nua-5

Collins, J. A. (2002). An international survey of the health Economics of IVF and ICSI. *Human Reproduction Update, 8*(3), 265–277.

Connell, J. (2011). *Medical tourism*. Wallingford: CABI International.

Davidson, C. M. (2010). Octomom and multi-fetal pregnancies: Why federal legislation should require insurers to cover in vitro fertilization. *William & Mary Journal of Women and the Law, 17*, 135–186.

Deonandan, R. (2015). Recent trends in reproductive tourism and International surrogacy: Ethical Considerations and challenges for policy. *Risk Management and Healthcare Policy, 8*, 111–119.

Edwards, W. (1989). *Modern Japan through its weddings*. Standford: Standford University Press.

Feldman, E. (1985). Medical ethics the Japanese way. *The Hasting Center Report, 15*(5), 21–24.

Ferraretti, A. P., Pennings, G., Gianaroli, L., Natali, F., & Magli, M. C. (2010). Cross-border reproductive care: A phenomenon expressing the controversial aspects of reproductive technologies. *Reproductive BioMedicine Online, 20*(2), 261–266.

Gerrits, T. (2018). Reproductive travel to Ghana: Testimonies, transnational relationships and stratified reproduction. *Medical Anthropology, 37*(2), 131–144.

Gunning, J. (2003). Regulation of assisted reproductive technology: A case study of Japan. *Medicine & Law, 22*(4), 751–761.

Hall, C. M., & Jenkins, J. (1995). *Tourism and public policy*. London: Routledge.

Harvey, V. (2016). Maternity and work: The impossible challenge of Japanese women. *Journal of Social Thought, 1*(1), 1–10.

Hayashitani, H., & Suzui, E. (2009). Problems of the couples experiencing infertility treatment and support measure [in Japanese]. *Kawasaki Medical Welfare Journal, 19*(1), 13–23.

Hibino, Y. (2016). Gestational surrogacy in Japan. In E. S. Sills (Ed.), *Handbook of gestational surrogacy: International clinical practice and policy issues* (pp. 174–180). Cambridge: Cambridge University Press.

Hibino, Y., & Shimazono, Y. (2014). Impact of egg donation deliveries from domestic and overseas sources on maternal care: A questionnaire survey of Japanese perinatal physicians. *Environmental Health and Preventive Medicine, 19*(4), 271–278.

Hibino, Y., Shimazono, Y., Kambayashi, Y., Hitomi, Y., & Nakamura, H. (2013). Attitudes towards cross-border reproductive care among infertile Japanese patients. *Environmental Health and Preventive Medicine, 18*(6), 477–484.

Hudson, N., & Culley, L. (2015). Cross-border reproductive travel. In N. Lunt, D. Horsfall, & J. Hanefeld (Eds.), *Handbook on medical tourism and patient mobility* (pp. 441–450). Northampton, MA: Edward Elgar Publishing.

Hudson, N., Culley, L., Blyth, E., Norton, W., Rapport, F., & Pacey, A. (2011). Cross-Border reproductive care: A review of the literature. *Reproductive BioMedicine Online, 22*, 673–685.

Ikemoto, L. C. (2009). Reproductive tourism: Equality concerns in the global market for fertility services. *Law and Inequality, 27*, 277–309.

Ikemoto, L. C. (2018). Reproductive tourism: Equity concerns in the global market for fertility services. In O. K. Obasogie & M. Darnovsky (Eds.), *Beyond bioethics: Towards a new biopolitics* (pp. 339–349). Oakland, CA: University of California Press.

Inhorn, M. C., Shrivastav, P., & Patrizio, P. (2012). Assisted reproductive technologies and fertility "tourism": Examples from global Dubai and the Ivy League. *Medical Anthropology: Cross-Cultural Studies in Health and Illness, 31*(3), 249–265.

Irizawa, H., & Gotoh, A. (2016). Discussion of current ethical issues regarding the use of assisted reproductive treatment in Japan. *Personalized Medicine Universe, 5*, 21–26.

Ito, M. (2015). The true cost of fertility treatment in Japan (2015, June 20). *The Japan Times*. Retrieved from http://www.japantimes.co.jp/life/2015/06/20/lifestyle/true-cost-fertility-treatment-japan/#.WTT-UZKGO70

The Japan Times. (2017). IVF accounts for 5% of babies born in Japan in 2015: Survey. *The Japan Times*. Retrieved from https://www.japantimes.co.jp/news/2017/09/12/national/social-issues/1-20-infants-born-vitro-fertilization-japan-survey/#.Wts1CdNua-5

Jordan, M., & Sullivan, K. (1998). Japan takes on a dim view of fertility treatments. July 5. *The Washington Post*. Retrieved from: https://search-proquest-com.ezproxy.lib.monash.edu.au/docview/408389236?accountid = 12528

King, R. B. (2009). Redefining motherhood: Discrimination in legal parenthood in Japan. *Pacific Rim Law & Policy Journal, 18*(1), 189–216.

Kisu, I., Banno, K., Mihara, M., Iida, T., & Yoshimura, Y. (2011). Current status of surrogacy in Japan and uterine transplantation research. *European Journal of Obstetrics & Gynecology and Reproductive Biology, 158*, 135–140.

Kubo, H. (2009). Epidemiology of infertility and recurrent pregnancy loss in society with fewer children. *Japan Medical Association Journal, 52*(1), 23–28.

Kumagai, F. (2015). *Family issues on marriage, divorce and older adults in Japan*. Singapore: Springer Science + Business Media.

Lee, F. R. (2005). Driven by costs, fertility clients head overseas. *The New York Times*. Retrieved from https://www.nytimes.com/2005/01/25/us/driven-by-costs-fertility-clients-head-overseas.html

Maeda, E., Ishihara, O., Saito, H., Kuwahara, A., Toyokawa, S., & Kobayashi, Y. (2014). Age-specific cost and public funding of a live birth following assisted reproductive treatment in Japan. *Journal of Obstetrics and Gynecology Research, 40*(5), 1338–1344.

Martin, J. L. (2015). *Reproductive tourism in the United States: Creating families in the mother country*. London: Routledge.

Matsubayashi, H., Hosaka, T., Izumi, S., Suzuki, T., & Makino, T. (2001). Emotional distress of infertile women in Japan. *Human Reproduction, 16*(5), 966–969.

Mayeda, M. (2006). Present state of reproductive medicine in Japan – ethical issues with a focus on those seen in Court cases. *BMC Medical Ethics, 7*(3), 3–19.

Moghimehfar, F., & Nasr-Esfahani, M. H. (2011). Decisive factors in medical tourism destination choice: A case study of Isfahan, Iran and fertility treatments. *Tourism Management, 32*(6), 1431–1434.

Mori, T. (1992). National regulation of and achievements in assisted reproduction in Japan. *Journal of Assisted Reproduction and Genetics, 9*(4), 293–298.

Morioka, Y., & Motegi, T. (2005). Medical ethics: Efforts of JAMS specialty societies in Japan. *Japan Medical Association Journal, 48*(4), 209–213.

National Institute of Population and Social Security Research (NIPSSR). (2015). *The 15th Japanese national fertility survey*. Tokyo: NIPSSR.

North, D. C. (1990). *Institutions, institutional change and economic Performance*. Cambridge: Cambridge University Press.

Oishi, A. S. (2008). Child care in a low birth society. In F. Coulmas, H. Conrad, A. Schad-Seifert, & G. Vogt (Eds.), *The demographic challenge: A handbook about Japan* (pp. 277–292). Leiden: Brill Academic Publishers.

Ooki, S. (2011). Effects of maternal age and fertility treatment on the increase in multiple births in Japan: Vital statistics, 1974-2009. *Journal of Epidemiology, 21*(6), 507–511.

Parasakou, A., & George, B. P. (2017). The market for reproductive tourism: An analysis with special reference to Greece. *Global Health Research and Policy, 2*(16), 1–12.

Peng, M. W., Sun, S. L., Pinkham, B., & Chen, H. (2009). The institution-based view as a third leg for a strategy tripod. *The Academy of Management Perspectives, 23*(3), 63–81.

Pennings, G. (2002). Reproductive tourism as moral pluralism in motion. *Journal of Medical Ethics, 28*, 337–341.

Pennings, G. (2004). Legal harmonization and reproductive tourism in Europe. *Human Reproduction, 20*, 3571–3572.

Pennings, G. (2009). The green grass on the other side: Looking at cross-border reproductive care. *Facts, Views and Vision in Obstetrics and Gynaecology, 1*(1), 1–6.

Powell, W., & Colyvas, J. (2008). Microfoundations on institutional theory. In R. Greenwood, C. Oliver, K. Sahlin, & R. Sunndaby (Eds.), *The Sage handbook of organizational institutionalism* (pp. 276–298). London: Sage.

Ramsey, J. R., Leonel, J. N., Gomes, G. Z., & Monteiro, P. R. R. (2011). Cultural intelligence's influence on international business traveler's stress. *Cross-Cultural Management: An International Journal, 18*(1), 21–37.

Raymo, J. M., & Iwasawa, M. (2008). Changing family life cycle and partnership transition – gender roles and marriage patterns. In F. Coulmas, H. Conrad, A. Schad-Seifert, & G. Vogt (Eds.), *The demographic challenge: A handbook about Japan* (pp. 255–276). Leiden: Brill Academic Publishers.

Retherford, R. D., Ogawa, N., & Sakamoto, S. (1996). Values and fertility change in Japan. *Population Studies, 50*(1), 5–25.

Saito, H., Jwa, S. C., Kuwahara, A., Saito, K., Ishikawa, T., Ishihara, O., … Irahara, M. (2017). Assisted reproductive technology in Japan: A summary report for 2015 by the ethics committee of the Japan society of obstetrics and gynecology. *Reproductive Medicine & Biology, 17*, 20–28.

Saito, Y., & Matsuo, H. (2009). Survey of Japanese infertile couples' attitudes toward surrogacy. *Journal of Psychosomatic Obstetrics & Gynecology, 30*(3), 156–161.

Scott, W. R. (1995). *Institutions and organizations* (Vol. 2). Thousand Oaks: Sage Publications.

Scott, W. R. (2003). Institutional carriers: Reviewing modes of transporting ideas over time and space and considering their consequences. *Industrial and Corporate Change, 12*(4), 879–894.

Scott, W. R. (2008). Approaching adulthood: The maturing of institutional theory. *Theory and Society, 37*(5), 427–442.

Shimazono, Y., & Hibino, Y. (2013). Japanese infertility patients' attitudes towards directed and non-directed oocyte donation: Analysis of a questionnaire survey and implications for public policy. *Asian Bioethics Review, 5*(4), 331–343.

Shinagawa, S. N. (1996). A short history of reproductive medical problems in Japan. *Eubios Journal of Asian and International Bioethics, 6*, 158–159.

Shirahase, S. (2000). Women's increased higher education and the declining fertility rate in Japan. *Review of Population and Social Policy, 9*, 47–63.

Shirai, C. (2010). Reproductive technologies and parent-child relationships: Japan's past and present examined through the lens of donor insemination. *International Journal of Japanese Sociology, 19*(1), 18–34.

Smith-Cavros, E. (2010). Fertility and inequality across borders: Assisted reproductive technology and globalization. *Social Compass, 4*, 466–475.

Sotoya, T. (2012). Ethical concerns of artificial insemination by donor in Japan. *Asian Bioethics Review, 4*(2), 135–142.

Spar, D. (2005). *The baby business: How money, science, and politics drive the commerce of conception*. Boston: Harvard Business Press.

Speier, A. (2016). *Fertility holidays: IVF tourism and the reproduction of whiteness*. New York: New York University Press.

Storrow, R. F. (2011). Assisted reproduction on treacherous terrain: The legal hazards of cross-border reproductive travel. *Reproductive BioMedicine Online, 23*, 538–545.

Suzuki, K., Hoshi, K., Minai, J., Yanaihara, T., Takeda, Y., & Yamagata, Z. (2006). Analysis of national representative opinion surveys concerning gestational surrogacy in Japan. *European Journal of Obstetrics & Gynecology and Reproductive Biology, 126*(1), 39–47.

Takeshita, N., Hanaoka, K., Shibui, Y., Jinnai, H., Abe, Y., & Kubo, H. (2003). Regulating assisted reproductive technologies in Japan. *Journal of Assisted Reproduction and Genetics, 20*(7), 260–264.

Tokuhiro, Y. (2010). *Marriage in contemporary Japan*. New York: Routledge.

Tsuge, A. (2005). How society responds to desires of childless couples: Japan's position on donor conception. *Meijigakuin University Journal, 35*, 21–34.

Tsuya, N. (2005). Shoshika to Jyosei-Gendaa Seisaku [Below-replacement fertility and gender policy in Japan]. In O. Hiroshi, & M. Atoh (Eds.), *Shoshika no Seisakugaku [policy studies on Below-Replacement fertility]* (pp. 157–187). Tokyo: Hara Shobou.

Utsunomiya, T. (2015). Reproductive medicine and medical tourism. In M. Cooper, K. Vafadari, & M. Hieda (Eds.), *Current issues and emerging trends in medical tourism* (pp. 123–138). Hershey, PA: IGI Global.

Viladrich, A., & Baron-Faust, R. (2014). Medical tourism in Tango paradise: The internet branding of cosmetic surgery in Argentina. *Annals of Tourism Research, 45*, 116–131.

Whittaker, A. (2010). Challenges of medical travel to global regulation: A case study of reproductive travel in Asia. *Global Social Policy, 10*(3), 396–415.

Whittaker, A., & Speier, A. (2010). "Cycling overseas": Care, commodification, and stratification in cross-border reproductive travel. *Medical Anthropology, 29*, 363–383.

Wu, A. K., Odisho, A. Y., Washington, S. L., Katz, P. P., & Smith, J. F. (2014). Out-of-pocket fertility patient expense: Data from a multi-center prospective infertility cohort. *Journal of Urology, 191*(2), 427–432.

Yamagata, Z., Hoshi, K., Hirata, S., & Takeda, Y. (2003). *Seishoku-hojo-iryo-gijutsu ni tsuiteno ishiki-chousa* (Report on Attitudes of General People toward ART) (in Japanese). Retrieved from http://www.mhlw.go.jp/shingi/2003/02/s0206-2 g.html

Yomiuri Shimbun. (2017). Sharp increase in children born through fertility treatments using donor eggs in Taiwan amongst Japanese couples. *Yomiuri Online*. Retrieved from https://yomidr.yomiuri.co.jp/article/20170508-OYTET50015/

Zanini, G. (2011). Abandoned by the state, betrayed by the church: Italian experiences of cross-border reproductive care. *Reproductive BioMedicine Online, 23*(5), 565–572.

Tourist behaviour through consumption in Melaka World Heritage Site

Nurul Diyana Md Khairi, Hairul Nizam Ismail and Syed Muhammad Rafy Syed Jaafar

ABSTRACT

The burgeoning popularity of World Heritage Sites (WHSs) has negatively impacted the sustainability of the historic environment of the sites due to the increasing number of tourist arrivals. Too many pressures from human activity, especially tourists could potentially damage the physical condition of the site and compromise its Outstanding Universal Value. Understanding tourist behaviour is one of the important aspects in managing a WHS through their consumption at the destination. The study aims to measure international tourist behaviour by integrating their non-spatial and spatial characteristics. A total of 384 Free Independent Tourists participated in the survey in Melaka WHS. Questionnaires, trip diaries, and tracking devices were the main instruments for capturing data. The evaluation of tourist behaviour shows that international tourists consume more spaces and time at primary elements such as historical attractions. However, they consume more money at secondary elements such as food and shopping. The finding is important for responsible authorities as it provides a different point of view in managing tourist behaviour. The results can be used to establish effective tools to develop a better visitor management plan for Melaka WHS.

1. Introduction

In reality, most locations, regardless of natural or cultural heritage, faced an increasing number of tourists after being designated as a World Heritage Sites (WHS) (Buckley, 2004; Shieldhouse, 2011). Like other WHSs, Melaka WHS is now faced with an increasing number of tourists following its WHS designation in 2008. Up until 2017, Melaka WHS still continues to receive a huge number of tourists each year. However, in many cases, the increase in tourists as a result of the WHS designation has also negatively affected the historic environmental sustainability of the designated sites (Kim, 2016). According to UNESCO, tourism is the second-most reported threat to WHS, next to pressure from development (UNESCO, 2016). Therefore, it is important to effectively manage tourists in order to sustain the WHS in the long term. Ertan and Egercioglu (2016) mentioned that most heritage cities are now experiencing uncontrolled development due to rapid economic growth, which result in the decay of historic cores. This phenomenon is hugely critical because a city's historic centre is the heart of its urban identity which once abandoned; the identity of the city diminishes as well. As mentioned by Ismail, Masron, and Ahmad (2014), there are always issues and challenges when managing such destinations. These include issues arising from the increase in tourists at the surrounding areas. As such, the physical capacity of the site could become strained from having to handle a larger number of tourists. These are among the current challenges faced by Melaka WHS in positioning itself as a major heritage destination in Malaysia.

The aim of the study is to adapt tourist behaviour so as to better plan and manage Melaka WHS as an urban heritage destination. This is done via the integration of tourist non-spatial (socio-demographic and travel behavioural pattern) and spatial characteristics (space, time, and money). These characteristics are the key influential factors in measuring tourist consumption. Tourist consumption will in turn help in understanding tourist behaviour in a more comprehensive manner. The incorporation of tourist behaviour via the integration of their non-spatial and spatial characteristics can help ensure proper space management and urban planning of a destination. Nevertheless, tourist spatial behaviour within heritage cities is not relatively well understood because only a few studies have addressed this phenomenon (Edwards & Griffin, 2013). In light of the above, the study argues that spatial aspects of tourists are an important criterion that needs to be considered, so as to understand tourist behaviour in a more comprehensive manner. The study also incorporated the value of money spent by tourists. This is framed in the context of space and time, which has not been previously studied. The tourism industry is now inundated with tourists of diverse profiles that choose to spend their time in different ways when visiting urban heritage destinations.

Kellner and Egger (2016) pointed out the importance of studying specific aspects of tourists, so as to understand their spatial-temporal behaviour. According to them, tourist movement within a space, the duration of time they spend at any given location, and the services they utilize should be the main focus of such a study. The results is valuable to all stakeholders in managing Melaka as a WHS. Based on Section 5.2.8 of the Conservation Management Plan (2011), a major threat to WHS is the pressures generated by the onslaught of mass tourism. This is evident in Melaka WHS where the key strength of the site is its tourism attraction and potential where if not managed properly, mass tourism could bring with it undue pressures and challenges that may undermine the long-term sustainability of the WHS. As seen today, Melaka is now facing tremendous growth in tourism activities since its establishment as a WHS in 2008. Nine years onward, the number of tourists is still on the rise. Teo, Khan, and Rahim (2014) also shared the same observation in that tourist interest to travel to Melaka has recently increased and is expected to continue to rise.

2. Literature review

2.1. Understanding tourist behaviour through their consumption

The overall analysis of tourism has tremendously contributed to the study of tourist behaviour (Pearce, 2006). Shaw, Agarwal, and Bull (2000) added that the study on tourism consumption and tourist behaviour is important. However, these aspects have often been neglected when it comes to tourism geography. It can be said that tourist behavioural patterns as well as demand have been studied widely. Ryan (2002) mentions that understanding the reason behind people visiting a destination, the link between various motivations, and the reasons why people are attracted to cities can help researchers understand tourist behaviour, particularly in an urban setting. Lew and McKercher (2006) also suggest that the way tourists behave and why they do certain things are mainly due to their own preference and priorities. Pearce (2005) also concludes that the study on tourist behaviour is important for the parties that make decisions concerning tourists, particularly at a tourism destination. Vu, Li, Law, and Ye (2015) mention that understanding tourist behaviour is important for tourism managers who engage in strategic planning and decision-making. The knowledge of which will help create a sustainable tourism industry. However, to gain full understanding of tourist behaviour, significant challenges needs to be addressed.

Understanding of tourist behaviour in terms of their non-spatial and spatial characteristics is relevant for gaining a better understanding of tourists that visit a heritage destination. The authorities and tourism players involved are highly dependent on sustainable development planning. The aim of which is to deliver a high-quality tourist experience. Tourist experience can be interpreted chronologically from their movement through space and time. Tourists move from one place to the other during a particular time interval (Tussyadiah & Fesenmaier, 2007; Xia et al., 2010). Tourist experience has an inevitable spatial-temporal dimension (Caldeira and Kastenholz, 2015). Therefore, urban destinations must accurately understand, facilitate and, to a certain extent, manage tourists' time-space activity. In doing so, a positive experience can be delivered (Edwards & Griffin, 2013). Edwards, Dickson, Griffin, and Hayllar (2010) conclude that an accurate understanding of tourist behaviour is crucial due to the perishable nature of the tourism industry.

One way to recognize tourist behaviour at a destination is through the way they consume the destination. This is due to the fact that capturing and measuring tourist behaviour is difficult and elusive. Tourists tend to change their behaviour over time (Mahika, 2011). However, tourist behaviour is constantly changing while their consumption is definite. Therefore, it is indeed easier to explore the patterns of tourism consumption than the behaviour that has produced them (Shaw et al., 2000). Understanding the city's function is seen as one of the way in describing tourist consumption in a destination. In the context of urban heritage tourism, the well-established systems approach, pioneered by Jansen-Verbeke (1986) is an ideal indicator. This approach defines the way tourists consume a destination where the inner-city environment is viewed as a 'leisure product'. The inter-relationship between the elements of inner-city tourism and its significance as a leisure product is illustrated in the model of Figure 1. The inner city comprises three main elements, which are primary, secondary, and additional elements, as classified by Jensen-Verbeke (1986). The 'primary' elements include a variety of facilities, which divide the inner city into an 'activity place' and a 'leisure' setting'. These facilities refer to the attractions and products that are available in urban heritage settings. On the other hand, facilities and services that are provided to support tourists in experiencing primary elements are called secondary elements. The facilities that make up this element include hotels, catering outlets, and shopping facilities.

Other elements of an inner city are additional elements. These consist of tourism infrastructure that provides ample experience for tourists when they visit the destination. Accessibility to and around the inner city (for example: signposts), accessibility and ease of parking, transport provision are some examples of this element. Tourist-specific services such as Tourist Information Centres (TICs), guides, maps, and information about things to see and do in the area (for example:

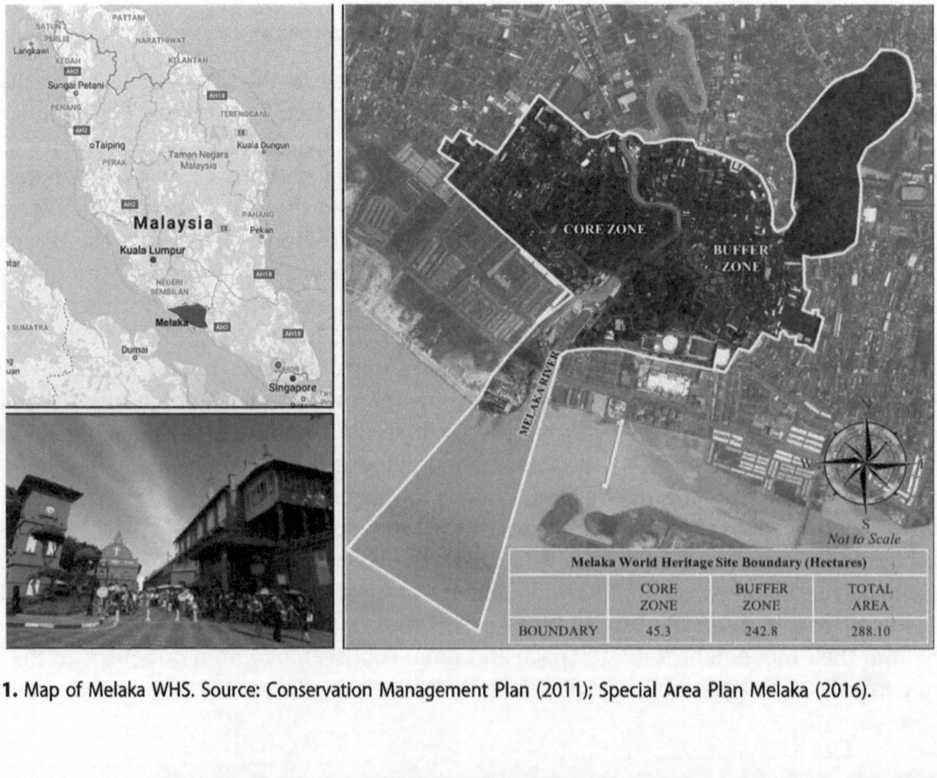

Figure 1. Map of Melaka WHS. Source: Conservation Management Plan (2011); Special Area Plan Melaka (2016).

promotional leaflets) are the other elements. Jansen-Verbeke (1986) claims that the elements of a leisure product are important, as they serve as 'pull factors' for tourist needs. These inner-city elements are present in most heritage destinations, especially in urban cities. Therefore, these elements are suitable indicators for measuring the consumption of tourists at a destination in terms of space, time, and money. The integration of non-spatial and spatial characteristics of tourists when visiting an urban heritage destination is important as it can explain the tourist consumption. The understanding of tourist consumption also helps in strengthening the knowledge on tourist behaviour which is crucial for tourism players and heritage managers. The determination of the actual consumption is basically based on the understanding of the non-spatial and spatial characteristics of the tourists.

2.2. Non-Spatial characteristics of tourist

Non-spatial characteristics of tourists can be categorized into two main groups, which are socio-demographic background and travel behavioural characteristics as indicated in Table 1. Dejbakhsh et al. (2011) stresses on the importance of determining tourist characteristics. Although tourists share common motivations when travelling, their actual behaviour at a destination mainly depends on these characteristics.

2.2.1. Socio-Demographic

Socio-demographic is associated with specific characteristics of tourists comprises country of origin, gender, and age (Debbage, 1991). Country of origin has been widely used in most studies related to tourist behaviour. There are also a few researches that have examined the spatial behaviour of tourists at a destination. These studies have also determined that this factor could also be influenced by cultural background differences (Pizam and Jeong, 1996). Previous research such as the study by Chu and Choi (2000) also indicate that cultural background, which is generated based on tourist country of origin, might have a significant effect on tourist behaviour. Scraton and Watson (1998) also claim that gender is also among the factors that may contribute to the difference in the spatial behaviour of tourists, especially when it comes to leisure activities.

Raun et al. (2016) propose that tourists can be segmented according to many different ways (for example: country of origin, length of stay, purpose of visit, mode of travel, gender, age), depending on the available data sources. In the case of tracking tourists, country of origin, length of stay, number of visits (first-time and repeat visitors), and the geographical locations of visits can all be considered (Kuusik et al., 2011; Tiru et al., 2010). McKercher and Lew (2003) added that there are important factors influencing the number and composition of tourists when analysing the country of origin. One of the most important is the distant decay principle, which states that more tourists come from nearby countries. Recent studies on people with disabilities has shown that some disabled tourists prefer to engage in smaller-scale activities (McKercher et al., 2003). Subsequently, the intensity of the relationship between individual fitness level and age will eventually influence the behaviour of the tourist.

Table 1. Non-spatial characteristics of tourists.

Characteristics		Variable (s)	
Non-spatial characteristics	Socio-demographic background	Country of origin	Pizam and Jeong (1996); Chu and Choi (2000); Scraton and Watson (1998); McKercher and Lew (2003); Tiru et al. (2010); Kuusik et al. (2011); Raun et al. (2016)
		Gender	Tiru et al. (2010); Kuusik et al. (2011); Raun et al. (2016)
		Age	Tiru et al. (2010); Kuusik et al. (2011); Raun et al. (2016)
	Travel behavioural characteristics	Length of stay	Lau and McKercher (2006); Pomfret and Bramwell (2016)
		Past-visit experience	Tideswell and Faulkner (1999); Kah et al. (2016); Smallwood et al. (2012)
		Travel Companion	Lehto et al. (2017); Baker and Crompton (2000)

2.2.2. Travel behavioural characteristics

The travel behavioural relate to the characteristics of a specific trip comprises length of stay, past-visit experience, and travel companion. Length of stay has not been much discussed in the literature on tourist spatial activities. Despite this fact, this variable has been widely used in the tourism industry and in numerous statistical analyses. Additionally, Pomfret and Bramwell (2016) point out that time budget will directly influence the number and range of activities that tourists participate in during their holidays. This is mainly because the amount of time that tourists have when going for holidays has already been decided before they arrive at the destination. Due to this, any modifications will be difficult to implement. Lau and McKercher (2006) agree that time schedule and the number of nights that tourists stay are the main elements that might affect their spatial behavioural pattern.

Past-visit experience also contributes to the difference in behavioural pattern. The combination of past experience and knowledge of tourism attractions could make tourists feel more familiar with the attractions visited (Tideswell and Faulkner, 1999). Apart from that, Kah et al. (2016) added that tourists tend to visit lesser attractions during their one-day visit of the destination. This holds true if the tourists had previously visited the same destination or city. Tourists would be less interested in visiting the same attractions once they have already become familiar with the attractions within that destination. Smallwood et al. (2012) observe that many scholars have studied tourist behaviour, especially first-time and repeat tourists, and how they arrived at their choice of destinations.

Travel companion also contributes to the difference in tourist behaviour at a destination. Tourists who travel individually will have a different range of spatial activities compared to the ones who travel in an organized group. Individual tourists are free to go with their own preferences. They can also choose the tourism attractions they want to visit within a destination. Lehto et al. (2017) suggest that tourists who travel with groups or with children are greatly impacted in terms of time spent and the number of activities they could undertake within a destination. Understanding the tourist satisfaction and intention to come back again would also help in understanding how tourists behave in and consume a destination. Baker and Crompton (2000) agreed with this statement. They postulate that behaviour predicts intention and corresponds to actual behaviour obtained from correct predictions of intentions. Reisinger and Turner (2003) further state that when experiences are met or are above expectations, tourists would feel enjoyment because they are satisfied with the destination. Additionally, Chi and Qu (2008) justify that tourist satisfaction is also related to loyalty (intention to come again). This enables the prediction of consumer behaviour, which is the key for business continuity.

2.3. Spatial characteristics of tourist

The spatial behaviour of tourists is of tremendous economic importance to a tourist destination (Modsching et al., 2006). The data considers how people move about or why they display certain movements while consuming activities. Tourist behaviour in a destination can differ within a space, even if their interests are identical, and driven by the same motivations (Walmsley, 2004). People perceive, think, and behave differently at various spatial and temporal scales. On the other hand, Wahab and Cooper (2001) argue that tourists need to competently and promptly explore a destination if they want to make their holidays more enjoyable and satisfying. Therefore, it can be said that tourists at a destination act differently in terms of their consumption. This is influenced by how the infrastructure and utilities are organized and provided.

There is a limited amount of research on the spatial characteristics of tourists. This is due to the over-dependency of ideas generated in the sociological and psychological aspects of tourism. Recently, many views on tourism demand have started to find their way in geographical literature. Lew and McKercher (2006) state that even though tourists might share the same motivations within a destination, the actual behaviour of tourists could still vary. This shows that it is not easy to analyse tourist activity, as tourists are heterogeneous and might come from different backgrounds and behave differently. Holden (2000) points out that it is important to determine the factors that may influence tourist behaviour. These factors include their demographic background, culture and lifestyle, level of education, as well as their attitudes towards the destination. He also adds that tourist experience could certainly vary when on holiday, as each individual has his or her own preferences and experience he or she wants to gain. Understanding the spatial behaviour of tourists is generally discussed in terms of space and time and serve as an important criterion that influences tourist behaviour. Therefore, it is important to acknowledge important factors relating to the spatial characteristics of tourists. These could be used to understand tourist behaviour.

2.3.1. Tourist space and time

Space can be defined as a reflection of human action and interaction within an area as well as the physical condition of human activity. Isaac (2008) stresses that the connections between different tourist elements, which the tourists establish, are a good indicator for understanding tourist space. Although each tourist element or attraction involves time spent, the fact that tourists might have their own time table should be considered. Tourists can create their own personal space through a combination of several attractions, facilities, and movement. This can also be analysed and explained using the time-space behaviour of tourists, which indicates the interaction between tourists and the space they use. Actual spaces are defined as areas that have a clear geographical boundary where different kinds of tourism activities could be accommodated within the area (Graham et al., 2016). Moreover, actual space is also seen as a functional tourism space, which includes attraction areas, identified as key to analysing tourist flow pattern. Shaw and Williams (2004) added that available attraction areas in urban area are not only spaces for certain users, but are open spaces that both local and tourists from the outside use.

There is an increasing interest among tourism researchers regarding the effects of time on the understanding of tourist behaviour. Participation time is an important factor in determining tourist behaviour (Lu et al., 2016). This is mainly because tourists might visit numerous attractions and take part in several activities, but many visits are done quickly. Time is one of the constraints that tourists face when travelling. Dickinson and Peeters (2014) argue that time is also important is shaping tourism consumption. This is mainly because the time for vacation is usually fixed and there is only a small chance of extending holidays. Nevertheless, McKercher and Zoltan (2014) argue that how tourists actually spend their time at a destination is highly dependent on how they value their time budget either as limited or extensive. This determines whether or not their time can be spent in many places. Pinjari and Bhat, (2011) assert that the location, time, types of activities, and duration of tourism activities pursued by an individual tourist can best be described based on activity-based approaches.

Some factors may also alter tourist behaviour when on site. These include the transportation system, changes in tourist preferences and taste, changes in activity pattern over time and location, as well as attributes provided by the destination. Buning and Gibson (2016) propose that activity duration as an important factor in activity participation. This might influence the travel behaviour of individual tourists. Recognizing the places that tourists visit in a destination in terms of places and times is crucial, especially for most sectors in the tourism industry. Modsching et al. (2006) also acknowledge that only little information on tourist actual behaviour can be found in regard to the activities they undertake and specific locations they visit. This is because the only information that public organizations have is the total number of nights that the tourists have stayed at the destination.

2.3.2. Tourist movement

One of the ways to understand tourist behaviour, particularly in urban tourism settings, is through the way they move within a destination. This is an important indicator that needs to be understood, as it could help in managing activities at a destination. The activities include planning, tour product development, transport, and attraction planning as well as accommodation development (McKercher and Lau, 2008). It can be said that the movement of tourists within space and time is much related to tourism. Tourists might experience attractions differently based on the differences in their movement pattern within a destination (Leung et al., 2011). Lew and McKercher (2006) also assert that spatial changes of individual tourists from one activity to another could also be described as movement patterns. Luberichs and Wachowiak (2010) point out that the spatial patterns of tourists are not randomly distributed. Therefore, these patterns can be used as a foundation to predict tourist behaviour within a destination. Due to this, time geography has been used in many studies as a method for analysing tourist movement in regard to space and time. McKercher et al. (2012) observe that there is a wide range of studies in tourism that have investigated movement patterns of tourists. In this context, movement patterns refer to the changes of activity consumption made by tourists from one location to another.

The spatial scale of tourist movement patterns can be described according to two main dimensions, which are the micro and the macro level. The changes in movement from one destination (originating region) to another destination, which may involve several to hundreds of kilometres, can be described as the macro level. The movement at the micro level is influenced by destination characteristics (for example: trip origins and destinations, locations of accommodation and attractions, transportation accessibility) and tourist characteristics (for example: time budget, tourists' motivations and interests, knowledge and information of destination, and emotional attachment) (Lew and McKercher, 2006). Tourists that move within a destination will incur changes in movement from one activity to another. This reflects the micro level (Lau and McKercher, 2006; Xia and Arrowsmith, 2005). However, the study of tourist movement pattern at the micro level has largely been neglected (McKercher and Lew, 2003) with many academicians having only studied movement patterns at the macro level in the last few decades (Lew and McKercher, 2002). Therefore, the study capitalizes on this knowledge gap as an opportunity to understand tourist movement at the micro level. The number of places that tourists could visit within a destination is unrestricted. McKercher and Lau (2008) suggest that mapping tourist movement could be complicated. The places that the tourists go to may not be in sequential order and might not even follow a logical pattern.

The movement pattern of tourists at micro level involves the use of public and private transportation networks, as well as 'on foot' transport. Lew and McKercher (2006) propose two forms of tourist movement analysis which are territoriality and linearity. Tourists usually take both territorial and linear paths during their trip, but not the 'no movement type'. Basically, accommodations such as hotels, or relatives or friends' home are the starting point in which movement occurs. Lew and McKercher (2006) indicate that tourists who are clustered at tourism nodes are the ones who like to spend most of their time near the hotel. They also prefer visiting attractions located near their hotel. McKercher and Lew (2003) assert that it is not only the location that plays an important role at the destination level, but the activities that tourist undertake also influence the distance taken. This greatly depends on time, money, and the willingness of the tourists to reach the destination.

2.3.3. Tourist mode of transport

The movement and behaviour of tourists are also influenced by mode of transport used within the destination. Based on previous literature, a close link between transport and tourism can be observed. In this case, long-haul transport has been given much more attention compared to the transportation network at the destination level (Prideaux, 2000). He also stresses that knowledge of the transportation network at the destination level plays an important role at the destination level. The inter-destination transportation network can be divided into four basic modes, which are automobiles, commercial company transportation, public transport services, and walking. However, the types of tourist transportation mode when travelling is highly influenced by travel distance and trip cost. Lumsdon (2006) concludes that buses and trains are the tourists' preferred mode of transportation when travelling, especially in an urban destination. Parahoo et al. (2014) claim that adequate provision of public transport has a positive influence on the perceived attractiveness of a city as a destination for tourists. On the other hand, Schofield and Thompson (2007) prove that there is a positive relationship between public transport satisfaction and the overall satisfaction experienced in cities. Martin et al. (2016) agree that understanding of the transport mode used by tourists during a visit to WHS is important. All stakeholders involved in such activity, mainly transport and destination policy makers can benefit from this knowledge.

Walking is another mode of transport that tourists choose when travelling within a destination. Walking offers freedom, flexibility, and cost effectiveness. Rahman et al. (2011) observe that tourist movement pattern in Melaka WHS consists of a lot of walking. This mode of transport is normally one of the best ways to experience the historic, cultural, and architectural aspects of a tourist city. Lo (2007) claims that the Free Independent Tourist (FIT) uses any transportation mode that is available in the destination as long as it is safe. As for Melaka WHS, heritage trails have become a broadly accepted way for promoting area history and enhancing tourism development (Ismail, 2012). These also promote the carbon-free tourist movement around the city centre. Van Cauwenberg et al. (2012) stresses that in order to promote walking as a medium of transportation within a neighbourhood, good access between shops and attractions should be provided. The availability of transport modes within a destination is also important. It is also important to know whether transport modes could alter the behaviour of tourists while visiting the attractions within a destination.

2.3.4. Tourist money expenditure

Tourist expenditure is the basic component of total tourism demand. According to the World Tourism Organisation, tourist expenditure refers to tourist total consumption expenditure during the trip and stay at a destination (Horner and Swarbrooke, 2016). On the other hand, Bernini and Cracolici (2016) suggested that tourist expenditure can be classified into two conditions, which are consumption and expenditure. Consumption can be explained from the tourists' purchase of goods and services. This purchase is not only for the tourists themselves but also for their friends and relatives. One example is the purchasing of gifts and souvenirs for family. On the other hand, expenditure is the amount of money spent within a destination. This is not only for the tourist himself or herself but also for his or her travel groups. This includes the parents who are responsible for paying for the goods their children buy. In light of this information, this study acknowledges tourism expenditure as a combination of these earlier definitions. In specific, tourist expenditure includes expenses and consumption of food and beverage, attractions (any attractions that involve fees), shopping (including souvenirs), transportation, and total expenditures during the trip.

Tourist expenditure has been studied widely but has not yet been applied in understanding the tourist spatial behaviour. Shoval and Isaacson (2010) observe that researchers might be interested in visitor money expenditure according to the time and space that he or she was exposed to during the visit. Therefore, the study considers money spent by tourists at each attraction visited. This is partly because the differences between the behaviours of tourists when experiencing the destination are projected from the way they spend their time, space, and money during the holidays. This area has become a major point of interest in the study of tourist behaviour. The pattern of tourist travel expenditure on foods and beverages, shopping, and transportation, as well as tourism products can shed light on their preferences when at a destination. The growing influence of tourism on the economy has been accompanied by an increased interest in understanding the factors that influence tourist expenditures (Seiler et al., 2002). Mok and Iverson (2000) point out that one of the key information in planning strategic facilities and amenities is the understanding of tourist expenditure patterns and the activities they perform during visitation.

3. Melaka world heritage site

The recognition of Melaka as a UNESCO World Heritage City in July 2008 has propelled it in becoming one of the most visited cities in Malaysia. One of the main reason Melaka WHS was selected as the study area is due to its popularity as a top heritage destination in Malaysia. Melaka WHS consistently receives a large number of tourist arrivals and tourist receipts from the year 2010 until 2016 as indicated in Table 2. These tourists generate complex movements within the WHS boundaries. For these reasons, the selection of Melaka WHS for the study is justified. Apart from that, Melaka WHS is seen as

Table 2. Melaka tourism basic data (2010–2016).

Year	Tourist arrival (million)	Tourist receipt (RM million)	RM spend/day	Average occupancy (%) OCC (%)	Average occupancy (%) No. of rooms	Average length of stay (nights)
2010	10.35	6.33	287.09	64.04	11,215	2.13
2011	12.17	7.64	294.87	63.45	11,215	2.15
2012	13.71	9.11	295.22	70.47	11,215	2.25
2013	14.31	10.99	337.00	70.11	11,215	2.28
2014	15.03	12.01	397.60	69.63	11,215	2.01
2015	15.74	16.76	488.53	65.59	11,215	2.18
2016	16.28	18.30	520.04	66.12	11,215	2.16

Source: Tourism Melaka (2017).

a popular destination and a remarkable example of historical colonial towns on the Straits of Melaka. The site also demonstrates a succession of historical and cultural influences. In terms of methodology, Melaka WHS was chosen because of its size; the core zone of the Melaka Historical City is only 45.3 hectares (15.7%) and is protected by 242.8 hectares of Buffer Zone.

The Melaka Historical City is located in the State of Melaka (Negeri Melaka), 125 kilometres south of Kuala Lumpur. Melaka is located on the south-western coast of Peninsular Malaysia. It takes around two hours of travel by road from Kuala Lumpur to Melaka and approximately three hours of travel by road to Singapore. The State of Melaka is basically divided into three main districts, which are Melaka Tengah, Alor Gajah, and Jasin. Geographically, Melaka WHS is small and compact and covers a total area of 288.10 hectares. Melaka WHS is also recognized as the oldest part of the Melaka Historical City, which includes the Melaka River, St. Paul's Civic Zone, and the Historic Residential and Commercial Zone. Essentially, Melaka WHS is divided into two main zones, which are the core zone and buffer zone, as indicated in Figure 1.

4. Research method

Quantitative research design was employed for the study, relying on a tracking tool and questionnaire. An accurate and precise measurement is provided using a descriptive research approach in explaining a particular phenomenon. The study focuses on the relationship between their non-spatial and spatial characteristics and how it is used in measuring the tourist consumption. Quantitative methods are used to construct the research instruments in measuring the tourists' consumption. Two instruments were used to track tourists consisting of the trip diaries and GPS Pal Columbia application. Questionnaire forms are used to gather information on tourist non-spatial data. The development of the survey instrument involves three stages, which lead to the selection of variables, the study area, and sampling method. FITs were selected as the main interest for the study, as these tourists are not restricted to timetables and are more flexible when exploring a destination. The actual data collection is conducted at selected budget hotels (starting point), which are determined during the pilot survey. The survey process involves two stages to acquire two different types of data (non-spatial and spatial data). The data collected were analyses through several descriptive and statistical analysis methods. Data validation is also conducted using open-ended questions to interview responsible authorities. This was done to validate the information on tourist behaviour and how it can assist responsible authorities in managing Melaka as a WHS.

5. Analysis and results

5.1. Tourists' profile

The overall profiles of international tourists were determined based on their non-spatial and spatial characteristics. These results are based on a survey of 384 respondents. In this survey, 92% (354 respondents) used trip-diaries and 8% (30 respondents) used GPS Pal application. The demographic and travel behavioural profile of respondents is shown in detail in Table 3. The spatial characteristics of respondents were gathered based on trip diaries and GPS Pal application. The tracking process was done in the first stage of the survey. This includes important information, which comprises preferred tourist attractions, tourist time and money expenditure, and tourist movement patterns.

5.2. Space, time, and money consumption by international tourists

The non-spatial and spatial characteristics of respondents are used in measuring the consumption of respondents during their one-day visitation. The study clarified that these characteristics are among the key factors that influence the consumption of tourists, particularly at a destination level. In this context, tourist consumption is better explained and comprehended using the Inner-City Tourism model by Jensen-Verbeke. This model consists of primary, secondary, and additional elements. Respondent consumption based on space, time, and money is outlined in Appendix G. The values are gathered and extracted from the trip diary and GPS tracking completed by the respondents. The values are further categorized and changed to PE (n), SE (n), and AE (n) and into means or

Table 3. Profile of the respondents.

Characteristics	Frequency	Percentage (%)
Gender		
Male	138	35.9
Female	246	64.1
Age Group		
20 and below	21	5.5
21–30	161	41.9
31–40	108	28.1
41–50	65	16.9
51 and above	29	7.6
Country		
United Kingdom (UK)	88	22.9
Netherlands	64	16.7
Germany	12	3.1
Sweden	4	1.0
Denmark	4	1.0
France	21	5.5
Ireland	5	1.3
Switzerland	5	1.3
Taiwan	8	2.1
China	31	8.1
Singapore	81	21.2
Thailand	11	2.9
Iran	15	3.9
Australia	25	6.5
New Zealand	10	2.6
Past-visit Experience		
First-time Tourist (FT-tourist)	237	61.7
Repeat Tourist (RP-tourist)	147	38.3
Number of Nights		
Two Nights	79	20.6
Three Nights	86	22.4
Four Nights	180	46.9
Five Nights and above	39	10.1
Travel Companion		
Partner	161	41.9
Individual	120	31.3
Family	15	3.9
Friends	88	22.9
Mode of Transport		
Walk only	286	74.5
Walk and Trishaw	40	10.4
Walk and Bicycle	58	15.1

averages so that tourist consumption can be understood in more detail. The formula shown below is used in developing a more comprehensive result that can further explain the spatial consumption of international tourists. The means of these variables are taken into account when explaining the results. The symbol (n) is the frequency for each element (primary, secondary, and additional) that the respondents consume. The percentage Y% is then obtained when each element is divided with the total number of elements and multiplied by 100. This value is then multiplied with the previous mean in order to get the mean for the primary, secondary, and additional elements for each category.

$$\text{Frequency } (n) \rightarrow \text{percentage } (\%) \rightarrow \text{mean } (\tilde{x}),$$

$$\text{PE, SE, AE} \rightarrow \frac{\text{PE, SE, AE}}{\text{TT}} \times 100\% = \text{Y\%} \rightarrow \text{Y\%} \times \text{mean} \rightarrow \text{mean of PE, SE, AE.}$$

5.2.1. Space consumption by respondents

The results in Table 4 show that the space consumed by respondents is highly influenced by their region, length of stay, past-visit experience, and travel companion. In terms of region, respondents from Europe ($\tilde{x} = 7.66$) and Oceania ($\tilde{x} = 7.52$) visited more primary elements compared to respondents from Asia ($\tilde{x} = 6.73$). On the contrary, respondents from Asia scored the highest mean ($\tilde{x} = 1.09$) in their consumption of secondary elements compared to the other group of respondents, which are the European ($\tilde{x} = 0.78$) and Oceania ($\tilde{x} = 0.88$) tourists. This result shows that respondents from Europe and Oceania prefer visiting historical attractions rather than secondary elements during their visitation. However, Asian respondents are more interested in secondary elements like shopping complexes and squares more than the primary elements.

Length of stay also influences the spatial consumption of respondents. Respondents who stayed for two nights ($\tilde{x} = 7.73$) and three nights ($\tilde{x} = 7.71$) visited more primary elements compared to the respondents who stayed for four nights ($\tilde{x} = 6.47$) and five nights ($\tilde{x} = 6.49$). The same goes for the secondary elements where respondents that stayed for two nights ($\tilde{x} = 1.41$) and three nights ($\tilde{x} = 1.12$) obtained higher mean scores compared to those that stayed for four ($\tilde{x} = 0.64$) nights and five nights ($\tilde{x} = 0.79$). This is mainly because respondents who stay for a shorter time have limited time budgets. This pushes them to visit as many attractions including primary and secondary elements within one day. This is in contrary to those who stay for a longer period, as they have plenty of time and their schedule is more organized. Therefore, they can visit a lesser number of attractions per day, as they have many days left to visit other attractions compared to respondents who stay for a shorter period. This result indicate that length of stay is an important determinant of tourist consumption as the number of possible experiences undertaken by the tourists is depending on their length of stay. This is also agreed by McKercher and Lau

Table 4. Space consumption of respondents.

Characteristic	Variable	*N*	Mean	PE (M)	SE (M)	AE (M)
Region	Europe	203	8.72	7.66	0.78	0.28
	Asian	146	7.97	6.73	1.09	0.14
	Oceania	35	8.63	7.52	0.88	0.23
Length of stay	2 Nights	79	9.24	7.73	1.41	0.10
	3 Nights	86	8.95	7.71	1.12	0.13
	4 Nights	180	7.39	6.47	0.64	0.28
	5 Nights	39	7.43	6.49	0.79	0.15
Past-visit experience	FT Tourist	237	9.57	8.40	0.81	0.36
	RP Tourist	147	6.47	5.42	1.05	0.01
Travel companion	Partner	161	8.76	7.52	0.98	0.25
	Individual	120	8.86	7.75	0.83	0.28
	Family	15	7.80	6.73	0.80	0.27
	Friends	88	7.34	6.36	0.91	0.07

Note: PE – primary element, SE – secondary element, AE – additional element, FT –first-timer, and RP – repeater.

(2008) where he stated that short-haul tourists and long-haul tourists are fundamentally different. This shows that length of stay is an important determinant of tourists' overall expenditure and consumption of local resources because the number of possible experiences that can be undertaken by tourists depends on their length of stay (De Menezes, Moniz, & Vieira, 2008).

In terms of number of visitations, FT tourists ($\tilde{x} = 8.40$) visited more primary elements compared to RP tourists ($\tilde{x} = 5.42$). For secondary elements, the RP tourists ($\tilde{x} = 1.05$) scored higher means compared to the FT tourists ($\tilde{x} = 0.81$). The finding also shows that FT tourists ($\tilde{x} = 0.36$) scored higher means in their consumption of additional elements compared to RP tourists ($\tilde{x} = 0.01$). This result indicates that most FT tourists tend to go to the TIC to get their initial information about Melaka WHS before they start their visit. Travel companion is the other factor that influences the spatial consumption of international tourists. The finding shows that respondents that travelled individually ($\tilde{x} = 7.75$) and with a partner ($\tilde{x} = 7.52$) visited more primary elements compared to those who travelled with family ($\tilde{x} = 6.36$) and friends ($\tilde{x} = 6.36$). This is mainly because a group of travel companions whether extended family, friends, or colleagues requires more time for planning and a stronger need for information than couples or singles. However, respondents who travelled with a partner ($\tilde{x} = 0.98$) and friends ($\tilde{x} = 0.91$) consumed secondary elements more than those who travelled individually ($\tilde{x} = 0.83$) and with family ($\tilde{x} = 0.80$).

5.2.2. Time consumption by respondents

Table 5 shows that the time consumed by respondents is highly influenced by gender, length of stay, past-visit experience, and travel companion. In terms of gender, female respondents spent more time at primary elements ($\tilde{x} = 174.12$), secondary elements ($\tilde{x} = 34.28$), and additional elements ($\tilde{x} = 5.91$) compared to male respondents. Length of stay is the other factor that influences the time consumed by respondents. Respondents who stayed for three nights ($\tilde{x} = 5186.77$) and two nights ($\tilde{x} = 174.38$) showed higher means in terms of time spent at primary and secondary elements compared to those who spent four nights ($\tilde{x} = 158.29$) and five nights ($\tilde{x} = 120.35$). This result clarifies that the respondents who stayed less than three nights consumed more time in Melaka WHS. The time consumed is reflected by the space consumption made by respondents who stayed for a shorter period, which shows that respondents who have limited time budget tend to go to as many attractions as they can. The findings also indicate that respondents who stayed for four nights ($\tilde{x} = 6.55$) scored the highest mean compared to the other groups. This is due to the fact that the majority of respondents who stayed for four nights are FT tourists. This is explained in more detail in Table 6.

In terms of past-visit experience, FT tourists ($\tilde{x} = 194.06$) spent more time at the primary elements compared to RP tourists ($\tilde{x} = 136.36$). However, RP tourists ($\tilde{x} = 37.42$) spent more time at the secondary elements compared to FT tourists ($\tilde{x} = 28.97$). The findings also show that there is a huge gap in time spent for FT tourists ($\tilde{x} = 8.25$) at additional elements compared to RP tourists ($\tilde{x} = 0.20$). Linking back to the previous finding, FT tourists visited more primary and additional elements, while RP

Table 5. Time consumption of respondents (minutes).

Characteristic	Variable	N	Mean	PE (M)	SE (M)	AE (M)
Gender	Male	138	199.93	167.68	28.41	3.84
	Female	246	214.3	174.12	34.28	5.91
Length of stay	2 Nights	79	223.81	174.38	46.80	2.63
	3 Nights	86	228.59	186.77	38.45	3.37
	4 Nights	180	188.57	158.29	23.73	6.55
	5 Nights	39	146.87	120.35	23.72	2.80
Past-visit experience	FT Tourist	237	231.28	194.06	28.97	8.25
	RP Tourist	147	173.98	136.36	37.42	0.20
Travel companion	Partner	161	217.06	176.16	34.84	6.06
	Individual	120	218.05	183.01	28.63	6.42
	Family	15	193.33	158.66	27.33	7.33
	Friends	88	186.08	151.53	33.07	1.48

Note: PE – primary element, SE – secondary element, AE – additional element, FT – first-timer, and RP – repeater.

Table 6. Cross-tabulation between length of stay and past-visit experience.

		Past-visit experience		
		FT tourists	RP tourists	Total
Length of stay	2	20	59	79
	3	30	56	86
	4	153	27	180
	5	34	5	39
	Total	237	147	384

Note: FT – First-timer, and RP – Repeat.

tourists visited more secondary elements. In regard to travel companion (refer to Table 4), the finding shows that respondents who travelled individually ($\tilde{x} = 183.01$) scored the highest mean of time spent at primary elements followed by respondents who travelled with a partner ($\tilde{x} = 176.16$), family ($\tilde{x} = 158.66$), and friends ($\tilde{x} = 151.53$). However, respondents who travelled with a partner ($\tilde{x} = 34.84$) spent more time at secondary elements followed by those with friends ($\tilde{x} = 33.07$), individually ($\tilde{x} = 28.63$), and with family ($\tilde{x} = 27.33$). For additional elements, the findings show that respondents who travelled with friends ($\tilde{x} = 1.48$) spent the least amount of time followed by respondents who travelled with family ($\tilde{x} = 7.33$), individual ($\tilde{x} = 6.42$) and with a partner ($\tilde{x} = 6.06$). In sum, the findings show that travelling in groups or with children highly impacts time spent and the intensity of pursuing different activities at a destination. This had also been highlighted by Lew and McKercher (2006) where organized groups are more restricted in their choice of transportation mode, destinations visited, expressions of interest, and time budget allocations.

5.2.3. Money consumption by respondents

Table 7 shows that the gender and region of the respondent greatly influences their money consumption. In terms of gender, female respondents ($\tilde{x} = 23.67$) spent more money at primary elements compared to male respondents ($\tilde{x} = 22.22$). Female respondents ($\tilde{x} = 117.84$) also spent more money at secondary elements compared to male respondents ($\tilde{x} = 94.68$). Female respondents ($\tilde{x} = 3.70$) spent more money on additional elements compared to male respondents ($\tilde{x} = 2.32$). This is mainly because female respondents tend to use the public transport or the trishaw more than the male respondents, which involve the use of money. In regard to region, respondents from Europe ($\tilde{x} = 23.59$) scored the highest mean in terms of money spent at primary elements followed by Oceania ($\tilde{x} = 22.71$) and Asian ($\tilde{x} = 22.64$) tourists.

Respondents from Asia ($\tilde{x} = 130.73$) proved to be the highest spender on secondary elements compared to respondents from Oceania ($\tilde{x} = 116.89$) and Europe ($\tilde{x} = 93.00$). Besides that, Asian ($\tilde{x} = 3.94$) respondents also spent more money on additional elements compared to respondents from Europe ($\tilde{x} = 2.81$) and Oceania ($\tilde{x} = 2.43$). The increasing purchasing power and spending behaviour of Asian and female tourists have made this demographic an attractive target for different industries. These results shows that money plays an important role in understanding the overall consumption that space and time might not able to capture. The study of tourist behaviour is basically on why people purchase certain tourism products, and how they make their decision when visiting a destination Therefore, it is proven that the determinants, which influence the purchase of tourism products, must be studied thoroughly to better understand tourist behaviour.

Table 7. Money consumption of respondents (Ringgit Malaysia).

Characteristic	Variable	*N*	Mean	PE (M)	SE (M)	AE (M)
Gender	Male	138	119.22	22.22	94.68	2.32
	Female	246	145.21	23.67	117.84	3.70
Region	Europe	203	119.39	23.59	93.00	2.81
	Asian	146	157.31	22.64	130.73	3.94
	Oceania	35	142.03	22.71	116.89	2.43

Note: PE – primary element, SE – secondary element, AE – additional element, FT – first-timer, and RP – repeater.

5.3. Tourist behaviour in Melaka WHS

Understanding tourist behaviour is better explained through their consumption. The evaluation of respondents' consumption of space, time, and money is used to understand tourist behaviour in a more comprehensive manner. Table 8 shows the tourist behaviour according to their consumption in terms of space, time, and money. The primary elements are divided into two, which are leisure setting and place of activity. The attractions are listed according to the frequency of visitation (N). Money spent is divided into four categories consisting of admission (AD), food (FD), shopping and souvenirs (SS), and public transport (PT). Time is presented in minutes.

Figure 2 shows the overall tourist behaviour according to categories comprising primary, secondary, and additional elements. In terms of space, 86.58% of the respondents visited primary elements, whilst 10.8% of respondents visited secondary elements. However, only a small portion of respondents (2.62%) visited additional elements. In terms of time, 82.05% respondents spent their time at primary elements, whilst 15.47% spent their time at secondary elements. Additionally, only 2.48% spent their time at additional elements, which basically refers to the TIC. With regard to money, there are four categories that are taken into consideration, i.e. admission, food, shopping, and souvenirs, as well as public transport. Since Melaka WHS offers a wide range of activities and services within its attractions other than for historic purposes, money plays an important role in explaining tourist behaviour. For the primary elements, respondents spent most of their money on shopping

Table 8. Overall tourist behaviour based on their consumption.

		Money spent (RM)				
Name of attraction	*N*	AD	FD	SS	PT	Time (min)
Primary Elements						
Leisure Setting						
Melaka River/Cruise	351	1545	1299	-	90	8805
Jonker Street	305	0	6907	2253	60	9805
Dutch Square and Queen Victoria Fountain	282	11	1384	2076	20	6965
The Stadhuys	254	1147	923	1221	965	6275
A' Famosa Fort	199	1086	40	796	30	5075
Melaka Sultanate Palace	107	194	-	10	0	3145
Dutch Graveyard	92	0	-	0	0	910
Taming Sari Tower	90	1660	134	1209	0	3360
Christ Church	86	405	0	0	0	1290
Proclaimation of Independence Memorial	79	142	0	0	0	1413
Makam Hang Kasturi	68	0	0	0	0	1295
St Francis Xaviers Church	61	0	0	0	0	900
Tokong Cheng Soon Teng	58	0	0	0	0	1165
Masjid Kampung Kling	55	0	0	0	0	1105
Makam Hang Jebat	46	0	100	200	0	970
Kuil Sri Poyattha	46	0	0	0	0	925
St Paul's Church	7	12	0	0	0	80
Activity Place						
Cheng Ho Cultural Museum	100	1720	0	0	0	2075
Galeri Seni Melaka	90	170	0	0	0	1645
Architecture Museum of Malaysia	89	0	0	0	0	1605
Hard Rock Cafe	77	0	710	4275	0	1800
Maritime Museum and Naval Museum	70	201	10	45	0	1880
Baba Nyonya Heritage Museum	69	504	0	0	0	1425
Melaka Stamp Museum	65	58	0	0	0	1260
Kite Museum	30	54	0	0	0	580
Melaka Islamic Museum	30	25	0	0	0	505
Secondary Elements						
Dataran Pahlawan Melaka	263	0	6018	2980	0	8555
Mahkota Parade	48	0	1375	2680	20	2205
Hatten Square	39	0	1805	3650	0	1730
Additional Elements						
Tourist Information Centre	85	0	0	0	0	2000

Note: AD – admission, FD – food, SS – shopping and souvenir, and PT – public transport.

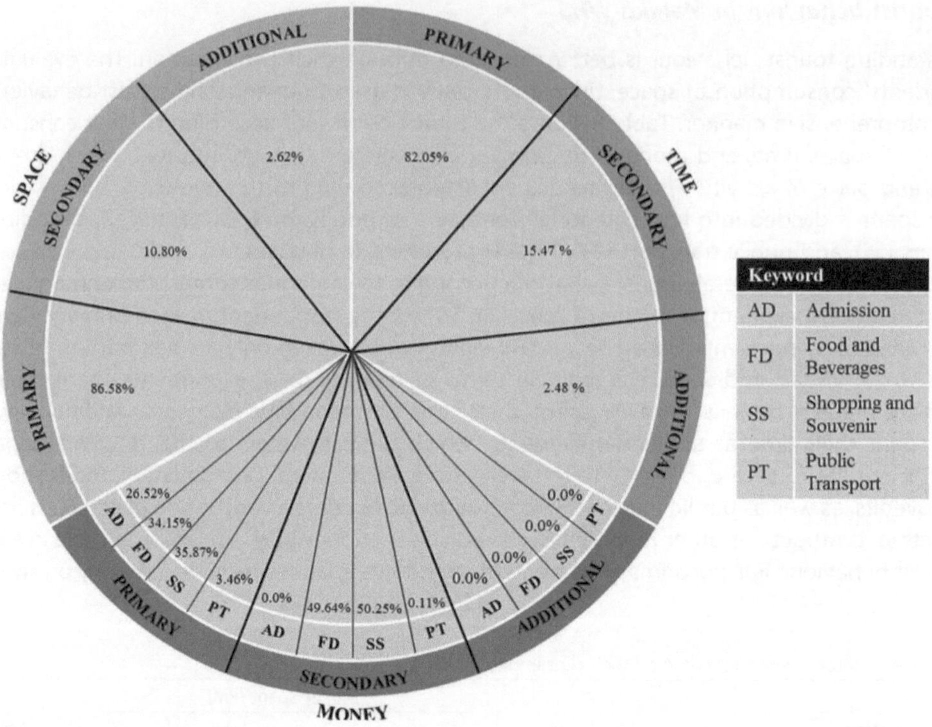

Figure 2. Space, time, and money consumption by respondents.

and souvenirs (35.8%) followed by food (34.15%), admission (26.52%), and public transport (3.46%). For secondary elements, respondents spent most of their money on shopping and souvenirs (50.25%) followed by food (49.64%) and public transport (0.11%). However, the respondents did not spend any money on additional elements, as there are no fees imposed to enter the TIC.

The overall findings on international tourist behaviour reveal that international tourists consume Melaka WHS according to the elements of the city (primary, secondary, and additional). Respondents are indeed visiting primary elements more than secondary and additional elements in terms of space and time consumption. However, when considering the money consumed by respondents based on different categories, the consumption is portrayed differently. Although respondents consumed more money on primary elements, further investigation shows that respondents consumed more money on food and beverages as well as shopping and souvenirs. However, these types of purchases are considered secondary element purchases.

6. Discussion

The aim of the study is to incorporate tourist behaviour to better plan and manage an urban heritage destination. Tourist behaviour can be determined from their consumption by integrating their non-spatial (socio-demographic and travel behavioural pattern) and spatial (space, time and money) characteristics. Apart from the construction of tourist profiles, the study also identified an existing relationship between all the variables of non-spatial and spatial characteristics of tourists. These variables were further integrated in measuring tourist consumption based on space, time, and money. Better understanding of the tourist consumption result in a more comprehensive understanding of tourist behaviour. The inter- and intra-relationship between both characteristics are used to measure tourist consumption while visiting Melaka WHS. The international tourist consumption of Melaka WHS is best described in terms of their spatial characteristics. It is determined that the consumption of tourists in terms of space, time, and money are highly influenced by their dynamic characteristics. However, dynamic characteristics are primarily associated with static characteristics. Therefore, both static and dynamic characteristics are important factors that need to be studied in order to better understand overall tourist behaviour.

The study suggests that to understand the spatial behaviour of tourists, focus must be placed not only on space and time, as suggested in previous literature, but also on money spent. Study by Henthorne (2000) on cruise expenditure behaviour in Ocho Rios, Jamaica also shows that the amount of time visitor spends at a destination has a substantial impact on the amount of money spent. Understanding money spent by tourists at each attraction or tourist spot can provide a new perspective for planners or tourism managers. This will help them understand how tourists consume cities, especially urban heritage destinations. Evaluation of tourist consumption based on their spatial and non-spatial behaviour is crucial for managing heritage sites. However, the management and conservation of the value of heritage attractions, especially a UNESCO city, is still far from promising. In line with the objective of the study, the information of tourist behaviour was further validated to the responsible authorities of Melaka WHS. Using this information, a proper visitor management should be implemented and provided by the local authorities to ensure the effective management of Melaka WHS in line with the growing number of tourist arrivals. This is crucial so as to maintain Melaka's status as a WHS and to protect its Outstanding Universal Value (OUV).

The study touches on Jansen-Verbeke's theory of the elements of an inner city. The model framework emphasizes the elements of an inner city that comprises primary, secondary, and additional elements. It can be noted that the division of elements in this model is crucial; however, this only explains tourist consumption in terms of time and space. Money should be considered as part of tourist consumption. Money consumption plays an important role. This is especially true in the case of tourists that spend more money on secondary elements even though they visited and spent more time at primary elements. Findings from the study also highlight that tourist behaviour can be easily understood through their consumption behaviour. This is mainly because any assessment of urban tourist activities will be influenced by the supply of services, attractions, and facilities at the destination. Overall, findings of the study provide implications on the theory of tourist behaviour and tourist consumption. The findings presented in the study correspondingly validate the importance of understanding both the non-spatial and spatial characteristics of tourists.

7. Implications and further research

In regard to tourist consumption, the study found that tourists consume Melaka WHS differently according to space, time, and money. This is mainly because each tourist has different preferences and purposes when visiting Melaka. The different categories of tourists also show that tourists visit places for different purposes. Some prefer walk-ins while others go for fully guided/tour packages. Still others might not have wanted to come to Melaka but the destination had already been pre-included in their itinerary. Tourists who purchase tickets and enter museums might be interested in heritage. However, there are tourists that come to Melaka but do not visit museums at all. Therefore, it can be said that it is difficult to determine the intention of tourists that come to Melaka. This has led to questions of whether the tourists are real tourists or vice versa. Even though tourist motivation was not taken into consideration in the study, the way in which tourists consume each element will eventually explain their internal behaviour. Hence, Melaka's tourism stakeholders should tailor their efforts and creativity in providing tourism services to satisfy the different needs and preferences of the different tourists in Melaka WHS. This will not only help enrich tourist experience and increase tourist satisfaction, but also encourage them to stay longer.

Apart from that, the State Government (JPBD) and local authorities (MBMB and PERZIM) should also control the development around the heritage city to conserve its heritage value. This means that they must not only focus on profits. The findings of the study echo this sentiment. It was found that repeated tourists spent more on secondary element purchases rather than primary elements. Therefore, it can be said that the aesthetic value of heritage attractions in Melaka WHS has started to fade. These attractions have changed the site from being a renowned heritage destination into just a conventional tourism destination. Tourists who visit Melaka WHS are not homogenous. This is a valuable source of information for Melaka's local authorities. The finding shows that international tourists consume Melaka WHS differently. Different tourists have different psychological and sociological needs and preferences that influence their spatial behaviour when visiting heritage destinations. To ensure that Melaka's tourism products and services fulfil the needs of different types of tourists, it is necessary to understand the different factors that influence tourist consumption. Such information can help the responsible authorities develop new tourist attractions.

The findings also provides valuable information for stakeholders and responsible agencies to prepare an effective visitor management plan. As Melaka WHS marks its ninth year on UNESCO's list, it thus becomes crucial for the site to have a proper visitor management plan so as to protect its Outstanding Universal Value (OUV). As a result, the carrying capacity of the site will be improved and better way finding, accessibility, and linkages within the heritage area can be provided. This way, the increased arrival of tourists can be managed properly. Good understanding of tourist movement and behaviour is one of the ways to develop a better visitor management system.

Disclosure statement

No potential conflict of interest was reported by the authors.

References

Baker, D. A., & Crompton, J. L. (2000). Quality, satisfaction and behavioural intentions. *Annals of Tourism Research, 27*(3), 785–804.

Bernini, C., & Cracolici, M. F. (2016). Is participation in the tourism market an opportunity for everyone? Some evidence from Italy. *Tourism Economics, 22*(1), 57–79.

Buckley, R. (2004). The effects of world heritage listing on tourism to Australian national parks. *Journal of Sustainable Tourism, 12*(1), 70–84.

Buning, R. J., & Gibson, H. J. (2016). The role of travel conditions in cycling tourism: Implications for destination and event management. *Journal of Sport & Tourism, 20*(3–4), 175–193.

Caldeira, A. M., & Kastenholz, E. (2015). Spatiotemporal Behaviour of the Urban Multi-Attraction Tourist: Does Distance Travelled From Country of Origin Make a Difference. *Tourism and Management Studies, 11*(1), 91–97.

Van Cauwenberg, J., Van Holle, V., Simons, D., Deridder, R., Clarys, P., Goubert, L., & Deforche, B. (2012). Environmental factors influencing older adults' walking for transportation: A study using walk-along interviews. *International Journal of Behavioural Nutrition and Physical Activity, 9*(1),85–96.

Chi, C. G. Q., & Qu, H. (2008). Examining the Structural Relationships of Destination Image, Tourist Satisfaction and Destination Loyalty: An Integrated Approach. *Tourism Management, 29*(4), 624–636.

Chu, R. K., & Choi, T. (2000). An importance-performance analysis of hotel selection factors in the Hong Kong hotel industry: A comparison of business and leisure travellers. *Tourism Management, 21*(4), 363–377.

Conservation Management Plan. (2011). *Conservation Management Plan and Special Area Plan 2011*. Melaka and George Town, Historic Cities of the Straits of Melaka: Government of Malaysia.

Debbage, K. G. (1991). Spatial behaviour in a Bahamian resort. *Annals of Tourism Research, 18*(2), 251–268.

Dejbakhsh, S., Arrowsmith, C., & Jackson, M. (2011). Cultural influence on spatial behaviour. *Tourism Geographies, 13*(1), 91–111.

De Menezes, A. G., Moniz, A., & Vieira, J. C. (2008). The determinants of length of stay of tourists in the Azores. *Tourism Economics, 14*(1), 205–222.

Dickinson, J. E., & Peeters, P. (2014). Time, tourism consumption and sustainable development. *International Journal of Tourism Research, 16*(1), 11–21.

Edwards, D., Dickson, T., Griffin, T., & Hayllar, B. (2010). Tracking the urban visitor: Methods for examining tourists' spatial behaviour and visual representations. In G. Richards & W. Munsters (Eds.), *Cultural tourism research methods* (pp. 104–114). Wallingford: Cabi.

Edwards, D., & Griffin, T. (2013). Understanding tourists' spatial behaviour: GPS tracking as an aid to sustainable destination management. *Journal of Sustainable Tourism, 21*(4), 580–595.

Ertan, T., & Egercioglu, Y. (2016). The impact of UNESCO world heritage list on historic urban city centres and its place in urban regeneration: The case of melaka, Malaysia and tire, Turkey. *Procedia-Social and Behavioural Sciences, 216*, 591–602.

Graham, B., Ashworth, G., & Tunbridge, J. (2016). *A geography of heritage*. London: Routledge.

Henthorne, T. L. (2000). An analysis of expenditures by cruise ship passengers in Jamaica. *Journal of Travel Research, 38*(3), 246–250.

Holden, A. (2000). Winter tourism and the environment in conflict: The case of cairngorm, Scotland. *The International Journal of Tourism Research, 2*(4), 247.

Horner, S., & Swarbrooke, J. (2016). *Consumer behaviour in tourism* (3rd ed.). London: Routledge.

Isaac, R. K. (2008). Master of Arts in Pilgrimage and Tourism. *Master of Arts in Pilgrimage and Tourism. Tourism and Hospitality Planning & Development, 5*, 73–76.

Ismail, H. N. (2012). *Heritage Awareness among Tour Operators in Heritage Sites: A Case Study of Rickshaw Pullers in Melaka*. 3rd International Conference on Business and Economic Research, Bandung, Indonesia.

Ismail, N., Masron, T., & Ahmad, A. (2014, December 9–11). Cultural heritage tourism in Malaysia: Issues and challenges. *Proceedings of the 4th International Conference on Tourism Research (ICTR)* (Vol. 12, pp. 1–8). Kota Kinabalu, Malaysia.

Jansen-Verbeke, M. (1986). Inner-city tourism: Resources, tourists and promoters. *Annals of Tourism Research, 13*(1), 79–100.

Kah, J. A., Lee, C. K., & Lee, S. H. (2016). Spatial–temporal distances in travel intention–behaviour. *Annals of Tourism Research, 57*, 160–175.

Kellner, L., & Egger, R. (2016). Tracking tourist spatial-temporal behaviour in urban places, a methodological overview and GPS case study. In A. Inversini & R. Schegg (Eds.), *Information and communication technologies in tourism 2016* (pp. 481–494). New York, NY: Springer International Publishing.

Kim, S. (2016). World heritage site designation impacts on a historic village: A case study on residents' perceptions of hahoe village (Korea). *Sustainability, 8*(3), 258.

Kuusik, A., Tiru, M., Ahas, R., & Varblane, U. (2011). Innovation in destination marketing. *Baltic Journal of Management, 6*(3), 378–399.

Lau, G., & McKercher, B. (2006). Understanding tourist movement patterns in a destination: A GIS approach. *Tourism and Hospitality Research, 7*(1), 39–49.

Lehto, X. Y., Fu, X., Li, H., & Zhou, L. (2017). Vacation benefits and activities: Understanding Chinese family travelers. *Journal of Hospitality & Tourism Research, 41*(3), 301–328.

Leung, D., Buhalis, D., & Law, R. (2011). E-tourism: Critical information and communication technologies for tourism destinations. In Y. Wang & A. Pizam (Eds.), *Tourism destination marketing and management: Collaborative strategies* (pp. 205–224). Wallingford: Cabi.

Lew, A., & McKercher, B. (2002). Trip destinations, gateways and itineraries: The example of Hong Kong. *Tourism Management, 23*(6), 609–621.

Lew, A., & McKercher, B. (2006). Modeling tourist movements: A local destination analysis. *Annals of Tourism Research, 33* (2), 403–423.

Lo, M. U. (2007). *A Comparative Study of Tourist Behaviour in Pleasure Travel: Group Package Tours and Free Independent Tours* (PhD thesis). International School of Management, Univeristy of South Australia, Australia.

Lu, J., Hung, K., Wang, L., Schuett, M. A., & Hu, L. (2016). Do perceptions of time affect outbound-travel motivations and intention? An investigation among Chinese seniors. *Tourism Management, 53*, 1–12.

Luberichs, J., & Wachowiak, H. (2010). Geographic Information System-Supported Segmentation Study of Visitors to Majorca Island. In S. C. Joseph (Ed.), *Advances in Hospitality and Leisure, Volume 6* (pp. 135–166). Warrington: Emerald Group Publishing Limited.

Lumsdon, L. M. (2006). Factors affecting the design of tourism bus services. *Annals of Tourism Research, 33*(3),748–766.

Mahika, E. C. (2011). Current trends in tourist motivation. *Cactus Tourism Journal, 2*(2), 15–24.

Martin, J. C., Marrero-Rodriguez, J. R., Moreira, P., Roman, C., & Santana, A. (2016). How Access Transport Mode to a World Heritage City Affects Visitors' Experienced Quality. *Tourism Economics, 22*(2), 207–226.

McKercher, B., & Lau, G. (2008). Movement Patterns of Tourists within a Destination. *Tourism Geographies, 10*, 355–374.

McKercher, B., & Lew, A. (2003). Distance decay and the impact of effective tourism exclusion zones on international travel flows. *Journal of Travel Research, 42*(2),159–165.

McKercher, B., Packer, T., Yau, M. K., & Lam, P. (2003). Travel Agents as Facilitators or Inhibitors of Travel: Perceptions of People with Disabilities. *Tourism Management, 24*(4), 465–474.

McKercher, B., Shoval, N., Ng, E., & Birenboim, A. (2012). First and repeat visitor behaviour: GPS tracking and GIS analysis in Hong Kong. *Tourism Geographies, 14*(1),147–161.

McKercher, B., & Zoltan, J. (2014). Tourist flows and spatial behaviour. In Lew, A. A., Hall, C. M., & Williams, A. M. (Eds.), *The Wiley Blackwell companion to tourism* (pp. 33–44). New York, NY: Wiley.

Modsching, M., Kramer, R., Gretzel, U., & Hagen, K. (2006). Capturing the beaten paths: A novel method for analysing tourists' spatial behaviour at an urban destination. In M. Hitz, M. Sigala, & J. Murphy (Eds.), *Information and communication technologies in tourism 2006* (pp. 75–86). Vienna: Springer.

Mok, C., & Iverson, T. J. (2000). Expenditure-Based segmentation: Taiwanese tourists to Guam. *Tourism Management, 21* (3),299–305.

Parahoo, S. K., Harvey, H. L., & Radi, G. Y. A. (2014). Satisfaction of tourists with public transport: An empirical investigation in Dubai. *Journal of Travel and Tourism Marketing, 31*(8),1004–1017.

Pearce, P. L. (2005). *Tourist behaviour: Themes and conceptual schemes*. New York, NY: Halsted Press.

Pearce, P. L. (2006). Australian tourism education: The quest for status. *Journal of Teaching in Travel and Tourism, 5*(3),251–267.

Pinjari, A. R., & Bhat, C. R. (2011). Activity-based travel demand analysis . *A Handbook of Transport Economics, 10*, 213–248.

Pizam, A., & Jeong, G. H. (1996). Cross-Cultural tourist behaviour: Perceptions of Korean tour-guides. *Tourism Management, 17*(4),277–286.

Pomfret, G., & Bramwell, B. (2016). The characteristics and motivational decisions of outdoor adventure tourists: A review and analysis. *Current Issues in Tourism, 19*(14),1447–1478.

Prideaux, B. (2000). The role of the transport system in destination development. *Tourism Management, 21*(1),53–63.

Rahman, J. I., Ismail, H. N., & Wai, C. L. (2011, September 19–22). Inquiry into Tourists' Movement Flow Pattern in the Melaka World Heritage Site: A Space Syntactic Analysis. Proceedings in the 11th International Congress of Asian Planning Schools Association (pp. 1475–1485). Tokyo, Japan.

Raun, J., Ahas, R., & Tiru, M. (2016). Measuring tourism destinations using mobile tracking data. *Tourism Management, 57*, 202–212.

Reisinger, Y., & Turner, L. W. (2003). *Cross-Cultural behaviour in tourism: Concepts and analysis*. London: Elsevier.

Ryan, C. (2002). Tourism and cultural proximity: Examples from New Zealand. *Annals of Tourism Research, 29*(4),952–971.

Schofield, P., & Thompson, K. (2007). Visitor motivation, satisfaction and behavioural intention: The 2005 Naadam festival, Ulaanbaatar. *International Journal of Tourism Research, 9*(5),329–344.

Scraton, S., & Watson, B. (1998). Gendered cities: Women and public leisure space in the 'postmodern city'. *Leisure Studies, 17*(2),123–137.

Seiler, V. L., Hsieh, S., Seiler, M. J., & Hsieh, C. (2002). Modeling travel expenditures for Taiwanese tourism. *Journal of Travel and Tourism Marketing, 13*(4),47–60.

Shaw, G., Agarwal, S., & Bull, P. (2000). Tourism consumption and tourist behaviour: A British perspective. *Tourism Geographies, 2*(3),264–289.

Shaw, G., & Williams, A. M. (2004). *Tourism and tourism spaces*. London: Sage.

Shieldhouse, R.G. (2011). *Ajagged path: Tourism, planning, and development in Mexican world heritage cities* (Ph.D. thesis). University of Florida, Gainesville, FL.

Shoval, N., & Isaacson, M. (2010). *Tourist mobility and advanced tracking technologies*. New York: Taylor and Francis.

Smallwood, C. B., Beckley, L. E., & Moore, S. A. (2012). An analysis of visitor movement patterns using travel networks in a large marine park, north-western Australia. *Tourism Management, 33*(3),517–528.

Special Area Plan Melaka (2016). *Special area plan Melaka historic cities of the straits of Melaka*. Melaka: JPBD.

Teo, C. B. C., Khan, N. R. M., & Rahim, F. H. A. (2014). Understanding cultural heritage visitor behaviour: The case of Melaka as world heritage city. *Procedia-Social and Behavioural Sciences, 130*, 1–10.

Tideswell, C., & Faulkner, B. (1999). Leveraging tourism benefits from the Sydney 2000 Olympics. *Pacific Tourism Review, 3* (3/4), 227–238.

Tiru, M., Kuusik, A., Lamp, M. L., & Ahas, R. (2010). LBS in marketing and tourism management: Measuring destination loyalty with mobile positioning data. *Journal of Location Based Services, 4*(2),120–140.

Tourism Melaka. (2017). *Melaka and Malaysia tourist arrival and tourist receipt (2000–2016)*. Melaka: Tourism Promotion Division.

Tussyadiah, I. P., & Fesenmaier, D. R. (2007). *Interpreting Tourist Experiences from First-Person Stories: A Foundation for Mobile Guides*. European Conference on Information Systems, St Gallen, Switzerland.

Tussyadiah, I. P., & Fesenmaier, D. R. (2007). Interpreting tourist experiences from first-person stories: A foundation for mobile guides. *ECIS*, 2259–2270.

UNESCO (2016). *UNESCO sustainable tourism toolkit*. Retrieved on 7 March 2016. Retrieved from http://whc.unesco.org/sustainabletourismtoolkit/guide

Vu, H. Q., Li, G., Law, R., & Ye, B. H. (2015). Exploring the travel behaviours of inbound tourists to Hong Kong using geo-tagged photos, *Tourism Management, 46*, 222–232.

Wahab, S., & Cooper, C. (2001). *Tourism in the Age of globalisation*. London: Routledge.

Walmsley, D. J. (2004). Behavioural approaches in tourism research. In Lew, A. A., Hall, C. M., & Williams, A. M. (Eds.), *A companion to tourism* (pp. 49–59). New York, NY: John Wiley and Sons.

Xia, J., & Arrowsmith, C. (2005, December 12–15). Managing Scale Issues in Spatio-Temporal Movement of Tourists Modelling. *Proceedings of International Congress on Modelling and Simulation* (pp. 162–170). Canberra.

Xia, J. C., Evans, F. H., Spilsbury, K., Ciesielski, V., Arrowsmith, C., & Wright, G. (2010). Market segments based on the dominant movement patterns of tourists. *Tourism Management, 31*(4),464–469.

Perceived pollution and inbound tourism for Shanghai: a panel VAR approach

Xu Xu and Markum Reed

ABSTRACT

Although there exists sizeable literature studying the impact of tourists' risk perceptions, less attention has been devoted to studying the difference between the impact of measured and perceived risks. We apply a panel vector autoregression (PVAR) to a dataset of inbound tourism in Shanghai, China to study the dynamic relationship between pollution and tourism. Particularly, we distinguish between measured pollution level and perceived pollution (measured by Google Trends search data) to test which one has the larger impact on tourists' decisions. Our results show that pollution (measured or perceived) can be a powerful deterrent to potential international tourists.

1. Introduction

Scholars have discussed the disincentive effects of the risks in tourist destination countries while neglecting the difference between perceived and measured risks. In this study, we attempt to measure these separate (yet interconnected) ideas. The particular risk we are looking at is pollution since it is often talked about and less empirically examined in the tourism literature. Pollution in China has been a growing concern for both residents in China and inbound tourists. This paper solely focuses on the latter. We attempt to disentangle the effects of measured and perceived pollution level on inbound tourist arrivals in Shanghai, China. The tourism-led growth hypothesis has been confirmed by various empirical studies, see Pablo-Romero and Molina (2013) and Brida, Cortes-Jimenez, and Pulina (2016) for a review of the literature. These studies press the importance of analysing the concerns of tourists for the industry not just at the national level and but also at the municipal level.

Fenichel, Kuminoff, Chowell, and Boni (2013) used Google Trends search data on 'swine flu' to measure people's concern over this issue. They find that Google Trends data are highly correlated with media attention but uncorrelated with the number of measured infected cases reported. One way the public responds to such risk is to skip their scheduled flights, worrying the high probabilities of catching the flu while flying. Interestingly, they find that Google Trends can help explain people forgoing their non-refundable flight tickets. Meanwhile, people show no response to measured cases reported. Results from Fenichel et al. (2013) alerted us that public attention and perception about certain risks may not accurately reflect the real scenario. It could be through over- or under-estimated or through miscommunication. Such perception and the choices made based on that perception may cause the unnecessary and additional individual, social, and economic loss.

Perceived pollution in this paper is defined following Xu and Reed (2017), as the monthly Google Trends interest in pollution. They use Google Trends search results to represent people's perception of pollution level in China and examine the causal effects of perceived pollution on tourist volume.

They find that perceived pollution does affect inbound tourism; however, they neglect to disaggregate the data. The difference between Xu and Reed (2017) and the current paper is that we look at a specific city, Shanghai and we filter search results by source country to gain a deeper understanding of how perceived pollution impacts tourism inflow. The higher public attention for Shanghai is based on search queries for 'Shanghai Pollution' not for China as a whole. Therefore, if we were to search other cities, such as Beijing, we would also see spikes but only related to Beijing's Pollution. The search results for a city are not related to one another directly. They are related indirectly. Based on the description of how Google Trends data are created, search query variation is based on how the key phrase has been searched since 2004. We also control for *measured pollution*, which is the amount of $PM_{2.5}$ reported for Shanghai.[1]

We use Shanghai as a case study to untangle the impact of perceived versus measured pollution levels on inbound tourism. With the arrival of over 5 million foreign tourists, Shanghai is the most popular tourist destinations in China according to the Yearbook of China Tourism Statistics (2015). The pollution level in Shanghai has attracted public concern in recent years, however, it is milder than the pollution in Beijing, another popular tourist destination in China. The annual mean concentrations of particulate matter ($PM_{2.5}$) for Beijing in 2014 was around 85, whereas Shanghai was around 56.[2] This is a considerable difference. However, what we propose is that pollution level does not have as much of an impact as people's perception of that pollution level. We also test whether concerns over pollution in Beijing would cause tourists to give up travelling to Shanghai. In other words, do concerns over air quality in Beijing or China affect inbound tourism to Shanghai?

We contribute to the literature in the following ways. First, we fill a gap in the literature by looking at how pollution affects international tourists' decisions based not only on measured pollution but also on their perception of the pollution. Second, we measure perceived pollution via Google Trends while previous studies have predominantly used survey data. Third, we adopt a panel vector autoregression (PVAR) analysis to control for tourists' source country fixed effects. Finally, although we study one particular type of risk in one city, a similar analysis can be applied to other types of concerns or regions of interest.

Our results show that people's perceptions about the pollution level have negative consequences for overall tourism. We also find that there are significant positive externalities associated with higher pollution in Beijing for Shanghai. This means that when people *perceive* higher pollution in Beijing, they will *change* their travel destination to Shanghai thereby increasing tourism in Shanghai. Furthermore, perceived pollution has a larger impact on inbound tourism than the measured pollution levels.

2. Literature review

Risk has long been identified as a major concern for travelling, especially for international tourism. Perceived risk acts as a key determinant of tourists travel decisions (Law, 2006; Lepp & Gibson, 2003; Maser & Weiermair, 1998; Sönmez & Graefe, 1998). Sönmez and Graefe (1998) identified and grouped risks associated with international travel into 10 categories. External events such as natural disasters, infectious diseases, and terrorist attacks have found to be detrimental to destination image and international travellers intentions (Law, 2006). Kozak, Crotts, and Law (2007) further examined two specific types of risks affecting destination choices of tourists: safety risks (e.g. health) and security risks (e.g. natural disasters).

Simpson and Siguaw (2008) suggested that the risk categories utilized in the tourism literature have two drawbacks. First, the categories are often too broad in nature. Second, such ad hoc categories are not generated from the perspective of current and potential travellers. This paper attempts to help resolve this deficiency in the literature by focusing on one specific type of risk: pollution. We also utilize online search data as a proxy for current and past perspectives of potential tourists.

Pollution has been considered as a kind of risk associated with travelling. The pollution level can potentially increase the likelihood of becoming sick while travelling abroad, which falls into the health-related risk category according to Sönmez and Graefe (1998). Using Thailand as a case study, Rittichainuwat and Chakraborty (2009) showed that 'polluted attractions' ranked top concerns for travellers. Also, Zhang, Zhong, Xu, Wang, and Dang (2015) found that haze pollution in Beijing has a considerable influence on domestic tourists' travel decisions using survey data. We further look at how pollution impacts international tourists decisions based not only on their perception of pollution but also on measured pollution level. Our findings concur with the results of Zhang et al. (2015). Moreover, we find that there is a potential substitution effect about higher polluted cities to relatively lower polluted cities.

Previous research on perceived risks has used questionnaire surveys as the main data source. For this paper, we collect Google Trends data to analyse perceived pollution levels following Xu and Reed (2017). Search engine data has gained popularity in the field of tourism study. Choi and Varian (2012) use Google Trends data to help predict tourists arrivals in Hong Kong. They find that search engine data has significant predictive power. Yang, Pan, Evans, and Lv (2015) further demonstrate the predictive ability of search engine data by comparing Google and Baidu. Gunter and Önder (2016) utilize the big data shrinkage methods to predict tourist arrivals to Vienna using Google Analytics data. Bangwayo-Skeete and Skeete (2015) and Li, Pan, Law, and Huang (2017) both construct composite search index to help forecast tourism demand. They use Google and Baidu Trends respectively. Park, Lee, and Song (2017) used Google Trends data to forecast tourists flow from Japan to South Korea. While a majority of the papers use search query includes common tourism-related keywords such as 'flight,' 'hotels' and 'trip,' our study focuses on one specific concern associated with travelling to China, 'Pollution'.

Finally, many studies focus on studying tourists' perceived travel risks; the difference between the impact of real risks and perceived risks has not been sufficiently addressed. In line with Fenichel et al. (2013), we focus on disentangling the effects of measured pollution level and perceived pollution level in Shanghai on inbound tourism. Tavitiyaman and Qu (2013) argued that one way to improve the destination image could be through decreasing the perception of risk. Our results further support this argument by demonstrating that perception matters just as much if not more than the risk itself.

3. Data

We collected monthly inbound tourism data (number of foreign tourists arrivals) for Shanghai from the *Shanghai Municipal Tourism Administration*[3] from December 2011 to October 2016. Figure 1 shows co-movements between the tourist source countries included in our dataset: Australia, Canada, Great Britain, and the United States. While examining the impact of perceived pollution on inbound tourism, we control for the measured level of pollution using $PM_{2.5}$. $PM_{2.5}$ form a haze when levels are high. These particulates make their way into people's respiratory tracts and reaching the lungs. It is often called the 'silent killer.' Exposure to air pollution was estimated by WHO to cause about 7 million deaths worldwide in 2012.[4] Recent research has emerged on pollution's debilitating impact on economies. The current research is looking at the damage done to the tourist industry of Shanghai.

Measured pollution ($PM_{2.5}$) data for Shanghai was made available through the *Mission China* air quality monitoring programme, collected by the U.S. Department of State. The monitoring programme for the main cities in China is intended to provide 'actionable health information related to air pollution' to US citizens living and travelling abroad. From Figure 2, we see that there is a significant variation in the amount of pollution in Shanghai over time; with relatively low pollution in the warmer months and higher pollution in the winter months. This is due to the power plant production to help cope with needed electrical usage in winter to produce heat in China. Furthermore, we see that there are often dips in total tourism[5] that also coincide with the winter months, but a slight peak during the Chinese New Year. However, Chinese New Year date often varies from year to year since it depends on the Lunar Calendar.

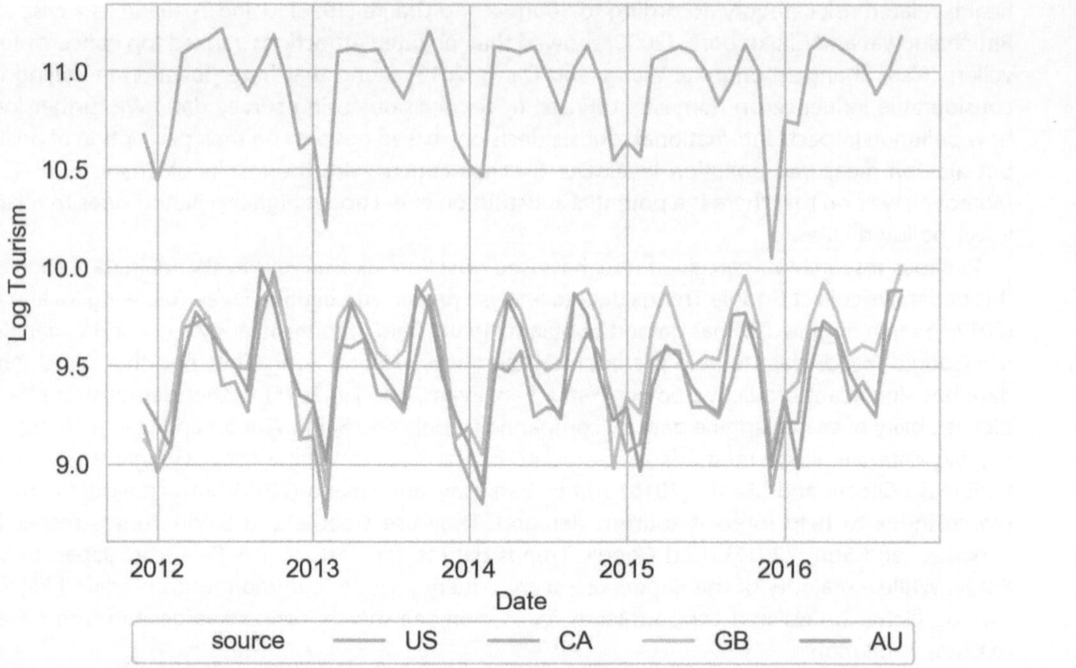

Figure 1. Log Monthly Inbound Tourism in Shanghai.

From Figure 2 we can gain an understanding through the smooth trend lines that tourism in China is on the rise, whereas pollution has been steadily falling. This should be good news for Shanghai. However, how does perceived pollution stack up? Does it show the same results as measured pollution? We seek to answer these questions in the paper.

Perceived pollution is defined using Google Trends query data. We are interested in keyword search results related to pollution in Shanghai. The more people who search for 'Shanghai Pollution' the higher the concern about pollution. The data is offered on a weekly basis and has been

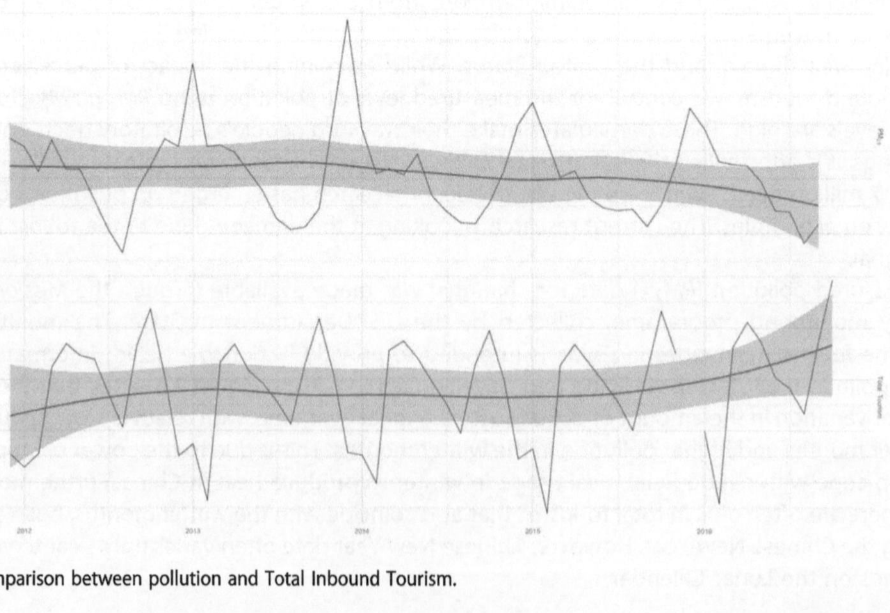

Figure 2. Comparison between pollution and Total Inbound Tourism.

aggregated to monthly for our analysis. For further discussion on the aggregation methodologies see Chamberlin (2010) and Xu and Reed (2017).

Pollution level (measured and perceived) varies as seen in Figure 3. Figure 3 depicts the co-movement between perceived pollution, measured using Google Trend data (PP^{bei}, PP^{china}, PP^{sha} represent perceived pollution level in Beijing, China, and Shanghai respectively.),[6] and measured pollution in Shanghai. Furthermore, we see similar movement between all four source countries implying that

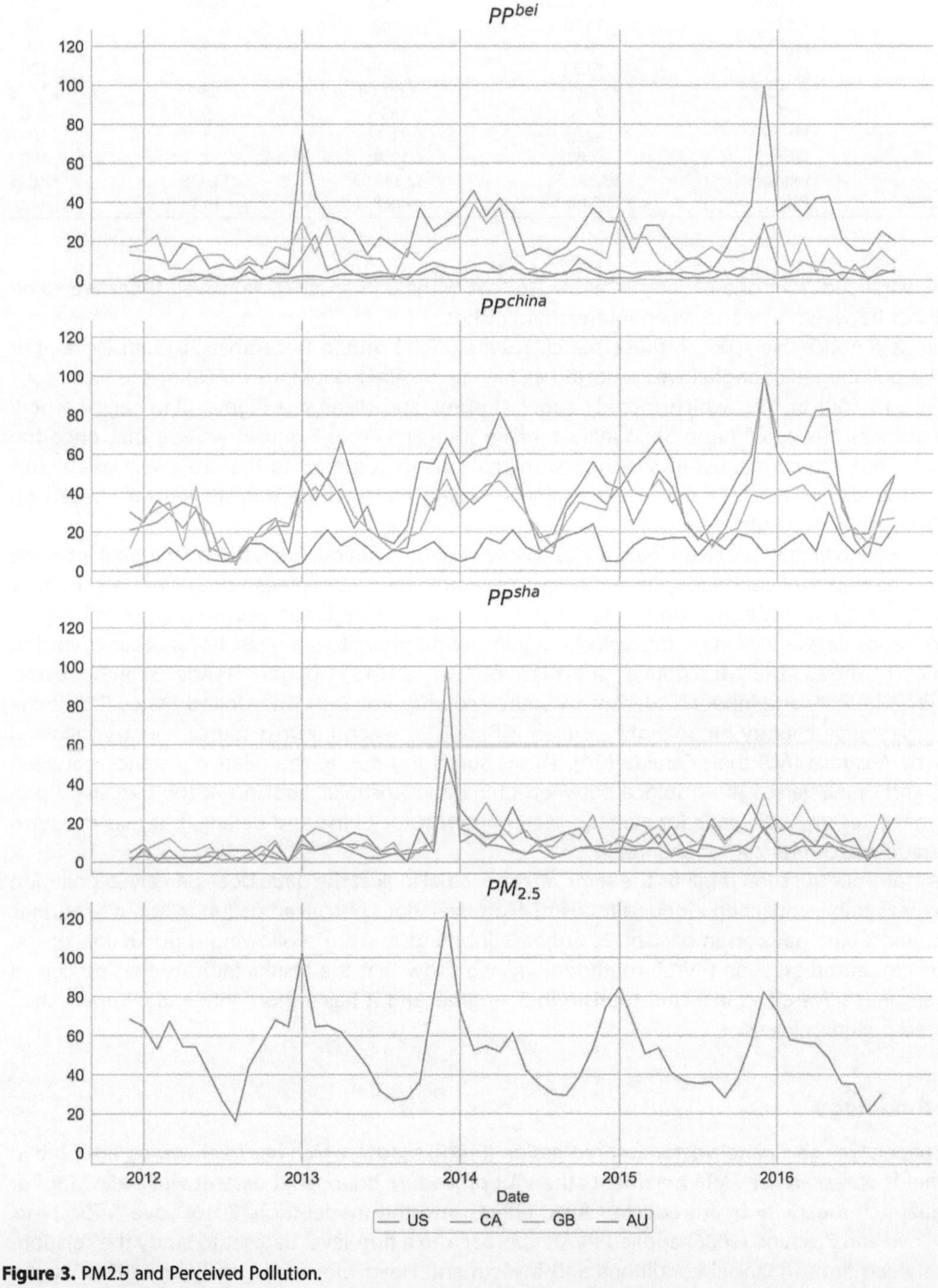

Figure 3. PM2.5 and Perceived Pollution.

Table 1. Descriptive Statistics.

Source	Variable	Mean	sd	Min	Max
AU	pp^{bei}	2.52	1.61	0	7
	pp^{china}	12.37	5.87	2	30
	pp^{sha}	11.58	9.28	0	53
	tourism	14,125.12	3424.70	5652	21,739
CA	pp^{bei}	4.98	4.95	0	29
	pp^{china}	32.63	17.20	5	100
	shangpol	8.12	12.71	0	100
	tourism	12,558.05	3775.32	6348	21,516
GB	pp^{bei}	14.93	8.62	0	39
	pp^{china}	29.41	11.76	3	54
	pp^{sha}	13.32	13.68	0	100
	tourism	14,892.61	3832.04	7592	22,858
US	pp^{bei}	24.68	16.74	5	100
	pp^{china}	42.25	21.77	7	100
	pp^{sha}	7.08	12.78	1	100
	tourism	53,331.56	12,340.03	13,436	75,015
Shanghai	$PM_{2.5}$	51.8	19.8	16.1	122

what is witnessed from one country is similar to that witness in another. However, there are some differences between how and when interest diminishes.

There is a noticeable spike in these search results (PP^{sha}) during December 2013. In December 2013 the pollution in Shanghai was reported as having an Air Quality Index (AQI) at the hazardous level (301 to 500) or 122, which forced school closures and disrupted flights. One can see how events such as this could harm Shanghai's tourism industry. From Figure 3 we see that once the spike occurred, search interest in Shanghai's pollution rarely returned to the pre-event levels. This anecdotal evidence indicates that there could be lasting harm to the tourism industry based on past high pollution events.

The following limitations to the data are as follows. We only run our analysis on Shanghai inbound tourism; due to lack of continuous (monthly) data available for other Chinese cities. We only use four English-speaking countries; we do this to control for potential language issues when searching for Google Trends data as well as for the lack of a significant number of search results for other countries.

Table 1 shows the descriptive statistics of $PM_{2.5}$, the Google Trends search results (PP^{bei}, PP^{china}, PP^{sha}) and inbound tourism by source country (*tourism*). The United States (US) is the top tourist source country for Shanghai; Britian (GB) falls in second tourist source country followed closely by Australia (AU) then Canada (CA). This is surprising due to the relative distance between the US and China versus the distance between China and Australia. Looking at the perceived pollution measures, the U.S. leads for average search interest for China and Beijing, but falls in fourth for average search interest for Shanghai.

Therefore, our question remains the same after an initial look at the data. Does perceived pollution have a measurable impact on inbound tourism? Moreover, does perceived pollution about Shanghai, Beijing, and China have an impact on Shanghai's inbound tourism? Following a discussion of the panel vector autoregression (PVAR) methodology, we show that the results indicate that perceived pollution does have affect inbound tourism in Shanghai, and it has a more influential impact than measured pollution levels.

4. Methodology

In this paper, we use a panel vector autoregression (PVAR) model, which has been widely adopted in many fields of research. PVAR implements the VAR procedure on a panel dataset while allowing for individual heterogeneity by introducing fixed effects into the model (Koh, 2016; Love & Zicchino, 2006). Love and Zicchino (2006) applied PVAR approach to a firm-level dataset to study the relationship between firms' financial conditions and investment. Head, Lloyd-Ellis, and Sun (2014) used a similar approach to examine the impact of city-specific income shocks on US housing markets. PVAR has been used less in the tourism literature. Mahadevan and Suardi (2017) used PVAR to study the impact of tourism growth on poverty and inequality. Antonakakis, Dragouni, Eeckels, and Filis (2017) re-examined the relationship between tourism and economic growth using PVAR. To the best of our knowledge, this is the first paper using PVAR to study the impact of pollution level, measured or perceived, on tourism. In this paper, we have monthly inbound tourist data from four source countries for less than five years. In other words, our panel dataset is relatively small in both time-series and cross-section dimensions. PVAR, compared to traditional VAR or panel data analysis, obtain efficiency gain from both dimensions. Also, using a PVAR approach can help account for the heterogeneity of each tourists' source country.

4.1. Panel vector autoregression

We adopt a $k-$variate PVAR of order p with the following representation as in Love and Zicchino (2006):

$$Y_{it} = Y_{it-1}A_1 + Y_{it-2}A_2 + \ldots + Y_{it-p}A_p + X_{it}B + u_i + e_{it} \quad i \in [1,\ N],\ t \in [1,\ T_i] \tag{1}$$

where Y_{it} is a vector of dependent variables; X_{it} represents the set of covariables; u_i stands for unobserved panel-specific fixed effects, and e_{it} is a vector of error terms. The matrices $A_1, A_2, \ldots, A_p$ and the matrix B are parameters to be estimated.

We utilize the Helmert transformation procedure to address the potential non-stationarity issues and to remove panel-specific fixed effects. The Helmert transformation is often preferred over the first-difference approach because it subtracts the mean of all future observations instead of subtracting the previous observation of a variable (Arellano & Bover, 1995). It also helps satisfy the orthogonality assumption between transformed variables and lagged regressors. Therefore, efficiency improvement of the estimation can be achieved by using system GMM instruments (Z_{it}) as proposed by Holtz-Eakin, Newey, and Rosen (1988).

The GMM estimator is given by

$$A = (\bar{Y}^{*'} Z\hat{W})Z'\bar{Y}^{*})^{-1}(\bar{Y}^{*'} Z\hat{W})Z'\bar{Y}^{*}) \tag{2}$$

where $\hat{W}$ is a weighting matrix that helps select maximum efficiency, the asterisk denotes the transformed Y variable (Hansen, 1982).

We can then continue to investigate the directions of Granger causality between the variables. The null hypothesis is that all estimated coefficients on the lag of variable n are jointly zero in the equation for variable m. Failing to reject the null hypothesis indicates that variable n Granger causes m. Such hypothesis testing can be implemented based on the GMM estimates of A and its covariance matrix (Granger, 1969).

4.2. Optimal moment and model selection

The model selection method used here is the consistent moment and model selection criteria (MMSC), proposed by Andrews and Lu (2001). MMSC for GMM models is based on the Hansen-J statistic of over-identifying restrictions (Hansen, 1982). There are three model and moment selection criteria: MMSC-BIC, MMSC-AIC, and MMSC-HQIC, which are analogous to the widely used model selection criteria Bayesian Information Criteria (BIC), Akaike Information Criteria (AIC), and Hannan-Quinn Information Criteria (HQIC).

Applying the MMSC to the GMM estimator in Eq (2), we will be able to select the pair of vectors $(p,\ q)$ that minimizes

$$MMSC_{BIC,n}(k, p, q) = J_n(k^2p, k^2q) - (|q| - |p|)k^2 ln(n) \tag{3a}$$

$$MMSC_{AIC,n}(k,p,q) = J_n(k^2p, k^2q) - 2k^2(|q| - |p|) \tag{3b}$$

$$MMSC_{HQIC,n}(p,q) = J_n(k^2p, k^2q) - Rk^2(|q| - |p|)lnln(n), \quad R > 2 \tag{3c}$$

where $J_n(k,p,q)$ is the statistic of over-identifying restriction for a PVAR specified in Eq (1).

4.3. Impulse response function and forecast-error variance decomposition

A stable PVAR model is invertible and has an infinite-order vector moving average (VMA) representation, which provides a familiar interpretation to estimated impulse response functions and forecast error variance decompositions. By rewriting the model as an infinite VMA, we can obtain the impulse-response function Φ_i, where Φ_i are the VMA parameters.

We can then orthogonalize the Φ_i using a matrix, P, to get the orthogonalized impulse responses, $P\Phi_i$, where $P'P = \Sigma$, which helps imposing identification restrictions on the system of dynamic equations. Confidence intervals can be then derived utilizing the asymptotic distribution of the PVAR parameters and the cross-equation error variance-covariance matrix, or be estimated using Monte Carlo simulation. See Lütkepohl (2005) for a complete discussion.

Also, the $h-$step ahead forecast-error has the following representation:

$$Y_{it+h} - E[Y_{it+h}] = \sum_{i=0}^{h-1} e_{i(t+h-i)}\Phi_i \tag{4}$$

where Y_{it+h} is an observed vector at period $t+h$ and $E[Y_{it+h}]$ is the h step ahead predicted vector made in period t. Similarly, each variable's contribution to the forecast-error variance is isolated using the orthogonalized shocks $e_{it}P^{-1}$, which have a covariance matrix I_k. Specifically, the $h-$step ahead forecast-error variance of variable n contributed by a variable m can be calculated as:

$$\sum_{i=0}^{h-1} \theta_{mn}^2 = \sum_{i=1}^{h-1} (i_n' P\Phi_i i_m)^2 \tag{5}$$

where i_s is the $s-$th column of I_k.

5. Results

In this paper we estimate the following PVAR model[7]:

$$BX_{ct} = \sum_{i=1}^{T} A_i X_{c,t-i} + F_c + \epsilon_{ct} \tag{6}$$

where $X_{ct} = [tourism_{ct}, PP_{ct}, PM_{ct}]'$ denotes the vector of the levels of inbound tourism (*tourism*), perceived pollution (*PP*) and measured pollution (*PM*) from each source country in the sample at each date. B and A_i are matrices of parameters, F_c is a vector of country fixed effects and $\epsilon_{ct} = [\epsilon_{ct}^{tourism}, \epsilon_{ct}^{PP}, \epsilon_{ct}^{PM}]'$ a vector of structural shocks.

The PVAR approach allows us to account for time-invariant characteristics intrinsic to each tourists source country which can be controlled for by using Helmert transformation procedure. We estimate the model by using 'GMM-style' instruments as proposed by Holtz-Eakin et al. (1988). This model has been widely employed in the economics literature; however, it has not been adopted in the tourism literature. This is a novel insight added by this paper.

Table 2 presents our model selection criteria proposed by Andrews and Lu (2001). From Table 2, we determine that we should use the first-order PVAR since the MBIC and MQIC are smallest.[8] We then fit a first-order PVAR model with the same specification of instruments as above using GMM estimation as shown in Table 3.

Table 2. Model Selection Criteria.

Lag	CD	J	J *p*-value	MBIC	MAIC	MQIC
1	0.956	114.110	0.002	−285.844	−35.890	−136.970
2	0.982	55.762	0.267	−210.874	−44.238	−111.624
3	0.992	20.209	0.736	−113.109	−29.790	−63.484

In Table 3, there are 222 observations included in the estimation. Our main variables of interest are the measured pollution ($PM_{2.5}$) and perceived pollution measured by Google Trends data (PP^j) where j represents Shanghai (PP^{sha}), Beijing (PP^{bei}), and China (PP^{china}). We are interested in the *tourism* column, with the number of inbound tourists being the explained variable. We find all variables except PP^{china} have significant impacts on inbound tourism. We believe that the insignificance of PP^{china} suggests tourists conduct a restrictive search when looking into pollution before they arrive. Lagged tourism is positively associated with current inbound tourism levels, similar to findings in the tourism demand literature, capturing a word-of-mouth effect, see Dogru, Sirakaya-Turk, and Crouch (2017) for a review.

Lagged $PM_{2.5}$ has a negative and significant impact on inbound tourism in Shanghai. Meaning, measured pollution in Shanghai has an adverse impact on tourism. Meanwhile, perceived pollution level in Shanghai (PP^{sha}) still exerts a negative and significant influence on inbound tourism after controlling for measured pollution level. This tells us that when a potential tourist searches 'Shanghai Pollution', they find that the results are not appealing to them; therefore, they will forego their trip to Shanghai and may decide to travel elsewhere.

Besides PP^{sha}, we also include two other variables: perceived pollution in China (PP^{china}) and in Beijing (PP^{bei}). The intuition is that when people make travel plans to China, they may have concerns about overall environmental quality in the country, as well as paying particular attention to their point of entry and other possible destination choices. We find interesting results showing that perceived pollution in China does not affect inbound tourism in Shanghai. On the other hand, if people worry about pollution in Beijing more, then we see an increase in tourism to Shanghai, e.g. we are witnessing a trade-off by tourists who decide go to Shanghai instead of Beijing due to (relatively) heavier pollution in Beijing. We are not arguing that higher perceived pollution level in Beijing is good news for Shanghai or saying there is any positive impact of it.

One limitation is that we do not know specific thresholds for 'unappealing' pollution. However, one would assume that any threshold made would be determined by an individual's current 'home' city or country level of pollution. Given WHO AAP data[9] we see that from all of our source countries, all cities within this database have lower yearly average levels of $PM_{2.5}$ than Shanghai. Therefore, we believe it is safe to assume that when individuals from these countries look at pollution levels in Shanghai, they see them as 'too polluted.' This observation gives more validation to our results of perceived pollution having a negative impact on tourist destination selection.

Table 3. PVAR Results.

	Tourism	$PM_{2.5}$	pp^{sha}	pp^{china}	pp^{bei}
$\boldsymbol{tourism_{t-1}}$	0.677***	0.211*	−0.099*	0.041	−0.119
	(0.059)	(0.086)	(0.041)	(0.096)	(0.073)
$\boldsymbol{PM_{2.5,t-1}}$	−0.091***	0.863***	−0.028	−0.048	0.002
	(0.020)	(0.056)	(0.024)	(0.037)	(0.025)
$\boldsymbol{PP^{sha}_{t-1}}$	−0.200***	−0.082	0.423	0.092***	0.114***
	(0.053)	(0.101)	(0.076)	(0.050)	(0.033)
$\boldsymbol{PP^{china}_{t-1}}$	−.006	0.155*	0.165***	0.562***	0.096**
	(0.045)	(0.073)	(0.037)	(0.063)	(0.034)
$\boldsymbol{PP^{bei}_{t-1}}$	0.316***	−0.349***	−0.137**	0.539***	0.521***
	(0.089)	(0.078)	(0.048)	(0.089)	(0.066)
N	222	222	222	222	222

Standard Errors in parentheses. *, **, *** denote significance at the 10%, the 5%, and the 1% levels.

Table 4. Granger Causality Results.

Null Hypothesis	*p*-value
$PM_{2.5}$ does not Granger Cause tourism	0.000
PP^{sha} does not Granger Cause tourism	0.000
PP^{china} does not Granger Cause tourism	0.176
PP^{bei} does not Granger Cause tourism	0.000

In order to help verify these finding we also use Granger Causality test to help determine the measured causal effects. Results from the Granger Causality tests in Table 4 show that $PM_{2.5}$, PP^{sha} and PP^{bei} can help explain inbound tourists volume in Shanghai. All variables of interest, except PP^{china} Granger-cause inbound tourism to Shanghai.

Since estimates of PVAR models are rarely interpreted by themselves, we look at the impact of exogenous changes in each endogenous variable to other variables in the PVAR system using impulse response functions (IRF) and forecast-error variance decompositions (FEVD) estimations. Prior to that, we first check the stability condition of the estimated PVAR. The resulting table and graph of eigenvalues confirm that the estimation is stable illustrated in Table 5 and Figure 4. Table 5 presents that the estimates satisfy the eigenvalue stability condition since the modulus of

Table 5. Stability Conditions.

Real Eigenvalue	Imaginary Eigenvalue	Modulus
0.807	0	0.807
0.695	−0.086	0.700
0.095	0.086	0.700
0.524	0	0.524
0.325	0	0.325

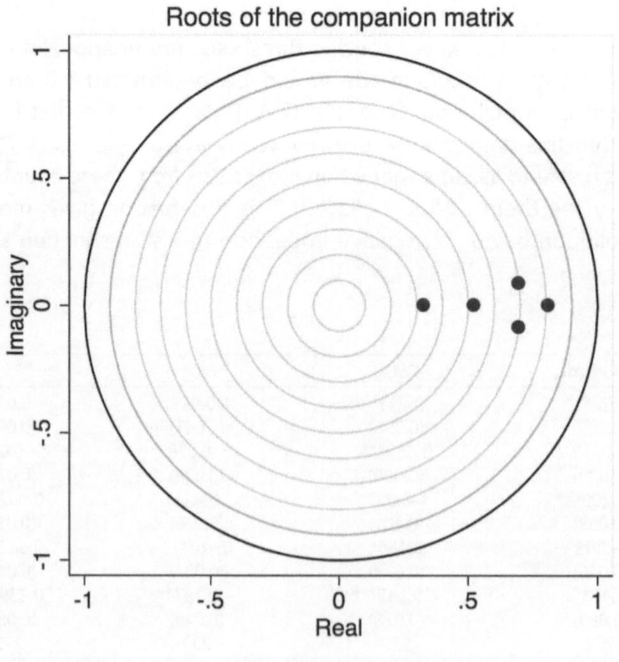

Figure 4. Stability Test.

Table 6. FEVD Estimate Results (response variable is *tourism*).

(1) Forecast Horizon	(2) tourism	(3) $PM_{2.5}$	(4) pp^{sha}	(5) pp^{china}	(6) pp^{bei}	(7) $pp^{combine}$
0	0	0	0	0	0	0
1	1	0	0	0	0	0
2	0.845	0.044	0.052	0.009	0.050	0.111
3	0.727	0.083	0.074	0.016	0.100	0.19
4	0.660	0.111	0.078	0.019	0.132	0.229
5	0.624	0.130	0.076	0.021	0.149	0.246
6	0.607	0.143	0.074	0.022	0.155	0.251
7	0.599	0.150	0.073	0.022	0.157	0.252
8	0.595	0.153	0.073	0.022	0.157	0.252
9	0.593	0.155	0.074	0.022	0.156	0.252
10	0.592	0.156	0.075	0.022	0.155	0.252

each eigenvalue is strictly less than 1. Figure 4 shows that none of the roots are outside of the unit circle indicating that the PVAR model is stable.

Based on the FEVD estimates in Table 6, we see that as much as 25 per cent of the variations in inbound tourists volume in Shanghai which can be explained by combined perceived pollution indicators, that is, the sum of the FEVD estimates for Shanghai (PP^{sha}), Beijing (PP^{bei}), and China (PP^{china}). The measured pollution level can explain as much as 15-16 per cent of the change in tourism. This shows that perceived pollution has a greater impact on inbound tourism than the measured pollution levels. Furthermore, based on the Impulse Response Function (IRF) plot, shown in Figure 5, we see that a positive shock on $PM_{2.5}$ or PP^{sha} leads to lower inbound tourism to Shanghai. It is also noteworthy that current shock in PP^{bei} has a positive impact on tourism mirroring our results from in Table 3.

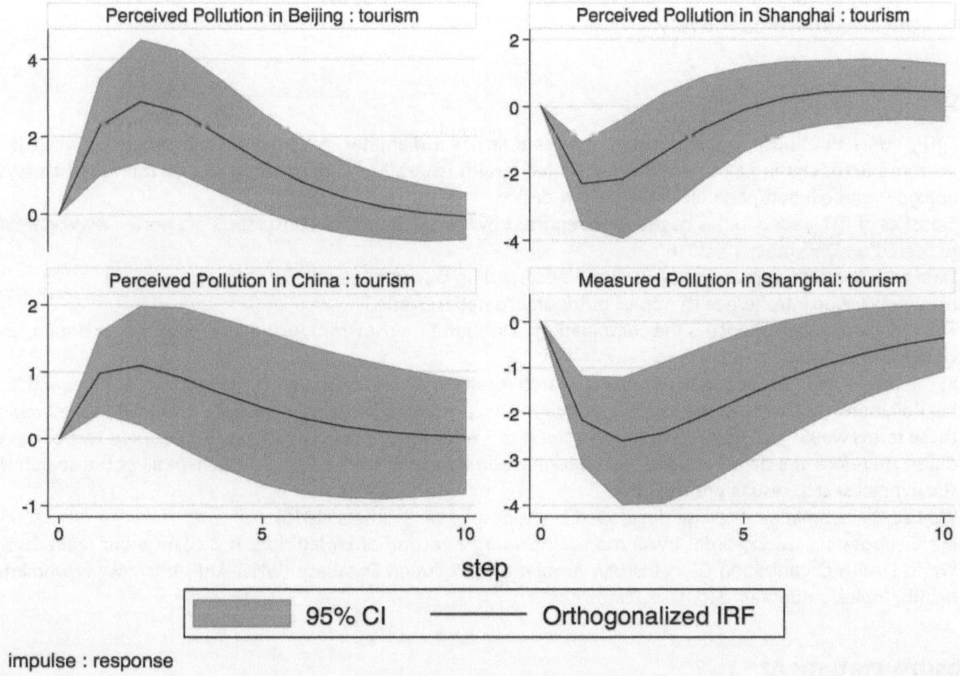

Figure 5. IRF Graph.

6. Conclusions

In this paper, we stress the power of perception, particularly the influence of perceived pollution on the tourist industry. We find that perceived pollution has a stronger impact on inbound tourism than measured pollution levels. The implications of this form of research have numerous benefits to the tourism industry.

We find that the perceived pollution level in one city could impact people's travel plans. In the future, it would be interesting to examine whether perceived pollution or other types of risks in one country could affect touring decisions to neighbouring countries. For example, if the pollution in Shanghai is too high for an individual, where would they decide to go instead? Does pollution in Shanghai have any impact on tourism in another city in China or Seoul, South Korea? Also, incorporating perceived pollution measures could potentially help predicting inbound tourism demand, although it is beyond the scope of this paper.

In the past, international attention has been focused on air pollution in China. Such focus treats all seasons and all cities in China homogeneously, and this has hurt the tourism industry. So how can we potentially combat the negative consequences of the perception of pollution? It may be up to central and local governments, along with media to help present the actual situation. On the one hand, China needs to tackle the pollution issue to make the country a more attractive destination. On the other hand, local government could help clarify the heterogeneous environmental quality across regions in China and build a better destination image through clear communication. Cities with cleaner air should emphasize it when promoting tourism. This could include clean city/region marketing campaigns. For example, 'Come to City X we are ranked cleanest air in Province Y.' This type of regional promotion could improve the regional economic development throughout China. Regions with lower pollution levels tend to have fewer factories and therefore are (sometimes) economically disadvantaged. It is also important to highlight seasons that are more suitable for tourism activities.

A note of caution, this does not mean hat cities/countries should promote unfounded pollution records to try to sway tourists. This research shows that tourists 'do their homework' before travelling. The best way to attract tourists would be to clean the air and promote factual evidence. This line of research could lay a good foundation for improving marketing communication messages to address tourists' concerns more effectively.

Notes

1. $PM_{2.5}$ refers to atmospheric particulate matter that has a diameter less than 2.5 micrometers, which is often responsible for causing air pollution and related health issues. More information can be found at https://www.epa.gov/pm-pollution/particulate-matter-pm-basics.
2. Statistics of $PM_{2.5}$ used in this paper were reported by the US Shanghai Consulate. Data portal: www.stateair.net. Accessed on December 12, 2016.
3. Data can be found at http://lyw.sh.gov.cn/. Accessed on December 2, 2016.
4. http://www.who.int/phe/health_topics/outdoorair/databases/en/
5. Total tourism in this instance is the combination of inbound tourism from Australia, Canada, Great Britain, and the United States.
6. In the case of the current paper, we look at search results for 'Pollution in China', 'Pollution in Beijing', and 'Shanghai Pollution'. For example, the top 3~5 search results presented by Google were the same. People searching for these terms would find similar links to only the first 3 webpages shown by Google. Subsequent webpages could differ. Therefore, the data from one search term could be considered a representative of all of the search results since initial search results are the same.
7. We use the estimation package developed by Abrigo, Love, & others (2015).
8. MAIC suggests a second-order PVAR model. Including a second-order lag does not change our main results.
9. World Health Organization Global Urban Ambient Air Pollution Database (WHO AAP) http://www.who.int/phe/health_topics/outdoorair/databases/cities/en/

Disclosure statement

No potential conflict of interest was reported by the author(s).

ORCID

Markum Reed http://orcid.org/0000-0002-0072-475X

References

Abrigo, Michael RM, & Love, Inessa. (2015). Estimation of panel vector autoregression in stata: A package of programs. *manuscript*.

Andrews, D. W., & Lu, B. (2001). Consistent model and moment selection procedures for GMM estimation with application to dynamic panel data models. *Journal of Econometrics*, *101*(1), 123–164.

Antonakakis, N., Dragouni, M., Eeckels, B., & Filis, G. (2017). The tourism and economic growth enigma: Examining an ambiguous relationship through multiple prisms. *Journal of Travel Research*, 0047287517744671.

Arellano, M., & Bover, O. (1995). Another look at the instrumental variable estimation of error-components models. *Journal of Econometrics*, *68*(1), 29–51.

Bangwayo-Skeete, P. F., & Skeete, R. W. (2015). Can Google data improve the forecasting performance of tourist arrivals? Mixed-data sampling approach. *Tourism Management*, *46*, 454–464.

Brida, J. G., Cortes-Jimenez, I., & Pulina, M. (2016). Has the tourism-led growth hypothesis been validated? A literature review. *Current Issues in Tourism*, *19*(5), 394–430.

Chamberlin, G. (2010). Googling the present. *Economic and Labour Market Review*, *4*(12), 56–95.

Choi, H., & Varian, H. (2012). Predicting the present with Google trends. *Economic Record*, *88*(s1), 2–9.

Dogru, T., Sirakaya-Turk, E., & Crouch, G. I. (2017). Remodeling international tourism demand: Old theory and new evidence. *Tourism Management*, *60*, 47–55.

Fenichel, E. P., Kuminoff, N. V., Chowell, G., & Boni, M. F. (2013). Skip the trip: Air travelers' behavioral responses to pandemic influenza. *PloS One*, *8*(3). https://doi.org/10.1371/journal.pone.0058249

Granger, C. W. J. (1969). Investigating causal relations by econometric models and cross-spectral methods. *Econometrica: Journal of the Econometric Society*, *37*, 424–438.

Gunter, U., & Önder, I. (2016). Forecasting city arrivals with Google analytics. *Annals of Tourism Research*, *61*, 199–212.

Hansen, L. P. (1982). Large sample properties of generalized method of moments estimators. *Econometrica: Journal of the Econometric Society*, *50*, 1029–1054.

Head, A., Lloyd-Ellis, H., & Sun, H. (2014). Search, liquidity, and the dynamics of house prices and construction. *The American Economic Review*, *104*(4), 1172–1210.

Holtz-Eakin, D., Newey, W., & Rosen, H. S. (1988). Estimating vector autoregressions with panel data. *Econometrica: Journal of the Econometric Society*, *56*, 1371–1395.

Koh, W. C. (2016). Oil price shocks and macroeconomic adjustments in oil-exporting countries. *International Economics and Economic Policy*, *14*, 1–24.

Kozak, M., Crotts, J. C., & Law, R. (2007). The impact of the perception of risk on international travellers. *International Journal of Tourism Research*, *9*(4), 233–242.

Law, R. (2006). The perceived impact of risks on travel decisions. *International Journal of Tourism Research*, *8*(4), 289–300.

Lepp, A., & Gibson, H. (2003). Tourist roles, perceived risk and international tourism. *Annals of Tourism Research*, *30*(3), 606–624.

Li, X., Pan, B., Law, R., & Huang, X. (2017). Forecasting tourism demand with composite search index. *Tourism Management*, *59*, 57–66.

Love, I., & Zicchino, L. (2006). Financial development and dynamic investment behavior: Evidence from panel var. *The Quarterly Review of Economics and Finance*, *46*(2), 190–210.

Lütkepohl, H. (2005). *New introduction to multiple time series analysis*. New York: Springer Science & Business Media.

Mahadevan, R., & Suardi, S. (2017). Panel evidence on the impact of tourism growth on poverty, poverty gap and income inequality. *Current Issues in Tourism*, *61*, 1–12.

Maser, B., & Weiermair, K. (1998). Travel decision-making: From the vantage point of perceived risk and information preferences. *Journal of Travel & Tourism Marketing*, *7*(4), 107–121.

Pablo-Romero, M. d. P., & Molina, J. A. (2013). Tourism and economic growth: A review of empirical literature. *Tourism Management Perspectives*, *8*, 28–41.

Park, S., Lee, J., & Song, W. (2017). Short-term forecasting of Japanese tourist inflow to South Korea using Google trends data. *Journal of Travel & Tourism Marketing*, *34*(3), 357–368.

Rittichainuwat, B. N., & Chakraborty, G. (2009). Perceived travel risks regarding terrorism and disease: The case of Thailand. *Tourism Management*, *30*(3), 410–418.

Simpson, P. M., & Siguaw, J. A. (2008). Perceived travel risks: The traveller perspective and manageability. *International Journal of Tourism Research*, *10*(4), 315–327.

Sönmez, S. F., & Graefe, A. R. (1998). Influence of terrorism risk on foreign tourism decisions. *Annals of Tourism Research*, *25*(1), 112–144.

Tavitiyaman, P., & Qu, H. (2013). Destination image and behavior intention of travelers to Thailand: The moderating effect of perceived risk. *Journal of Travel & Tourism Marketing, 30*(3), 169–185.

Xu, X., & Reed, M. (2017). Perceived pollution and inbound tourism in China. *Tourism Management Perspectives, 21*, 109–112.

Yang, X., Pan, B., Evans, J. A., & Lv, B. (2015). Forecasting Chinese tourist volume with search engine data. *Tourism Management, 46*, 386–397.

Zhang, A., Zhong, L., Xu, Y., Wang, H., & Dang, L. (2015). Tourists' perception of haze pollution and the potential impacts on travel: Reshaping the features of tourism seasonality in Beijing, China. *Sustainability, 7*(3), 2397–2414.

Profiling the elite middle-age Chinese outbound travellers: a 3rd wave?*

Jigang Bao, Xin Jin and David Weaver

ABSTRACT

With inductive analysis of relevant statistics and literature, this paper reveals an elite middle-age Chinese cohort of significant size, with ample buying power and considerable discretionary time to support the outbound travel of themselves and their children. This middle age group has experienced a time of dramatic change in recent years causing a transformation of traditional livelihood activities and lifestyles. Associated changes in wealth and values contribute to overseas travel proclivities that could entail both positive and negative impacts for both host and guest. The depiction of the current status and potential development in Chinese outbound travel suggests that this group could possibly form a '3rd wave' of Chinese outbound tourists. The segmentation based on generation cohort and social status is particularly suitable in helping the tourism industry service providers in identifying market characteristics for marketing, product development and comprehensive destination development.

1. Introduction

The increase in Chinese outbound trips from 10 million in 2000 to 128 million in 2015 (United Nations World Tourism Organisation [UNWTO], 2016) has attracted considerable international attention, with associated 2015 expenditures of US$292 billion continuing China's pattern of double-digit growth every year since 2004 (UNWTO, 2016). Facilitating factors have included a range of economic, social, and political stimuli (Jin & Wang, 2016), and notably the Approved Destination Status system that has incrementally resulted in the opening of most foreign countries to the Chinese leisure market (Arita, Edmonds, La Croix, & Mak, 2011). From a Mainland China perspective, trips to Hong Kong and Macau – the special administrative regions of China – are regarded as outbound travel in both literature and statistics, and account for most such activity. However, the Chinese market has emerged as a significant source market for many other short and long-haul destinations, indicating a transition from Burton's 'industrialising' to 'industrialised' tourism phase, where widespread participation in domestic tourism is augmented by growing participation in international travel (Burton, 1995).

The enthusiastic attitude of Chinese to travel abroad is almost certain to continue as China moves into its 'post-industrial' stage, notwithstanding the anticipated economic disruptions that typically accompany the transition from a manufacturing model to one of services-dominated product and business innovation (World Tourism Cities Federation [WTCF], 2015). Concurrently, the Chinese

*Authors contribute equally to the paper.

public has become increasingly attached to travel as its multiple benefits to quality of life and well-being become more apparent (Chen, Lehto, & Cai, 2013), and its links to the Chinese Dream more salient (Weaver, 2015). An important challenge for stakeholders is to accurately identify the Chinese market segments most likely to experience sustained growth and to exhibit patterns of high expenditure. Indeed, a large body of literature has explored the overall Chinese outbound tourism market with regard to travel motivations and barriers (e.g. Sparks & Pan, 2009), expectations of destinations (e.g. Li, Lai, Harrill, Kline, & Wang, 2011), destination image and positioning (e.g. Hsu & Song, 2012), service quality, shopping management, and unethical practices (e.g. Zhang, Heung, & Yan, 2009), but these studies tend to assume a market homogeneity that generates generic rather than segment-specific findings and implications (Jin & Wang, 2016).

There is a neglect of age-based segmentation in literature and statistics always depict Chinese outbound tourists as mostly young, with about two-thirds born after 1980 (WTCF, 2015). General research and marketing, accordingly, tend to focus on the youthful so-called '2nd wave' of Chinese tourists (Arlt, 2013). This emphasis contrasts with neighbouring East Asian countries which are well into a post-industrial phase of tourism articulation. For example, Japanese after age 60 have a higher travel propensity than those in any other age group for both sexes (Sakai, Brown, & Mak, 2000). Studies in other countries (e.g. US and Canada) indicate that the better educated seniors have more money, are predisposed to recreational spending, and tend to travel farther from home if their health allows (Zimmer, Brayley, & Searle, 1995).

Moreover, depending on destination and tour purposes, the number of older Chinese outbound travellers could already exceed that of younger Chinese outbound travellers. For example, the number of above 40 year-old Chinese holiday tourists to Australia has exceeded that of under-40 Chinese tourists since 2012. Also, in 2016, the above-40 Chinese visiting friends and relative (VFR) tourists to Australia surpassed 1.3 million, which was 2.3 times of the figure of under-40 VFR tourists (TRA, 2017).

Yet, little is discussed regarding this older Chinese outbound traveller segment in the literature. Thus, through inductive analysis of secondary data, this paper aims to inform researchers of the shift of older Chinese outbound tourists from a position of latency to salience as a key market segment based on their potential expenditure. An inductive approach is also employed by Cheng and Foley (2017) in connection with the post-1980s Chinese generation to inform subsequent empirical research, and as such provides a helpful precedent. 'Older Chinese tourists' mainly refer to Sun and Wang (2010)'s 'beginning of economic reform' generation (born 1956–1975) in their study of changing values among different Chinese age groups. Prospects of the segment and key sub-segments with respect to magnitude, group characteristics and dynamics, and buying power, are discussed. It also considers the attendant implications for travel product development, destination marketing, and sustainable tourism development. While older travellers can currently be regarded as a latent 'niche market', a niche in the Chinese population can still be very big and influential.

Common sense steps in segmenting markets (Dolnicar, 2007) were followed in this research, entailing (1) selection of the segmentation criteria (in this case, age/generation), (2) profiling the attendant segments by identifying personality characteristics that characterize each group while distinguishing them from other groups, and (3) assessing the pragmatic usefulness of the segments and formulating relevant targeted marketing activities.

2. Chinese generations and outbound tourists

2.1. Generation and generation theory in tourism research

A generation is an 'aggregate of all people born over roughly the span of a phase of life who share a common location in history and, hence, a common collective persona' (Strauss & Howe, 1997, p. 61). The inner beliefs of a generation arguably remain consistent as its members age together (Li, Li, & Hudson, 2013). Each generation may also develop long-lasting core values, derived from societal events, which significantly influence their life and provide vital cues for lifelong consumer and other behaviour (Muller, 1991; Schewe & Noble, 2000). Continuity theory suggests that individual travel patterns observed in a generation are difficult to change, especially because these patterns often emerge at a relatively early age (Filep, Hughes, Mostafanezhad, & Wheeler, 2015; Pilcher, 1994),

Numerous studies in tourism and other disciplinary areas use generation theory in their studies and many focus on a particular generation cohort because of their alleged behavioural consistencies, such as cohort comparison between baby boomers and the silent generation (Lehto, Jang, Achana, & O'Leary, 2008), tourism and generation Y (e.g. Benckendorff, Moscardo, & Pendergast, 2010; Vijaya, 2010). Other studies do not specifically use the term 'generation cohort' but focus similarly on a particular age group (e.g. senior travel, backpackers, grey nomads). With regard to senior travel, much research has focused on traveller motivations, constraints and behaviour. Motivations include opportunities for social interaction, physical exercise, learning, excitement and nostalgia (e.g. Guinn, 1980), as well as socialization and exposure to novice situations (Hagan & Uysal, 1991). Health, financial limitations, lack of time and information are some of the important constraints (McGuire, Dottavio, & O'Leary, 1986). Factors such as income, education, rural residency, willingness to spend money on recreation, and health status influence the choice of destination. Education, providing an awareness of other cultures, history and geography, is an important factor influencing travel behaviour of seniors (Zimmer et al., 1995).

2.2. Chinese generations

The fact that the Chinese belong to a collectivist culture where certain impulses are widely shared and replicated (Michailova & Hutchings, 2006) endorses the broad prospect of generational fidelity. Because China has experienced its own unique social and political trajectory since the mid-twentieth century, the use of conventional Western generational distinctions (i.e. silent generation – prior to 1945, baby boomers – 1946–1964, generation X 1965–1980, generation Y 1981–1990, and the millenniums – after 1991) is inappropriate in Chinese contexts. Simple use of decades to define group cohorts (e.g. post 60s, 70s, 80s, post 90s) is also inappropriate since these do not align with the critical events that have influenced China's generational dispensations. Reflecting China's exceptionalism, Erickson's (2009) modified Western framework categorizes Chinese generations after 1949 as 'boomers' (born from 1946–1960/64), 'generation X' (born from 1961/65 to 1979), and 'generation Y' (born from 1980 to 1995). More indigenous still are Egri and Ralson (2004) and Sun and Wang (2010). The former distinguish 'consolidation' (1951–1960), 'cultural revolution' (1961–1970) and 'social reform' (1971–1975) generations, while the latter define four Chinese older generations (COG) as 'great leap forward' (born before 1941), 'cultural revolution' (1941–1956), 'the beginning of economic reform' (1957–1972), and 'societal transition' (1973–1977).

We adopt Sun and Wang (2010) framework (and respectively label the four generations as COG1-4), since it emphasizes the formative events that influence generational cohorts as their members progress from adolescence to adulthood (Mannheim, 1953; Schuman & Scott, 1989), a cohort being a group of people within a delineated population who experience the same significant social event within a given period of time (Pilcher, 1994). Both studies concur that Chinese born between 1955 and 1975 are more open to change and self-enhancement (Egri & Ralson, 2004), and hold on to more modern values such as individualism and secular orientation (Sun & Wang, 2010).

2.3. Younger and older Chinese outbound tourists

According to World Tourism Cities Federation 2015 report, the majority of outbound mainland Chinese travellers over the past 10 years are younger (with those born after 1980s and 1990s accounting for 68%), are childless (35.3%) or have very young children only (59.25%). Corroboratively, Tuniu.com, the second largest online travel platform in China, reported that those between 26–35 accounted for 36% of their 2015 sales and those above 56 just 18% (Tuniu.com, 2016). These younger and allegedly more independent and sophisticated tourists are those referred to by Arlt (2013) as the 2nd wave of Chinese outbound tourists, juxtaposed against the supposedly unsophisticated and older package tourists of the 1st wave that have dominated the industrializing and industrialized tourism stages and embody the classic mass tourism 'environmental bubble' (Cohen, 1972).

The 2nd wave, more prominent since 2005, is thought to be 'mostly affluent travellers' with experience of overseas education who are no longer 'hostage of the tour guide' and are 'latte-drinking', 'avant-garde', 'self-organising' and 'English speaking' (Arlt, 2013). They are believed to take time in their trips for more in-depth experiences and to travel further afield in search of 'off-the-beaten-track' destinations and activities. This pattern underpins recent segmentation studies that focus specifically on Chinese youth travellers (Liu & Ryan, 2011) and (mostly young) backpackers to Macau (Ong & du Cros, 2012), or that otherwise implicate a prevalence of younger Chinese travellers among outbound film tourists to Korea (Kim, 2012), casino visitors (e.g. Wong, Fong, & Liu, 2012); and self-drive tourists (Wu & Pearce, 2014).

Internationally, middle age tourists constitute a large market that has been well researched in countries such as the United States, Canada and Australia with large 'baby boomer' populations. Investigations of comparable Chinese travellers, however, are lacking. Some historical indication is provided by Jang, Yu, and Pearson (2003), who found that earlier Chinese travellers to the United States were mostly business or VFR travellers and middle-aged males. The business travellers were mainly from managerial and executive positions while a quarter of the VFR travellers were retirees visiting their children or relatives; neither group were active leisure participants (Jang et al., 2003). Other studies, while not focusing on older cohorts, have included them in their samples, usually as a relatively small group (e.g. Sparks & Pan, 2009, Jin, Wu, Becken, & Ding, 2016).

The existence of the 1st (older, unsophisticated) and 2nd (younger, more sophisticated) waves, though lacking rigorous empirical support, is well accepted in industry and academia alike (Li, 2016). A critical issue, however, is the emphasis on younger travellers in segmentation studies, as secondary data in certain destinations indicate the rise of older Chinese outbound tourists in the recent years. Data from Australia are used here as evidence for the alleged 3rd wave, as statistics from Tourism Research Australia, the peak national tourism research body, are reliable and well-respected, and correlate well with the purpose of the paper. These data (Figure 1) indicate that older Chinese tourists (40+) have exceeded younger Chinese tourists (under-40) in recent years. While the number of business tourists remain stable in the last decade for both groups, there is a clear and significant increase of older (40+) Chinese tourists travelling to Australia for holiday and VFR purposes. Chinese older holiday tourists (40+) increased from about 1.28 million in 2005 to 1.94 million in 2016,

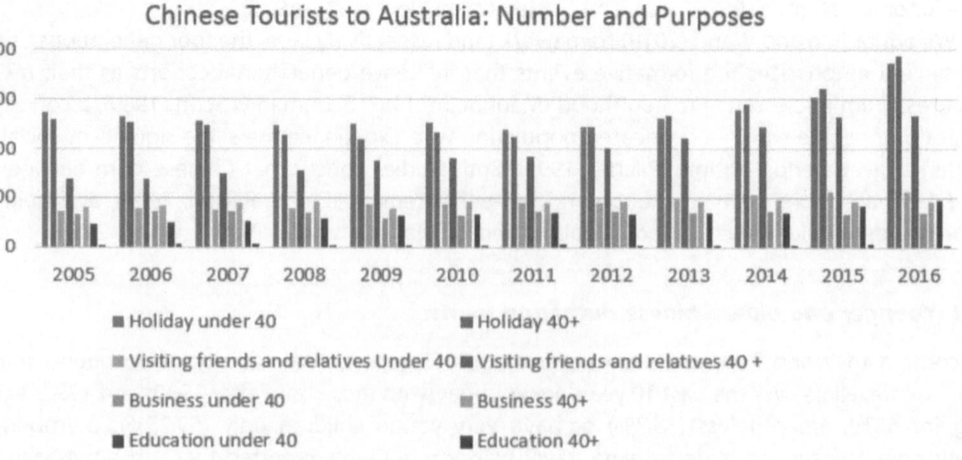

Figure 1. Older (40+) and younger (under 40) Chinese tourists to Australia: Number and trip purpose.

surpassing the number of the younger group (under 40) since 2012. Chinese older VFR tourists (40+) increased more dramatically from 695,000 in 2005 to 1.33 million in 2016, whereas VFR markets for younger groups do not see significant growth.

Top foreign destinations for Chinese outbound tourists include Thailand, South Korea, Japan, Vietnam, the United States, Singapore, Russia, Australia, Indonesia and Malaysia (China National Tourism Adminstration [CNTA], 2016). Statistics form TRA, reported in Figure 1, indicate that older Chinese visitors are on the rise. Data broken down by age and visitation purpose from other key destinations are not available for this paper; however, our contention is that the older Chinese travellers, and perhaps specific segments thereof, constitute a relatively neglected travel market that merits more concerted attention by destination managers, marketers and planners as well as entrepreneurs.

The next sections of this paper outline the profiles and characteristics of the older Chinese outbound tourists, by, first of all, distinguishing between 'senior' and 'middle aged' cohorts and identifying an elite component of the middle age cohort (Sellick, 2004, as 40–59). Then, based on literature that discusses older Chinese tourists with regard to psychological characteristics such as travel motives, risk perceptions, and cognitive age, and demographics (e.g. Hsu, Cai, & Wong, 2007), the paper discusses external conditions (e.g. societal progress, personal finance, time, and health, family support and responsibility) that support the long term growth of this market. The final sections raise questions with regards to internal desires of the cohort and the implications for both themselves and receiving destinations.

3. Older Chinese outbound tourists – identifying an elite group

3.1. Identifying an 'elite' component

Not all older Chinese have the resources or inclination for outbound travel, so we use 'education' and 'urban dweller' as two variables that facilitate identification of segments more conducive to such activity. Years of schooling and educational level, beyond their associations with higher income, are useful to designate social and economic status, with the better educated situated at the top of the hierarchy (Davey, De Lian, & Higgins, 2007). Concurrently, city dwellers, with disproportionately high education levels, have been identified as the group benefitting most from recent economic growth (Tang & Parish, 2000). Table 1 establishes a broad demographic context by situating all Chinese born between 1956 and 1975 by education level and urban/rural status. Subsequently, it isolates middle-aged (COG3-4) Chinese urban dwellers with at least Tertiary (2–3 year) educational attainment, who account for 6.1%% of older Chinese residents and 20.4% of city dwellers in this age cohort (born between 1956 and 1975) but a disproportionately higher percentage of disposable income.

Table 1. Number of tertiary graduates born between 1956 and 1975 residing in cities as of 2015.

	General population information			40+ graduates living in cities = 23.86 million					
Year of birth	Total population (millions)	Living in cities (millions)	Living in towns (millions)	Tertiary (2–3 year) graduates subtotal (millions)	Male %	University (4 year) subtotal (millions)	Male %	Post graduates subtotal (millions)	Male %
1975–1971	124.8	37.8	26	5.1	53%	3.38	57%	0.48	63%
1970–1966	105.6	32.4	21.2	3.7	55%	2.69	60%	0.32	68%
1965–1961	78.8	24.3	15	3.1	58%	1.99	66%	0.25	76%
1960–1956	81.3	22.4	15	1.9	61%	0.87	67%	0.08	79%
Sub-Total	265.7	116.9	77.2	13.8		8.93		1.13	

Total 23,864,518 graduates born between 1956 and 1975 reside in cities, accounting for 6.11% of population of the age group and 20.4% of city dwellers in the age group.

Source: The sixth national population census, National Bureau of Statistics of the People's Republic of China, 2015.

As indicated, the absolute total number of tertiary graduates born between 1956 and 1975 residing in cities exceeds 23.8 million. Males are disproportionately represented, especially at the highest levels of education and among older members of the group. Understanding the motivations of these elite middle-aged urban-dwelling Chinese to travel overseas and the constraints for travelling (such as family duties, financial resources, time) would provide significant implications to the patterns and growth of the Chinese outbound travel market.

3.2. Relevant formative (defining) events

The formative years for common generation identity and shared values are believed to be roughly from 17 to 25 years of age (Mannheim, 1953; Pilcher, 1994). Based on literature discussing Chinese generations in tourism and sociological studies (e.g. Cheng & Foley, 2017; Sun & Wang, 2010; Van Norden, 2015), Table 2 sums up the major social events that shaped generation core values and highlights in bold years of age of the elite middle age groups at the time of these major social events, thereby influencing their characteristics and behaviour.

The elite middle-age cohort, accordingly, was born after the radical agricultural and industrial reform of the Great Leap Forward (1958–1961) which resulted in disastrous economic disruptions which induced the Great Famine of 1959–1961. They were also not significantly influenced by the most tumultuous years of The Cultural Revolution (1966–1969), which plunged China into disastrous political and social upheavals, but were influenced by the social recoveries that succeeded this disruption. Notably, they were the first generation to enter universities by passing entrance examinations, a system regarded as just and equal for attaining status in the social hierarchy. Previously, the exams were suspended and universities were close during the Cultural Revolution (Davey et al., 2007). This generation is the last generation born and raised in the era of ubiquitous material scarcity, and therefore has less experience and memory of real suffering from extreme poverty or political turmoil than their elders. These memories therefore have less direct impact on their psychological and behavioural patterns in politics and everyday life, although memories and traditions passed on from the previous generations could modify their perceptions and behavioural patterns.

3.3. Wealth accumulation and the buying power of the elite middle-aged group

The middle-aged elite generation came of age in tandem with China's two decades of rapid economic development, and therefore this is the generation with real economic power. In particular, this elite group has accumulated significant wealth through the possession of housing, dramatic increases in household income, and participation in financial markets (Meng, 2007). In 1993, Deng Xiaoping made the famous South China tour to affirm China's open policy and economic reform. Housing reform began in 1994 and ended in the early 2000s, first by selling public sector housing to their occupants (mainly social and political elites) at a highly subsidized price and then by

Table 2. Major events and the age of the above 40s at the time of the events/year

Indicative age of different age groups in a year of major event					
Year of birth:	1956	1960	1965	1970	1975
1951–1961 – Great leap forward and famine	4				
1966 – Beginning of cultural revolution	10	6	1		
1976 – End of cultural revolution	***20***	***16***	11	6	
1979 – first year to resume entrance examination to universities/colleges, beginning year to implement the one child policy	***23***	***19***	14	9	4
1989 – Tiananmen square incident	33	29	***24***	***19***	14
1993 – Deng Xiaoping South China tour – start of the economic taking off	37	33	28	***23***	***18***
2003 – GDP per capita exceed CNY10,000	47	43	38	33	28
2010 – China becomes world 2nd largest economy	54	50	45	40	35
2015 – Year of the census	59	55	50	45	40

Source: compiled by the authors. Formative years (17–25) are italicized in the table.

opening the housing market to all buyers (Liu, Winter, & Zilibotti, 2015). Between 1995 and 2002, households who owned their homes increased from 37% to 78% and urban average real household total wealth increased more than four-fold (Meng, 2007). Research shows that privatization of public housing gave the elites a sudden windfall of private housing wealth as they received larger and better housing prior to the reform and then higher price subsidies during the reform. Since 2002, elite wealth accumulation expanded through investment-oriented house purchases, induced by rising real estate value as well as land-use and mortgage reforms. Ownership of more than one property in particular has contributed to asset accumulation. The fact that the elite group has multiple house ownerships reflects research findings that disposable income is not a full proxy for accumulated wealth, as there is no long term relationship between housing price and income (Guo & Huang, 2010). In contrast, the following generation (born after the 1980s) have accumulated their wealth mainly by first sharing houses with their elders (Rosenzweig & Zhang, 2014).

The elite cohorts, in addition, were the first generation to be affected by the one-child policy, implemented from 1979, which was more strictly enforced in urban settings; in rural areas, a second child was generally allowed five years after the first child was born especially when the first child is a girl (Hesketh, Lu, & Xing, 2005). This contributed further to wealth accumulation by increasing the saving rate (Choukhmane, Coeurdacier, & Jin, 2013). The active 2nd wave outbound Chinese travellers, born in the 1980s and 1990s, are the children of this generation whose decisions about travelling overseas for education or pleasure are mainly made and financed by their parents, thus indicating a secondary or indirect form of tourism momentum. Statistics indicate that over 901,000 Chinese students studied abroad on a student visa in 2015 and every year about 523,000 new Chinese students go overseas for tertiary education (Chinese Ministry of Education, 2016). According to UNWTO criteria, these individuals qualify as tourists. As to the question of whether this elite group has sustainable financial resources to support their own travelling after they support their children's education, the above review on the nature of wealth accumulation and the following discussions indicate sufficient resources to additionally fund such activity.

3.4. Changing family obligations

Chinese culture has traditionally placed great importance on family obligations, including a belief in the need for children to repay parents for their efforts in raising the children, a willingness to sacrifice for the family (including funding of their overseas education), and respect for the authority of the family. These obligations are rooted in Confucian teachings, a patriarchal clan system, and the long agrarian tradition of Chinese society (Qi, 2015). However, the attainment of wealth and property through individual initiative in free-market economies has allowed traditional obligations to be integrated with more individualistic desires (Fuligni & Zhang, 2004), both impulses of which are accommodated in contemporary discourses about the Chinese Dream (Wang, 2014). As the only generation that experienced the rapid transition from controlled economy to market economy, the elite middle age cohort has attitudes regarding family obligations different from their parents and children.

Consisting mostly of one-child parents, the elite group has paid great attention to their children's education success, which is seen as bringing honour and respect to their families (Chao & Tseng, 2002). They still uphold family obligations to take care of their elders, but such traditions could decline as a result of long-term exposure to a market economy. Children with more education and greater economic resources request and use more formal age support services (Sun, 2002). The ongoing re-interpretation and renegotiation of filial piety reflects China's transition to the individualization of society in which self-interest paradoxically drives behaviour for care of the elderly (Qi, 2015). As a result, the elite cohort may not envisage a governmental role in caring for their parents, but simultaneously do not expect the same care-giving from their children given their own accumulated wealth and various urban welfare schemes. With their children growing up and leaving the household, the group is therefore attaining considerable discretionary time and income for outbound travel and other leisure activity.

3.5. Behavioural patterns in an era of changing lifestyles and social values

Witnesses to rapidly improving social values and lifestyles in their prime years (18–38), members of COG3-4 transitioned from a monotonous and naïve childhood that gave them an 'idealistic' complex (Zhang, 2009) to a privileged adulthood of consumption where increasing prosperity and the questioning of old value systems were both taken for granted. An emerging habit of independent thinking after the Cultural Revolution is associated with the emphasis on 'scientific training' experienced by graduates of higher education. The elite group harbours memories of The 1989 Tiananmen Square incident, with some serving as active participants in the protest or advocates of the student movement. The incident itself symbolized profound social change – although the Communist political leadership was still firmly entrenched, the incident brought a conscious recognition to the regime in power that China needed to develop a pragmatic plan to relate more positively with the rest of the world (especially the West) and to shift away from a rigid demand economy toward a 'manageable' market driven economy (Ralston, Holt, & Terpstra, 1995). The South China tour in 1993 conducted by Deng Xiaoping, the supreme leader at the time, symbolized this pragmatic mentality and initiated the development that propelled China's economy into the world's second biggest.

With China thus integrating into the world community and Western managers and entrepreneurs importing Western business systems, values, expectations, and behaviour patterns previously unfamiliar to the Chinese, members of the elite group have become more open to new methods and processes, embracing organizational changes, and accepting greater risks through individual decision making. Compared to their elders, they are more likely to regard 'self-development' as the most important thing in life, and are more individualistic and secular (Sun & Wang, 2010). Yet, and paradoxically, some conventions remain influential (Ralston et al., 1995). For example, the traditional emphasis on informal reciprocal obligations (*guanxi*) persists as a key factor in interpersonal relationships. Yet, research is lacking as to how changing value structures that amalgamate Eastern and Western impulses may impact on psychological and behavioural patterns with regard to travel decisions of themselves or their children.

A related factor is the extent to which the elite middle age group had been substantially influenced by outside pop culture during China's transition from a controlled to market-oriented economy (e.g. Zhou, 2015). In the early 1990s, there was an overwhelmingly unidirectional flow of Asian popular culture to the mainland – a survey in 1993 indicated that over 85% of young people were substantially influenced by Hong Kong and Taiwan pop culture (Gold, 1993). This offered an outlet for individuality, subverting the authorities without confronting the social mechanics of domination per se. Sentiments expressed in foreign pop culture as to directions for personal development were certainly assimilated, an example being 'Go With Your Feelings', a popular song by the Taiwanese singer Su Rui (Gold, 1993). More recently, with the economic growth of the Mainland quickly surpassing those of Hong Kong and Taiwan, mainland Chinese pop culture has superseded those of Hong Kong and Taiwan in influence. Concurrently, social and political stability have been emphasized as necessities for economic and national development. What implications this second dichotomy/hybridisation of external/internal pop culture influence has on travel psychology and patterns is also unclear.

Because of the Cultural Revolution and concomitant destruction of associated tangible and intangible cultural heritage, members of this elite group might have experienced a lacuna in their formative years through lack of exposure to traditional ideologies such as Confucianism. However, these have recrudesced substantially. A renewed interest in Confucianism has originated inside various walks of life without evident initiative by the central and local authorities, yet, from the late 1980s the promotion of Confucianism has been increasingly orchestrated by the government (Holbig & Gilley, 2010). With the opportunities for self-development and actualization, 'investment' in oneself and one's children has been the main driving force. How these are reflected in the tourism sector is not clear. Relevant research would answer a call by Arlt (2013) to explore whether 'investment' in personal prestige, self-esteem, social and cultural capital and learning, has been the main driving force of Chinese outbound tourism.

4. Agenda for future research

This analysis of statistics and literature reveal an elite middle age Chinese cohort of significant size and with ample buying power to support the outbound travel of themselves and their children. However, the extent to which this group distinctly contributes and will contribute to Chinese outbound travel, as VFR, business package or other types of tourist, and especially during trips to long-haul destinations such as Europe, Australia, and the US, is unclear. Within-group consistency with regards to ideology, value of life, and fictions about other countries is probable, but actual consumption and behavioural patterns are less certain and remain unexplored. Personal, psychological, marketing, and product/experiences related factors that shape the demand and behaviour of this cohort are largely under-researched.

For example, how do members decide which destinations to visit or what activities to participate in? To date, the majority of mainland Chinese travellers have stayed within East or Southeast Asia before venturing further, but will they choose 'mainly the most famous sights in the most famous cities in major destinations' or will they choose destinations based on childhood/early adulthood interests in culture and destinations shaped by revolutionary developments in Chinese society and economy? The higher market elite middle age group has already been known to visit remote destinations in pursuit of exotic activities, but what about the other elite middle age tourists? Will they do the same? How do their travel experiences shape their interests and increase their confidence/efficacy in travelling? Their spending patterns and information search behaviour requires further empirical research.

Personal, psychological and behavioural factors also warrant more investigation. Chinese outbound tourism is induced not only by economic development, but also by psychological impulses implicating self-reinvention – to explore if 'being Chinese' means something new. Outbound trips could be used as a way to develop and affirm the tourists' self-identity, with recent studies exploring the Chinese 'gaze' on host destinations. For example, female Chinese felt confused by Macau's city presentation of its urban image due to Chinese cities' being fast urbanized and thus contrasted with the perceived lack of change in Macau's downtown urban development (Zhang & Hitchcock, 2017). Research indicates that self-referencing and other values shape Chinese tourists' perceptions during outbound travel but to what extent classical Chinese culture may be wholly relevant in the future is questioned (Sun, Zhang, & Ryan, 2015). This elite middle age cohort grew up with the concept of 'learning to the West' in China's transition into a more open and capitalistic society, yet, with Chinese cities increasingly becoming leaders in urbanization, what constitutes 'modernism' in the eyes of this elite Chinese cohort, what would be their 'gaze' on their overseas trips, and what constitutes 'otherness', are all timely and thought-provoking research questions.

Understanding the mindset and expectations of this cohort and what impacts their behaviour seem particularly relevant, as the cohort is currently in key positions in the workplace, and their mindset will shape future trends in China's values and socioeconomic development. Moreover, while older Chinese, in concert with longstanding societal norms, are actively involved in taking care of their grandchildren, their interaction with 'foreign' destinations may have an influence on their attitude toward traditional Chinese educational ideology. An important implication for countries such as Australia which heavily emphasize their protected areas and other natural attractions is whether the elite middle age Chinese visitors who are increasingly attracted to such places will accept the Western 'scientific' approach to product interpretation, or prefer the Eastern 'aesthetic' approach that privileges story-telling, poetry and metaphor in a meta-narrative of human-nature oneness and quest for harmony (Xu, Cui, Ballantyne, & Packer, 2013).

Studies indicate that contact between people from different cultural backgrounds (via tourism or other channels) may lead to positive attitude change, friendship and respect building (e.g. Pizam, Uriely, & Reichel, 2000), whereas other research indicates outcomes of tension, misunderstanding, stereotyping, irritation and ethnocentrism (e.g. Reisinger & Turner, 2002). The differences might be context-based and attributable to quality and quantity of the contact, cultural distance, and the degree of 'interculturalness' (e.g. Pizam et al., 2000; Reisinger & Turner, 2002). Urry (2002) famously posits the 'tourist gaze' as a dynamic and constructed context of social contact that is influenced by social group and historical period. Host gaze is an equally important reciprocal context. Chinese outbound tourists in diverse outbound destinations have been consistently reported as displaying poor behaviour. Chan (2006), for example, associated Chinese tourists to Vietnam with 'litter', 'haggle a lot', 'spit', 'bargain too much', 'trouble-making' and 'difficult to satisfy'. Conscious of possible damage to image and concomitant soft power aspirations, the Chinese National Tourism Administration even launched a national educational campaign to promote appropriate behaviour. The poor behaviour might be from tourists of a newly rich but relatively less well-off segment. Some of these behavioural problems might be 'culturally justifiable' (Moufakkir, 2011), and research finds that residents' empathy for tourists could be positively mediated by high cultural distance – the higher the distance, the more sympathy the residents could have towards the tourists (Moufakkir, 2011). Yet, research is warranted as to the cultural experience and behaviour of the middle class elite cohort.

This phenomenon calls for more research on examining the 'Chinese tourist gaze', as well as its connection with the 'demonstration effect', and whether these in tandem construct new ways of tourist self-reflection. The Chinese self could be shaped through interaction with host people. Chinese tourists might 'copy' or at least become conscious of the behavioural patterns of the host community, as social behaviour of people from a collectivistic culture (such as China) could be strongly influenced by in-group and outgroup distinctions (Triandis, 1989). While the aforementioned exposure to and influence of Western culture (often via Hong Kong and Taiwan) during puberty and early adulthood indicates receptiveness to this effect, at least in highly developed Western destinations, the rise of mainland China economic influence and self-confidence may serve as a mitigating factor, and also potentially as a cause of seemingly arrogant behaviour. With the elite group entering middle age, their mindset may be more fossilized, and how this reconciles with the sense of freedom and critical thinking that are legacies of external (e.g. Hong Kong and Taiwan) pop culture is unclear. The gaze of the older Chinese tourists and the perceived demonstration effect could offer an opportunity to understand identity formation and self-positioning of this elite group in an increasingly globalized world.

In addition, although within-group consistency is probable with regard to formative events, ideology, tradition, and value of life, a disparity in wealth status, political influence, and variety of subcultures may exist within this elite middle age group due to regional differences within China. Within-group differences may also manifest as a gap in economic status, influence of external culture, attitude on self and family development, and on values and behavioural states and patterns. For example, the fictions about other countries within the group would be largely individualized, affected by their experiences with regional difference as a contributing factor. Currently, a large percentage of Chinese outbound tourists are from the more commercialized and industrialized eastern coastal regions, followed by tourists from the fast-growing middle inland areas, and less from western interior regions or the economically struggling northeast. Future research is warranted into whether regional economic differences, political influence, subcultures, and formative experiences equate with potentials of generating outbound travels within this group and their travel behaviours.

Literature indicates that recent segmentation studies focus specifically on Chinese youth travellers. With the attendant discussion of the 2nd wave of Chinese outbound tourists and foreign independent travel, destinations are getting 'China ready' but with too much focus perhaps on assisting independent travellers. The China National Tourism Administration reported that more than 57.3 million outbound tourists were organized by travel agencies in 2016, among which about 45 million travelled to foreign countries, 9 million to Hong Kong and Macau, and 3 million to Taiwan (CNTA, 2017). Although the percentage of older Chinese travellers among the above figures is unknown, destination marketing organizations may need to collaborate more closely with travel agencies to further assist the group travellers and respond to the needs and wants of older elite travellers accustomed to high standards of service and experience. Based on the discussions, a

Psychology – gaze, otherness, authenticity, demonstration effect	Marketing – information search, destination and activity choices	Product/experiences – travel type, travel companion, experiences, spending patterns
• Is there a 'demonstration effect' for the Chinese middle age tourists? • What are the 'demonstration effects'? – what do they learn from their overseas trips? • If and how will the 'demonstration effect' make changes to their value systems or behavioral patterns? • What do they look for, 'modernity' or 'otherness'? • What are considered as 'otherness' and 'authenticity' to these tourists? • What are the impacts of their trips on their personal life? • What are their attitudes twoards diversity, flexibility and social issues in tourism?	• How do they decide on destinations? • Do they pick the mainly the most famous sights in the most famous cities in major destinations? • Will they pick destinations based on their childhood or early adulthood interests in culture and regions? • What activities give them a sense of 'self-reinventing' and self-actualization? • What marketing strategies and techniques are most effective to this group? • What are their information search patterns? • Will social media be effective in marketing to this cohort?	•What are their spending patterns? •Would they continue the current spending pattern which favors shopping rather than accommodations and activities? •What's their consumption priority? •How do travel experiences shape their interests and efficacy in overseas travel?

Figure 2. Future research questions.

conceptual map is outlined below with relevant questions which academics and destination managers and marketers need to explore to attract, satisfy and engender loyalty among elite middle aged Chinese tourists (Figure 2).

5. Conclusion

This paper identifies an elite middle age group of Chinese outbound tourists, a possible 3rd wave, by depicting their current status and potential development for the future. Segmenting a large market using distinctive generation cohort characteristics is plausible theoretically and practically, and compatible with China's collectivist inclinations. This segmentation is particularly suitable in helping the tourism industry service providers in identifying market characteristics for marketing, product development and comprehensive destination management. This elite middle age group has experienced a time of dramatic change causing a transformation of their traditional livelihood activities and lifestyles. Associated changes in materials wealth and values contribute to overseas travel proclivities that could entail both positive and negative impacts for both host and guest. More research is warranted to understand the travel urge, and generate experiences and services beneficial to the segment, so that diverse outbound destinations can achieve economic benefits from their presence.

Disclosure statement

No potential conflict of interest was reported by the authors.

References

Arita, S., Edmonds, C., La Croix, S., & Mak, J. (2011). Impact of approved destination status on Chinese travel abroad: An econometric analysis. *Tourism Economics, 17*(5), 983–996.

Arlt, W. G. (2013). The second wave of Chinese outbound tourism. *Tourism Planning & Development, 10*(2), 126–133.

Benckendorff, P., Moscardo, G., & Pendergast, D. (Eds.). (2010). *Tourism and generation Y*. Walingford: CABI.

Burton, R. (1995). *Travel geography* (2nd ed.). London: Longman.

Chan, Y. W. (2006). Coming of age of the Chinese tourists the emergence of non-Western tourism and host-guest interactions in Vietnam's border tourism. *Tourist Studies, 6*(3), 187–213.

Chao, R., & Tseng, V. (2002). Parenting of Asians. *Handbook of Parenting, 4*, 59–93.

Chen, Y., Lehto, X. Y., & Cai, L. (2013). Vacation and well-being: A study of Chinese tourists. *Annals of Tourism Research, 42*, 284–310.

Cheng, M., & Foley, C. (2017). Understanding the distinctiveness of Chinese Post-80s tourists through an exploration of their formative experiences. *Current Issues in Tourism*, 1–17.

China National Tourism Adminstration. (2016). *Annual report of China outbound tourism development*. Beijing: China Tourism Academy.

China National Tourism Adminstration. (2017). Retrieved from http://www.cnta.gov.cn/zwgk/lysj/201711/t20171108_846343.shtml

Choukhmane, T., Coeurdacier, N., & Jin, K. (2013). *The one-child policy and household savings* (SSRN Scholarly Paper No. ID 2341051). Rochester, NY: Social Science Research Network. Retrieved from https://papers.ssrn.com/abstract=2341051

Cohen, E. (1972). Toward a sociology of international tourism. *Social Research, 39*(1), 164–182.

Davey, G., De_Lian, C., & Higgins, L. (2007). The university entrance examination system in China. *Journal of Further and Higher Education, 31*(4), 385–396.

Dolnicar, S. (2007). Market segmentation in tourism. In A. G. Woodside & D. Martin (Eds.), *Tourism management: Analysis, behaviour, and strategy* (pp. 129–146). Wallingford: CABI.

Egri, C. P., & Ralston, D. A. (2004). Generation cohorts and personal values: A comparison of China and the United States. *Organization Science, 15*(2), 210–220.

Erickson, T. (2009). *Generations in China* [online]. Retrieved from http://blogs.harvardbusiness.org/erickson/2009/03/generations_in_china.html

Filep, S., Hughes, M., Mostafanezhad, M., & Wheeler, F. (2015). Generation tourism: Towards a common identity. *Current Issues in Tourism, 18*(6), 511–523.

Fuligni, A. J., & Zhang, W. (2004). Attitudes toward family obligation among adolescents in contemporary urban and rural China. *Child Development, 75*(1), 180–192.

Gold, T. B. (1993). Go with your feelings: Hong Kong and Taiwan popular culture in greater China. *The China Quarterly, 136*, 907–925.

Guinn, R. (1980). Elderly recreational vehicle tourists: Motivations for leisure. *Journal of Travel Research, 19*(1), 9–12.

Guo, F., & Huang, Y. S. (2010). Does 'hot money' drive China's real estate and stock markets? *International Review of Economics & Finance, 19*(3), 452–466.

Hagan, L., & Uysal, M. (1991). *An examination of motivations and activities of pre-retirement (50–64) and postretirement (65–98) age group for a touring trip*. Proceedings of the 22nd Annual Conference of the Travel and Tourism Research Association, Long Beach, CA. pp. 381–386.

Hesketh, T., Lu, L., & Xing, Z. W. (2005). The effect of China's one-child family policy after 25 years. *The New England Journal of Medicine, 353*, 1171–1176.

Holbig, H., & Gilley, B. (2010). Reclaiming legitimacy in China. *Politics & Policy, 38*(3), 395–422.

Hsu, C. H., Cai, L. A., & Wong, K. K. F. (2007). A model of senior tourism motivations—Anecdotes from Beijing and Shanghai. *Tourism Management, 28*(5), 1262–1273.

Hsu, C. H., & Song, H. (2012). 'Projected images of major Chinese outbound destinations'. *Asia Pacific Journal of Tourism Research, 17*(5), 577–593.

Jang, S., Yu, L., & Pearson, T. (2003). Chinese travellers to the United States: A comparison of business travel and visiting friends and relatives. *Tourism Geographies, 5*(1), 87–108.

Jin, X., & Wang, Y. (2016). Chinese outbound tourism research: A review. *Journal of Travel Research, 55*(4), 440–453.

Jin, X., Wu, L., Becken, S., & Ding, P. (2016). How do worry, self-efficacy, and coping interact? Examining Chinese tourists to Australia. *Journal of China Tourism Research, 12*(3–4), 374–393.

Kim, S. (2012). A cross-cultural study of on-site film-tourism experiences among Chinese, Japanese, Taiwanese and Thai visitors to the Daejanggeum Theme Park, South Korea. *Current Issues in Tourism, 15*(8), 759–776.

Lehto, X. Y., Jang, S., Achana, F. T., & O'Leary, J. T. (2008). Exploring tourism experience sought: A cohort comparison of baby boomers and the silent generation. *Journal of Vacation Marketing, 14*(3), 237–252.

Li, X. (2016). *Chinese outbound tourism 2.0.* Boca Raton, FL: Apple Academic Press, Inc.

Li, X., Lai, C. T., Harrill, R., Kline, S., & Wang, L. (2011). When east meets west: An exploratory study on Chinese outbound tourists' travel expectations. *Tourism Management, 32*(4), 741–749.

Li, X., Li, X., & Hudson, S. (2013). The application of generational theory to tourism consumer behavior: An American perspective. *Tourism Management, 37*, 147–164.

Liu, G., & Ryan, C. (2011). The role of Chinese students as tourists and hosts for overseas travel. *Asia Pacific Journal of Tourism Research, 16*(4), 445–464.

Liu, L., Winter, C., & Zilibotti, F. (2015). *Housing and wealth accumulation in urban China, before and after the 1994 housing reform.* Retrieved from https://www.econ.uzh.ch/dam/jcr:00000000-57f0-1d93-0000-000058e18ee1/housing_and_wealth_feb2015.pdf

Mannheim, K. (1953). *Essays on sociology and social psychology.* New York, NY: Oxford University Press.

McGuire, F., Dottavio, D., & O'Leary, J. (1986). Constraints to participation in outdoor recreation across the life span: A nationwide study of limitors and prohibitors. *The Gerontologist, 26*(5), 538–544.

Meng, X. (2007). Wealth accumulation and distribution in urban China. *IZA discussion paper series.* Retrieved from http://ftp.iza.org/dp2553.pdf

Michailova, S., & Hutchings, K. (2006). National cultural influences on knowledge sharing: A comparison of China and Russia. *Journal of Management Studies, 43*, 383–405.

Moufakkir, O. (2011). The role of cultural distance in mediating the Host Gaze. *Tourist Studies, 11*(1), 73–89.

Muller, T. E. (1991). Using personal values to define segments in an international tourism market. *International Marketing Review, 8*(1), 49–70.

Ong, C. E., & du Cros, H. (2012). The Post-Mao Gazes: Chinese backpackers in Macau. *Annals of Tourism Research, 39*(2), 735–754.

Pilcher, J. (1994). Mannheim's sociology of generations: An undervalued legacy. *The British Journal of Sociology, 45*(3), 481–495.

Pizam, A., Uriely, N., & Reichel, A. (2000). The intensity of tourist–host social relationship and its effects on satisfaction and change of attitudes: The case of working tourists in Israel. *Tourism Management, 21*(4), 395–406.

Qi, X. (2015). Filial obligation in contemporary China: Evolution of the culture-system. *Journal for the Theory of Social Behaviour, 45*(1), 141–161.

Ralston, D. A., Holt, D. H., & Terpstra, R. H. (1995). The impact of culture and ideology on managerial work values: A study of the United States, Russia, Japan, and China. *Academy of Management Proceedings, 1995*(1), 187–191.

Reisinger, Y., & Turner, L. W. (2002). Cultural differences between Asian tourist markets and Australian hosts, part 1. *Journal of Travel Research, 40*(3), 295–315.

Rosenzweig, M., & Zhang, J. (2014). *Co-residence, life-cycle savings and inter-generational support in Urban China* (Working Paper No. 20057). National Bureau of Economic Research.

Sakai, M., Brown, J., & Mak, J. (2000). Population aging and Japanese international travel in the 21st century. *Journal of Travel Research, 38*(3), 212–220.

Schewe, C. D., & Noble, S. M. (2000). Market segmentation by cohorts: The value and validity of cohorts in America and abroad. *Journal of Marketing Management, 16*(1e3), 129–142.

Schuman, H., & Scott, J. (1989). Generations and collective memories. *American Sociological Review, 54*(3), 359–381.

Sellick, M. C. (2004). Discovery, connection, Nostalgia. *Journal of Travel & Tourism Marketing, 17*(1), 55–71.

Sparks, B., & Pan, W. (2009). Chinese outbound tourists: Understanding their attitudes, constraints and use of information sources. *Tourism Management, 30*(4), 483–494.

Strauss, A., & Howe, N. (1997). *The fourth turning: An American prophecy.* Broadway, New York: Bantam Doubleday Dell Publishing Group.

Sun, R. (2002). Old age support in contemporary urban China from both parents' and children's perspectives. *Research on Aging, 24*(3), 337–359.

Sun, J., & Wang, X. (2010). Value differences between generations in China: A study in Shanghai. *Journal of Youth Studies, 13*(1), 65–81.

Sun, M., Zhang, X., & Ryan, C. (2015). Perceiving tourist destination landscapes through Chinese eyes: The case of South Island, New Zealand. *Tourism Management, 46*, 582–595.

Tang, W., & Parish, W. L. (2000). *Chinese urban life under reform: The changing social contract.* Cambridge: Cambridge University Press.

Tourism Research Australia. (2017). *International visitor survey data.* Unpublished data.

Triandis, H. C. (1989). The self and social behavior in differing cultural contexts. *Psychological Review, 96*(3), 506–520.

Tuniu.com. (2016). *Chinese outbound tourists behaviour analysis*. Retrieved from http://science.china.com.cn/2018-02/05/content_40216059.htm

United Nations World Tourism Organisation. (2016). UNWTO tourism highlights 2016 edition. Retrieved from http://www.e-unwto.org/doi/pdf/10.18111/9789284418145

Urry, J. (2002). The tourist gaze (Theory, culture and society series).

Van Norden. (2015). Five colossal events changed China forever. Retrieved from http://nationalinterest.org/feature/5-colossal-events-changed-china-forever-13046

Vijaya, R. M. (2010). Medical tourism: Revenue generation or international transfer of healthcare problems? *Journal of Economic Issues*, *44*(1), 53–70.

Wang, Z. (2014). The Chinese dream: Concept and context. *Journal of Chinese Political Science*, *19*, 1–13.

Weaver, D. (2015). Tourism and the Chinese dream: Framework for engagement. *Annals of Tourism Research*, *51*, 54–56.

Wong, I. A., Fong, H. I. V., & Liu, M. T. (2012). Understanding perceived casino service difference among casino players. *International Journal of Contemporary Hospitality Management*, *24*(5), 753–773.

World Tourism Cities Federation. (2015). Retrieved from https://www.ipsos.com/sites/default/files/2016-06/021.1-Chinese-Outbound-Tourist.pdf

Wu, M. Y., & Pearce, P. (2014). Chinese recreational vehicle users in Australia: A netnographic study of tourist motivation. *Tourism Management*, *43*, 22–35.

Xu, H., Cui, Q., Ballantyne, R., & Packer, J. (2013). Effective environmental interpretation at Chinese natural attractions: The need for an aesthetic approach. *Journal of Sustainable Tourism*, *21*, 117–133.

Zhang, Z. (2009). Bearing witness: Chinese urban cinema in the era of the 'transformation'. *Shanghai University Journal: Social Science Edition*, *4*, 96–113.

Zhang, H. Q., Heung, V., & Yan, Q. (2009). Play or not to play—An analysis of the mechanism of the zero-commission Chinese outbound tours through a game theory approach. *Tourism Management*, *30*(3), 366–371.

Zhang, Y., & Hitchcock, M. J. (2017). The Chinese female tourist gaze: A netnography of young women's blogs on Macao. *Current Issues in Tourism*, *20*(3), 315–330.

Zhou, X. Y. (2015). Rock music in contemporary China: An ideological arena. *Journal of Literature and Art Studies*, 5(12), 1146–1151. Retrieved from http://www.davidpublisher.org/Public/uploads/Contribute/567a39b9a2774.pdf

Zimmer, Z., Brayley, R. E., & Searle, M. S. (1995). Whether to go and where to go: Identification of important influences on seniors' decisions to travel. *Journal of Travel Research*, *33*(3), 3–10.

Structural relationships among mega-event experiences, emotional responses, and satisfaction: focused on the 2014 Incheon Asian Games

HakJun Song, MinCheol Kim and Yunseon Choe

ABSTRACT

In this study, the relationships among the experiences of visitors to the 2014 Incheon Asian Games, their emotional responses, and their satisfaction are explored by employing a theoretical background derived from Pine and Gilmore's experience economy theory. It was found that the visitors' satisfaction with the Incheon Asian Games was mainly affected by their emotional responses of pleasure and arousal. Furthermore, their entertainment, educational, and esthetic experiences were positive antecedents of pleasure and arousal. The results also showed that their educational and esthetic experiences positively influenced dominance, although the construct of dominance was not a significant antecedent of satisfaction.

Introduction

Mega-events are defined as 'large-scale cultural (including commercial and sporting) events, which have a dramatic character, mass popular appeal, and international significance' (Roche, 2000, p. 1). They have generally created effective economic benefits, social impacts, and cultural effects in host regions and nations (Chaohui, Lin, & Qiaoyun, 2012), including the improvement of hospitality facilities, accommodations, infrastructure, and amenities; tourism-related training; and international prominence (Dansero & Puttilli, 2010). Mega-events seem to provide visitors with many advantages in regard to the experience economy, including entertainment, education, escape from daily routine, and appreciation of the tourism industry (Li, Huang, & Cai, 2009). However, little research has recognized the important role of experience in the context of the mega-event especially for mega-sports events. Specifically, a paucity of studies exists to identify the important relationships among the mega-event experience, and an individual's emotional responses and satisfaction (Song, Lee, Park, Hwang, & Reisinger, 2015).

Several scholars (e.g. Andersson, 2007; Getz, 2010; Oh, Fiore, & Jeoung, 2007) have stated that the concept of experience can become a vital part of understanding mega-event visitors' behaviours. Other scholars (e.g. Andersson, 2007; Pine & Gilmore, 1998; Pine & Gilmore, 1999) have emphasized two important elements related to hosting successful mega-events: whether valuable experiences and emotions are offered to visitors and whether the visitors are satisfied with the experience-based attractions available during the mega-events. Therefore, the present research tries to extend previous experience economy studies (Pine & Gilmore, 1998) by focusing on the 2014 Incheon

Asian Games South Korea (2014) in order to determine whether the visitors' satisfaction was impacted by their emotional responses. In addition to providing the findings of this study, this paper will suggest successful strategies by which to attract more visitors, provide memorable experiences, and increase visitor satisfaction.

Mega-event experience and emotional response

In order to better understand tourists, researchers have applied the experience economy concept to tourism studies (Hayes & MacLeod, 2007; Hosany & Witham, 2010; Mehmetoglu & Engen, 2011; Oh et al., 2007; Quadri-Felitti & Fiore, 2012; Yuan & Wu, 2008). Some researchers (Getz, 2010; Li et al., 2009) have emphasized that festivals or mega-events, particularly in the various tourism sectors, need to use the concept of experience to increase the level of visitor satisfaction. Emotional response has also considered as a significant construct which can affect the satisfaction level of visitors at festivals and mega-events.

The results also showed that emotional value was positively influenced by the festival programme and natural environments. Lee and Kyle (2013) investigated festival visitors' emotional responses, particularly based on love, joy, surprise, fear, anger, and sadness. Song, Ahn, and Lee (2015) found that the experience economy performed an important role in regard to explaining the 2012 Yeosu Expo visitors' emotions. Based on previous studies, it can be assumed that the 4Es (i.e. entertainment (ENTE), educational (EDUE), escape (ESCE), and esthetic (ESTE) experience) provide a significant theoretical framework for hosting a mega-event to attract spectators and event participants.

Emotional response and satisfaction

Oliver (1993) studied several key elements that required for an emotional response related to satisfaction: pleasure, surprise, interest, and hostility. Giese and Cote (2000) concluded that consumer

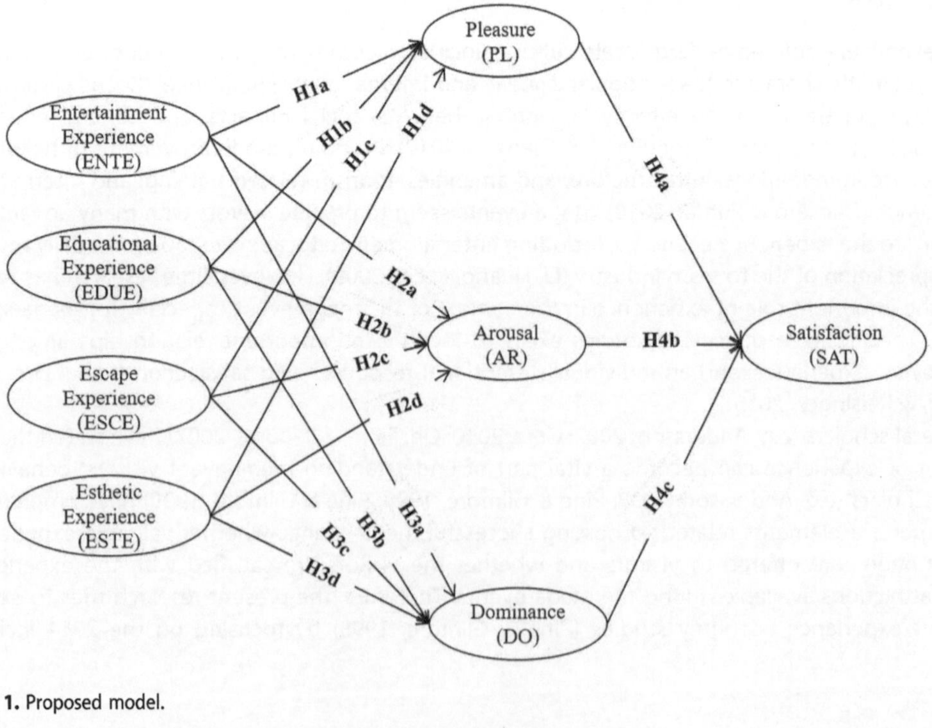

Figure 1. Proposed model.

satisfaction is explained by an individual's response (emotional or cognitive). Izquierdo and Samaniego (2011) found that a positive museum activity produced positive emotions, developed more pleasurable user experiences, and provided a greater degree of satisfaction in a study that identified relationships among visitor emotions, satisfaction, and future intentions.

Method

This study provides a framework for a research model for hypotheses testing as shown in Figure 1. The present study includes four sections: (1) mega-event experiences at the Incheon Asian Games, (2) the Pleasure-Arousal-Dominance (PAD) Emotional State Model, (3) satisfaction, and (4) demographic information. This study developed a preliminary construct measurement based on an intensive review of the available literature and then adapted experience economy to the mega-event framework (Mehrabian & Russell, 1974; Oh et al., 2007; Pine & Gilmore, 1998; Song, Choi, & Lee, 2011). The measurement items, except for the demographic information, were rated on

Table 1. Confirmatory factor analysis for the measurement model.

Constructs	Factor loading	*t*-value	Cronbach's alpha
F1: Entertainment Experience (ENTE)			0.864
The Incheon Asian Games was amusing to watch	0.838		
The Incheon Asian Games was very entertaining	0.874	13.502	
The entertaining activities of the Incheon Asian Games were captivating	0.774	9.833	
F2: Educational Experience (EDUE)			0.815
The experience of the Incheon Asian Games increased my knowledge, and I learned a lot from it.	0.685		
The experience of the Incheon Asian Games inspired my curiosity and interest in learning new things	0.857	7.779	
The experience of the Incheon Asian Games was highly educational for me	0.778	7.256	
F3: Escape Experience (ESCE)			0.879
Through the Incheon Asian Games I felt placed in a completely different world	0.869		
Through the Incheon Asian Games I completely forgot about my daily life	0.887	11.696	
Through the Incheon Asian Games I completely escaped from reality	0.773	10.178	
F4: Esthetic Experience (ESTE)			0.910
I would feel good just being around the setting of the Incheon Asian Games	0.817		
The setting of the Incheon Asian Games was interesting	0.899	11.121	
The setting of the Incheon Asian Games provided pleasure to my senses	0.924	11.698	
F5: Pleasure (PL)			0.909
I felt 'joyful' when I visited the Incheon Asian Games	0.881		
I felt 'cheerful' when I visited the Incheon Asian Games	0.877	15.843	
I felt 'happy' when I visited the Incheon Asian Games	0.805	15.494	
F6: Arousal (AR)			0.872
I felt 'excited' when I visited the Incheon Asian Games	0.805		
I felt 'astonished' when I visited the Incheon Asian Games	0.834	11.714	
I felt 'elated' when I visited the Incheon Asian Games	0.825	11.975	
F7: Dominance (DO)			0.877
I felt 'confident' when I visited the Incheon Asian Games	0.825		
I felt 'flexible' when I visited the Incheon Asian Games	0.895	10.750	
I felt 'proactive' when I visited the Incheon Asian Games	0.808	8.950	
F8: Satisfaction (SAT)			0.923
My expectations were satisfied by my visit to the Incheon Asian Games	0.865		
My feelings about the Incheon Asian Games are positive	0.921	11.260	
Overall, I was satisfied with my visit to the Incheon Asian Games	0.912	15.349	

Measurement Model	χ^2	S-B χ^2	*df*	Normed S-B χ^2	NFI	NNFI	CFI	RMSEA
	308.686	267.441	224	1.194	0.833	0.959	0.967	0.036
Suggested value*				≤3	≥0.9	≥0.9	≥0.9	≤0.08

Note: Merdia's normalized = 11.046.
*All standardized factor loadings are significant at $p < 0.001$.

a 5-point Likert scale that ranged from 1 (strongly disagree) to 5 (strongly agree). The measurements for the mega-event experiences consisted of 12 items. A unipolar PAD scale that consisted of nine items was employed to measure positive emotional response related to visitors' experiences at the Incheon Asian Games. This scale was designed by Mehrabian and Russell (1974) and Tsaur, Chiu, and Wang (2007). Three items were adapted from Oliver (1997) to measure visitors' satisfaction.

Since the majority of the visitors at the Incheon Asian Games preferred to watch ball games, an on-site survey of Korean visitors was held at two main sites: the Samsan World Gymnasium for basketball games and the Incheon Asiad Main Stadium for football games. A research team, composed of eight trained members, conducted the self-administered questionnaire survey from 24 September 2014 to 3 October 2014. A face-to-face survey and convenience sampling method were used for the data collection process. The field researchers briefly explained the purpose of the survey and asked whether the participants were available for the survey. Upon consent, the survey questionnaires were distributed to the participants. A total of 420 questionnaires were collected. Of these questionnaires, 18 uncompleted questionnaires were excluded from the study sample. Therefore, a total number of 402 questionnaires were selected to be included in the final sample of this study.

Results

The results of the measurement model showed acceptable values for all of the fit indices (see Table 1), indicating that the covariance matrix of the data and the proposed measurement model were similar (i.e. S-B $\chi^2 = 267.441$ (df= 224); Normed Fit Index (NFI) = 0.833; Non-normed Fit Index (NNFI) = 0.959; Comparative Fit Index (CFI) = 0.967, and Root Mean Square Error of Approximation (RMSEA) = 0.036). All of the latent variables showed Cronbach's alpha coefficients for reliability that were higher than 0.7, the minimum value (Nunnally & Bernstein, 1994). In the measurement model, a satisfactory value for the convergent validity was confirmed because the value of the standardized factor loadings for each measurement item was larger than the criteria value of 0.5 (i.e. they ranged from 0.685 to 0.924). The convergent validity was also checked because the average variance extracted (AVE) was greater than the minimum value of 0.5 and the composite reliability values were above the minimum value of 0.7 (Hair, Black, Babin, Anderson, & Tatham, 2006) (see Table 2).

Table 2. Results of measurement model.

Construct	ENTE	EDUE	ESCE	ESTE	PL	AR	DO	SAT
ENTE	0.688	0.368 (0.607)	0.211 (0.459)	0.399 (0.632)	0.413 (0.643)	0.410 (0.640)	0.196 (0.443)	0.354 (0.595)
EDUE	0.077	0.603	0.288 (0.537)	0.433 (0.658)	0.444 (0.666)	0.454 (0.674)	0.426 (0.653)	0.304 (0.551)
ESCE	0.091	0.074	0.713	0.440 (0.663)	0.406 (0.637)	0.376 (0.613)	0.353 (0.594)	0.239 (0.489)
ESTE	0.081	0.064	0.088	0.776	0.563 (0.750)	0.498 (0.706)	0.444 (0.666)	0.442 (0.665)
PL	0.089	0.068	0.084	0.078	0.731	0.752 (0.867)*	0.387 (0.622)	0.588 (0.767)
AR	0.096	0.081	0.100	0.091	0.099	0.675	0.387 (0.622)	0.587 (0.766)
DO	0.075	0.069	0.086	0.077	0.073	0.081	0.712	0.310 (0.557)
SAT	0.092	0.072	0.092	0.082	0.086	0.095	0.078	0.809
CR	0.869	0.819	0.881	0.912	0.891	0.862	0.881	0.927

Notes: ENTE = Entertainment Experience; EDUE = Educational Experience; ESCE = Escape Experience; ESTE = Esthetic; Experience; PL, pleasure; AR, arousal; DO, dominance; CR = composite reliability; * = Highest correlations between pairs of constructs.

The values of average variance extracted are along the diagonal; Squared correlations among latent constructs are above the diagonal; Correlations among latent constructs are in the parenthesis; Standard errors among latent constructs are below the diagonal.

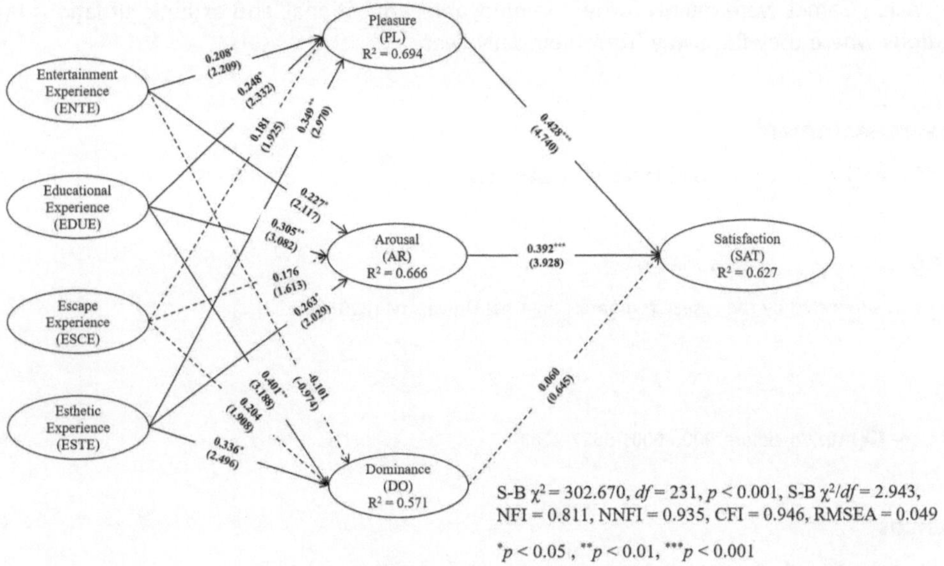

Figure 2. Final model.

The proposed structural model showed a marginal fit to the sample data (i.e. S-B $\chi^2 = 302.670$ (df = 231), $p < 0.001$, NFI = 0.811, NNFI = 0.935, CFI = 0.946, and RMSEA = 0.049 (Figure 2)). In terms of the hypothesis testing for the research model, the esthetic, educational, and entertainment experiences had positive effects on pleasure. Arousal was positively influenced by entertainment, educational, and esthetic experiences. Moreover, the educational and esthetic experiences had positive effects on dominance. Among the latent variables related to emotional responses, pleasure and arousal had significant effects on satisfaction. Overall, the four constructs (i.e. entertainment, educational, escape, and esthetic experiences) were generally found to play important roles in explaining visitors' emotional responses to the Incheon Asian Games.

Conclusion

The results of the current study provide empirical evidence of the experience economy. Within the context of a mega-event, entertainment, educational, and esthetic experiences affected the emotional responses of pleasure and arousal, while dominance was influenced by educational and esthetic experiences. The results also showed that pleasure most significantly influenced satisfaction related to visiting the Incheon Asian Games, followed by arousal. This study showed that the visitors' pleasure and arousal should be the primary marketing precursors to satisfactory visitor experiences. The study results are consistent with Lee, Lee, Lee, and Babin's (2008) findings, which stated that emotions are responsible for the mediation between event experiences and visitor satisfaction. Moreover, the study findings are in accordance with previous research that showed that satisfaction is not affected by dominance (Robert & John, 1982; Sherman, Mathur, & Smith, 1997). Russell and Pratt (1980) also stated that, because dominance is primarily based on an individual's cognitive interpretation, the dimension has low predictive values, which preclude it from being used when examining affective responses.

Interestingly, this study showed that escape experiences did not have any statistical effect on emotional response. It seems that the result of this study can be derived because the experience has relatively less influence on the emotional response rather than the other experience components of experience economy. The study results imply that the emotional responses of the visitors to the

Incheon Asian Games were mainly formed in enjoyable, educational, and esthetic situations instead of situations where they felt away from their daily lives.

Disclosure statement

No potential conflict of interest was reported by the authors.

Funding

This work was supported by the research grant of Pai Chai University in 2018.

ORCID

Yunseon Choe http://orcid.org/0000-0001-5527-2263

References

The 2014 Incheon Asian Games. (2014). *Overview*. Retrieved from http://www.incheon2014.kr/asiadStory/incheonAsiad/overview/index

Andersson, T. D. (2007). The tourist in the experience economy. *Scandinavian Journal of Hospitality and Tourism*, *7*(1), 46–58.

Chaohui, W., Lin, L., & Qiaoyun, X. (2012). Impact of tourists' perceived value on behavioral intention for mega events: Analysis of inbound and domestic tourists at Shanghai World Expo. *Chinese Geographical Science*, *22*(6), 742–754.

Dansero, E., & Puttilli, M. (2010). Mega-events tourism legacies: The case of the Torino 2006 Winter Olympic Games – a territorialisation approach. *Leisure Studies*, *29*(3), 321–341.

Getz, D. (2010). The nature and scope of festival studies. *International Journal of Event Management Research*, *5*(1), 1–47.

Giese, J. L., & Cote, J. A. (2000). Defining consumer satisfaction. *Academy of Marketing Science Review*, *4*(2), 1–24.

HairJr, J. F., Black, W. C., Babin, B. J., Anderson, R. E., & Tatham, R. L. (2006). *Multivariate data analysis* (6th ed.). Upper Saddle River, NJ: Pearson Prentice Hall.

Hayes, D., & MacLeod, N. (2007). Packaging places: Designing heritage trails using an experience economy perspective to maximize visitor engagement. *Journal of Vacation Marketing*, *13*(1), 45–58.

Hosany, S., & Witham, M. (2010). Dimensions of cruisers' experiences, satisfaction, and intention to recommend. *Journal of Travel Research*, *49*(3), 351–364.

Izquierdo, C. C., & Samaniego, M. J. G. (2011). Strengthening members' relationships through cultural activities in museums. *Journal of Leisure Research*, *43*(4), 560–588.

Lee, J. J., & Kyle, G. T. (2013). The measurement of emotions elicited within festival contexts: A psychometric test of a festival consumption emotions (FCE) scale. *Tourism Analysis*, *18*(6), 635–649.

Lee, Y. K., Lee, C. K., Lee, S. K., & Babin, B. J. (2008). Festivalscapes and patrons' emotions, satisfaction, and loyalty. *Journal of Business Research*, *61*(1), 56–64.

Li, M., Huang, Z., & Cai, L. A. (2009). Benefit segmentation of visitors to a rural community-based festival. *Journal of Travel & Tourism Marketing*, *26*(5-6), 585–598.

Mehmetoglu, M., & Engen, M. (2011). Pine and Gilmore's concept of experience economy and its dimensions: An empirical examination in tourism. *Journal of Quality Assurance in Hospitality & Tourism*, *12*(4), 237–255.

Mehrabian, A., & Russell, J. A. (1974). *An approach to environmental psychology*. Cambridge, MA: MIT Press.

Nunnally, J. C., & Bernstein, I. H. (1994). *Psychometric theory*. New York, NY: MGraw-Hill.

Oh, H., Fiore, A. M., & Jeoung, M. (2007). Measuring experience economy concepts: Tourism applications. *Journal of Travel Research*, *46*(2), 119–132.

Oliver, R. L. (1993). Cognitive, affective, and attribute bases of the satisfaction response. *Journal of Consumer Research*, *20* (3), 418–430.

Oliver, T. A. (1997). *Satisfaction: A behavioral perspective on the consumer*. Boston, MA: McGraw-Hill.

Pine, B. J., & Gilmore, J. H. (1998). Welcome to the experience economy. *Harvard Business Review*, *76*(4), 97–105.

Pine, B. J., & Gilmore, J. H. (1999). *The experience economy: Work is theatre and every business a stage*. Boston, MA: Harvard Business School Press.

Quadri-Felitti, D., & Fiore, A. M. (2012). Experience economy constructs as a framework for understanding wine tourism. *Journal of Vacation Marketing*, *18*(3), 3–15.

Robert, D., & John, R. (1982). Store atmosphere: An environmental psychology approach. *Journal of Retailing*, *58*, 34–57.

Roche, M. (2000). *Mega-events and modernity: Olympics and expos in the growth of global culture*. London: Routledge.

Russell, J. A., & Pratt, G. A. (1980). A description of the affective quality attributed to environments. *Journal of Personality and Social Psychology, 38*(2), 311–322.

Sherman, E., Mathur, A., & Smith, R. B. (1997). Store environment and consumer purchase behavior: Mediating role of consumer emotions. *Psychology & Marketing, 14*(4), 361–378.

Song, H. J., Ahn, Y. J., & Lee, C. K. (2015). Structural relationships among strategic experiential modules, emotion and satisfaction at the Expo 2012 Yeosu Korea. *International Journal of Tourism Research, 17*(3), 239–248.

Song, H. K., Choi, Y. J., & Lee, C. K. (2011). A study of festival visitors' loyalty based on experience economy: The case of Boryeong Mud festival. *Korean Journal of Tourism Research, 25*(6), 179–198.

Song, H. J., Lee, C. K., Park, J. A., Hwang, Y. H., & Reisinger, Y. (2015). The influence of tourist experience on perceived value and satisfaction with temple stays: The experience economy theory. *Journal of Travel & Tourism Marketing, 32*(4), 401–415.

Tsaur, S. H., Chiu, Y. T., & Wang, C. H. (2007). The visitors behavioral consequences of experiential marketing: An empirical study on Taipei Zoo. *Journal of Travel & Tourism Marketing, 21*(1), 47–64.

Yuan, Y. H., & Wu, C. (2008). Relationship among experiential marketing, experiential value, and customer satisfaction. *Journal of Hospitality and Tourism Research, 32*(3), 387–410.

Sustainable Asian Destinations

When destination attractiveness shifts in response to climate change: tourists' adaptation intention in Taiwan's Kenting National Park

Wei-Ching Wang, Chung-Hsien Lin, Wen-Bor Lu and Su-Hsin Lee

ABSTRACT

We tested a structural model that integrates protection motivation theory with the individual's (farmer's) climate change adaptation process. The model helps us grasp the effects of climate change perception and hypothetical shifts in destination attractiveness, as well as threat and coping appraisals in light of tourists' adaptation intentions in coastal destinations. We collected a total of 333 on-site valid questionnaires from domestic tourists at Kenting National Park in Taiwan and analysed the structural relationships in the aforementioned study constructs using structural equation modelling. Our findings show that when tourists have a higher level of perception regarding global climate change, they believe that destination attractiveness will decrease. When presented with scenarios of hypothetical shifts in destination attractiveness, tourists with higher levels of adaptation intention will perceive greater levels of threat to tourism behaviours, as well as higher effectiveness in adaptive measures. We confirmed that the proposed theoretical framework for tourists' adaptation intention toward climate change is useful; the framework also sheds light on tourists' acknowledgement of hypothetical alterations in destination attractiveness caused by climate change, in addition to their psychological adaptations. We discuss theoretical and practical implications.

Introduction

As climate change awareness grows, tourism systems stakeholders' perception toward adapting to climate change has become an important research topic (Hopkins, 2014; Morrison & Pickering, 2013). Climate change, as defined by the Intergovernmental Panel on Climate Change (IPCC) refers to 'any change in climate over time, whether due to natural variability or as a result of human activity' (IPCC, 2007, p. 871). The IPCC defines adaptation as the 'adjustment in natural or human systems in response to actual or expected climatic stimuli or their effects, which moderates harm or exploits beneficial opportunities' (IPCC, 2007, p. 869). In Southeast Asian countries like Taiwan, Japan, and Korea, the most popular tourist and outdoor recreation destinations are coastal and marine environments (United Nations World Tourism Organization, 2013). Extreme weather events, along with major changes in climate frequency and intensity, have been observed in Taiwan (Hong, 2012; Yeh, 2014; Zhuo & Lu, 2014). Such events in the near future, such as rising sea levels and seawater temperature, could impact the country's coastal regions (Dai, 2002; Lee, 2008).

For coastal destination management organizations (DMOs), understanding tourists' adaptation behaviour in response to climate change is becoming increasingly important (Martinez et al., 2014; Schliephack & Dickinson, 2017). Weather is a situational variable that can influence tourists' choice of destination (Woodside & Lysonski, 1989). Furthermore, climate change may cause tourists' behaviours to shift, like by substituting their tourist destination for another one, engaging in other activities, or reducing the length of their trip (Dawson, Havitz, & Scott, 2011). In the existing tourism literature, psychological factors influencing tourists' adaptation to climate change have been neglected, and our understanding of their adaptation behaviour remains limited (Cocolas, Walters, & Ruhanen, 2016; Jopp, DeLacy, & Mair, 2010). The impact of climate change on the sustainable development of numerous tourist destinations has been evaluated to understand tourists' adaptive behaviours (Gössling, Scott, Hall, Ceron, & Dubois, 2012).

By reviewing the literature on climate change in other fields, we can see that researchers have applied the individual's (farmer's) climate change adaptation process (Bryant et al., 2000) or protection motivation theory (PMT) (Lam, 2015) to investigate people's adaptive behaviours and intentions. The form of PMT proposed by Rogers (1975) states that when individuals receive information on threats (e.g. climate change), they go through the cognitive mediating process before engaging in adaptive/maladaptive coping actions, which are then determined by their adaptation intention (i.e. protection motivation). Conversely, intention can be predicted using threat appraisals and coping appraisals. Dang, Li, and Bruwer (2012) successfully applied PMT to model the farmer's adaptation intention toward climate change. In their vulnerability assessment of coastal tourism, Moreno and Becken (2009) revealed that most studies on the individual's adaptive behaviours to climate change fail to consider climate change's impact on attributes of concern (e.g. destination attractiveness in tourism systems), which in turn affects the system's sensitivity (e.g. destination attractiveness in tourism systems).

Bryant et al. (2000) suggested that the adaptation process toward climate change includes characteristics of stress (e.g. climate change perception), the system's scale (e.g. destination, region, country), and other components that could affect the final adaptive response (e.g. crop insurance, crop choice). Bryant et al. (2000) believe that a better understanding can be gleaned by examining the relationships among factors that influence adaptation intention. Lam's (2015) study on the farmer's adaptation intention toward climate change has verified this view. However, this research on the process of adaptation intention does not consider the individual's psychological and cognitive factors (Bujosa, Riera, & Torres, 2015; Landauer, Haider, & Pröbstl-Haider, 2014).

We believe that the current understanding of tourists' adaptation intention toward climate change is limited for two reasons. First, tourism researchers have not yet adequately expressed the importance of tourists' adaptation intention process toward climate change; some evidence suggests that adaptation intention could encourage adaptive behaviours (Horng, Hu, Teng, & Lin, 2014). The notion of adaptation intention must be expanded to other systems (e.g. tourism systems) and contexts (e.g. coastal destinations) to deepen our understanding of the individual's adaptation behaviour (Gössling et al., 2012; Moreno & Becken, 2009). Second, past studies have conducted scenario analyses to explore whether hypothetical shifts in destination attractiveness that are induced by climate change within tourism systems contribute to tourists' climate change adaptation. Findings indicate that coastal destination attributes have been among the determining factors of tourists' climate change adaptation (Bujosa et al., 2015). However, these studies have called for further research on coastal destinations in different climate zones for a broader understanding of the effects of destination-managed realignment for tourist responses (Schliephack & Dickinson, 2017).

PMT and the individual's climate change adaptation processes have been examined separately. In light of our knowledge, we believe that PMT and the abovementioned process have not yet been integrated. The use of a comprehensive framework in this study is desirable for the following reasons. First, the characteristics of systems in an individual's process of adapting to climate change, and the cognitive mediating factors of PMT, can mutually compensate for the respective shortcomings in these two theories. Second, past studies on tourist behaviour commonly use the influential relationship among different psychological factors to predict future behaviour (e.g. Fishbein & Ajzen, 1975; Woodside & Lysonski, 1989). This will address concerns raised previously by numerous researchers. Therefore, we believe that the formation of tourists' adaptation intention regarding destinations must be determined by the characteristics of stress (i.e. climate change perception), the features of tourism systems (i.e. hypothetical shifts in destination attractiveness), and the cognitive mediating process (i.e. threat and coping appraisals) (Wang, 2015).

Past research has revealed how tourism stakeholders recognize the potential impacts of climate change and their adaptation strategies (Hopkins, 2014; Morrison & Pickering, 2013). Previous studies have also enhanced our understanding of the factors that influence adaptation intention toward climate change (Cocolas et al., 2016; Dawson et al., 2011; Jopp et al., 2010). Few studies have explored the structural relationships among tourists' adaptation intention, climate change perception, and hypothetical shifts in destination attractiveness, as well as threat and coping appraisals. Moreover, the climate change adaptation process may occur on different scales, including the individual, regional, national and international levels (Bryant et al., 2000). The adaptation process for each scale has its own qualities. Hence, the aforementioned structural relationships must be tested at a given destination. In this study, we focus on individual coastal destinations in order to expand our understanding of tourists' climate change adaptation intention and the underlying factors. We propose a structural model that integrates PMT with the individual's climate change adaptation process in order to examine the impact of climate change perception, hypothetical shifts in destination attractiveness and threat and coping appraisals on tourists' adaptation intention (see Figure 1).

The contributions of this study are two-fold. First, this study expands the literature on tourists' climate change adaptation intention, thereby clarifying the formation process of their adaptation intention by providing empirical evidence for relationships among the concerned factors. Second, this study clarifies the roles played by a hypothetical shift in tourism systems' destination attractiveness and the psychological factors of threat and coping appraisals in tourists' adaptive behaviours to climate change. The results can serve as a reference for coastal DMOs in the formation and design of strategies for economic, social and environmental sustainable development.

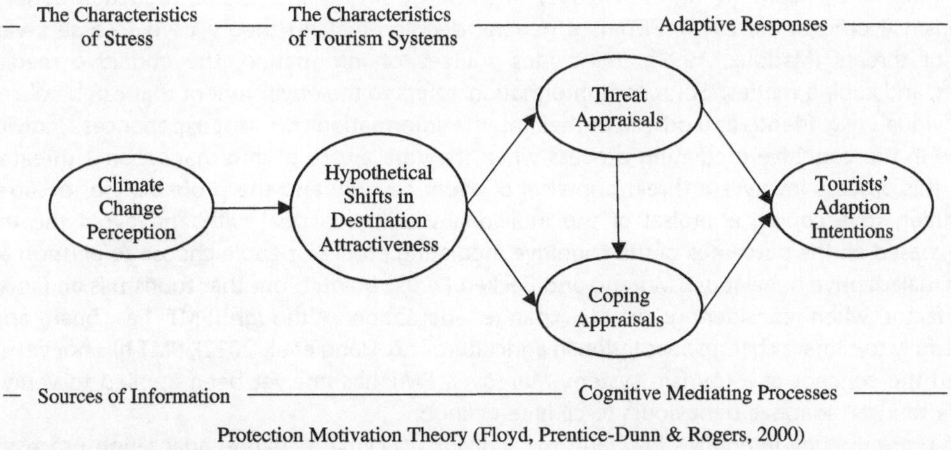

Figure 1. Proposed theoretical model for the study. Revised from farmer's climate change adaptation process (Bryant et al., 2000). PMT (Floyd et al., 2000).

Literature review

An integrated framework of tourists' climate change adaptation process

Previous studies on adaptation to climate change in the tourism industry have mostly stressed strategies (e.g. Schliephack & Dickinson, 2017; Scott, de Freitas, & Matzarakis, 2009), measures (e.g. Bujosa et al., 2015), and influencing factors of adaptation (e.g. Cocolas et al., 2016). These studies have rarely discussed the formation of tourists' adaptation intention. A literature review of research in other fields indicates that past research has already explored the farmer's adaptation intention toward climate change using PMT. Bryant et al. (2000) reviewed the literature on the farmer's adaptation intention toward climate change in Canada, then collated and proposed a process for farmers' adaptation intention toward climate change, with the following components: (1) characteristics of stress; (2) features of agricultural systems; (3) scales; (4) adaptive responses. 'Characteristics of stress' refers to climatic disruptions to which agricultural systems are sensitive, long-term changes in average climatic variables or elements of extreme weather events (e.g. type, size, frequency). 'Features of agricultural systems' refers to factors that can influence the sensitivity, adaptive capacity and responses of agricultural systems (e.g. local biophysical, cultural, and economic aspects). 'Scale' refers to the scale at which adaptive responses to climate change are analysed (i.e. farms and regional, national and international levels). 'Adaptive responses' refers to the different types of adaptations, including autonomous responses (e.g. delaying travel time to destinations and purchasing insurance) and conscious adaptation (e.g. introducing new technology, enhancing public facilities, and other public policy interventions).

The adaptation intention process described above is similar to past studies on climate change, which have mostly focused on physical environments and resources, but have not acknowledged the importance of cognitive factors in the adaptation process (Bujosa et al., 2015; Jopp et al., 2010). This process has increasingly received the attention of tourism researchers. Dawson et al. (2011) verified the importance of cognitive factors and found that, in poor snow conditions, highly committed skiers were more likely to change their skiing behaviour than less committed skiers.

PMT has previously been used as a socio-cognitive model to investigate individual adaptive behaviours and is among the main research theories on health protection behaviours (Floyd, Prentice-Dunn, & Rogers, 2000). Originally established to explain the connection between appeals to fear and attitude change, PMT proposes that people alter their attitudes out of fear to avoid the consequences of a threat or to prevent the occurrence of a threatening event (e.g. disease) (Rogers, 1975). Aside from PMT's application in health risks, it has also been used to explain other behaviours such as consumer marketing communication (Cismaru & Lavack, 2006), farmer's adaptation behaviours to climate change (Dang et al., 2012), and energy saving and carbon reduction behaviours in tourism (Horng et al., 2014). PMT is a general decision-making theory used to assess various types of threats (Maddux, 1993); it includes sources of information, the cognitive mediating process, and coping modes. 'Sources of information' refers to the environment that can be observed by individuals (e.g. friends and others) or their internal information (e.g. past experiences). Individuals engage in the cognitive mediating process when they are aware of information on a threatening event; this process involves a threat appraisal of event severity and the probability of occurrence, in addition to a coping appraisal of the individual's ability to deal with and avoid the threat. Finally, based on the outcomes of the cognitive mediating process, people choose to perform adaptive or maladaptive behaviours. Moreno and Becken (2009) pointed out that tourism is an indispensable factor when considering climate change adaptation. Although PMT has been applied successfully to climate change adaptation in agriculture (e.g. Dang et al., 2012), PMT has not yet incorporated the concept of a tourism system. Moreover, PMT has not yet been applied to verify and explore tourists' adaptive behaviours to climate change.

The present study integrates PMT and the individual's climate change adaptation process, and presents a framework for tourists' adaptation process. This framework is composed of the following:

(1) characteristics of stress/sources of information (e.g. climate change perception); (2) features of systems (e.g. a shift in destination attractiveness); (3) cognitive mediating process (e.g. threat and coping appraisals); (4) adaptive responses/coping behaviour (tourist adaptation intention). Integrating the two theories allows a simultaneous consideration of the potential transformations induced by climate change in light of the qualities of tourism systems and possible cognitive mediating factors that these changes may generate from tourists. This helps us understand the formation process of tourists' adaptation intention toward climate change. We conducted a literature review for the study constructs of this integrated framework.

Tourist adaptation intention (adaptive responses/coping behaviour)

Both PMT and the farmer's climate change adaptation processes mention concepts that are similar to adaptation intention. For example, according to the adaptation process proposed by Bryant et al. (2000), adaptive responses include autonomous responses, such as the coping measures of individual tourists (e.g. delaying travel time to a destination, purchasing insurance) and conscious adaptations, such as government or public policy interventions for climate change. Similar to autonomous responses, in the form of PMT proposed by Floyd et al. (2000), 'coping behaviors' refers to the individual's selection of adaptive responses. Within the context of this study, 'adaptation intention' is a type of autonomous response or coping behaviour that refers to an individual's autonomous choice to conduct tourism behaviours involving climate change adaptation.

Reser, Bradley, Glendon, Ellul, and Callaghan (2012) defined 'adaptation' as the adjustments and adaptive responses or actions generated after individuals go through internal assessments, judgements, and processing when faced with environmental changes and threats. Many scholars have found that when investigating an individual's behavioural occurrence/changes, behavioural intention can be used to predict the actual behavioural occurrence. If behavioural intention can be measured appropriately, the results obtained can approximate most actual behaviours (Fishbein & Ajzen, 1975). Woodside and Lysonski (1989) defined 'behavioural intention' as the individual's subjective judgement of the probability that he/she would engage in a given behaviour, thus reflecting his/her willingness or desire to invest effort in that behaviour. Within the context of this study, 'tourists' adaptation intention' refers to the potential adjustments made by individuals to their activities or behaviour during travel, in order to adapt to the probability of environmental changes in a destination caused by climate anomalies. A review of past literature related to climate change adaptation in the field of tourism indicates that most scholars accept that adaptation is a multidimensional concept. The substitution behaviour theory has been used to explain adaptation, divided into psychological and behavioural aspects (Dawson et al., 2011). 'Psychological adaptation' encompasses internal psychological processes (risk assessment, motivation response, coping decisions) that occur when individuals are faced with threats. These threats often relate to the harm caused by climate change to tourism systems (e.g. destination attractiveness) or people's direct experience of extreme weather in their daily lives. 'Behavioral adaptation' refers to behavioural changes in tourists caused by climate change, including activity, spatial, or temporal substitution.

Climate change as a multidimensional notion has been explored by scholars from three angles: (1) technology, (2) management, and (3) behavioural adaptations (Scott et al., 2009). 'Technological adaptation' refers to coping methods for climate change and technology innovations to address vulnerabilities; it encompasses the costs and complexity of adaptive technologies and often requires government intervention (Jopp et al., 2010). Studies have found that tourists have opposing views of adaptive technologies used by destination operators. For example, in response to beach restoration in a coastal landscape, some tourists did not feel that beach replenishment was incongruous with the original landscape and were only concerned with the beach's recreational value (e.g. sunbathing); other tourists were aware of the incongruity but could understand that the hotel was coping with the possible erosion of the beach owing to climate change, and thus were in favour of beach restoration measures (Buzinde, Manuel-Navarrete, Yoo, & Morais, 2010). 'Management adaptation' indicates attempts by stakeholders in tourism systems to alter their marketing methods, like reducing or increasing demand during specific tourism periods or re-directing tourists to different destinations and participate in different tourism activities (Schliephack & Dickinson, 2017). 'Behavioral adaptation' relates to tourism activities in which tourists decide to participate or the destinations they visit. However, behavioural adaptation is frequently influenced by technological and management adaptation (Jopp et al., 2010). Dawson et al. (2011) found that highly-involved tourists were more likely to engage in activity and temporal substitution; less involved tourists would engage in spatial substitution and choose a different skiing destination. Researchers have focused on different behavioural constructs, and climate change adaptation is commonly measured using multidimensional approaches.

Threat appaisals and coping appraisals (the cognitive mediating process)

Threat and coping appraisals are key psychological factors that influence adaptation intention toward climate change. They are also the main adaptation appraisal in the cognitive mediating process of PMT (Floyd et al., 2000). Scholars have defined 'threat' as an emotion generated when individuals face danger, or a motivation that drives them to avoid certain things (Tolman, 1923). In PMT, this concept is regarded as a relational construct that can evoke an individual's responses to threatening stimuli and self-protection. It is necessary to formulate adaptation strategies for specific attributes of destinations that are undergoing alterations (Bujosa et al., 2015; Wall, 1998). As destination attractiveness might undergo changes because of the impacts of climate change, these changes can be viewed as threat in tourists' adaptive behaviours (Schliephack & Dickinson, 2017). 'Threat appraisal' refers to the trade-off between the severity of the potential threat and vulnerability if the individual does not employ any protective measures. Within the context of this study, 'threat appraisal' refers to tourists' perceived severity of the impact or interference by climate change on their tourism activities, and the probability or inconvenience of being affected. 'Coping appraisal' is the individual's assessment of his/her ability to cope with or avoid threats, which includes response efficacy and self-efficacy. 'Response efficacy' refers to the individual's ability to successfully reduce the threat if recommended measures to improve the situation are adopted (Rogers, 1975). Within the context of this study, 'response efficacy' refers to tourists' willingness to follow recommendations to reduce or alter climate change in their tourism activities. Horng et al. (2014) recommended that by saving energy and water when travelling, tourists can reduce the threat of climate change. 'Self-efficacy' refers to the perceived ability of an individual to use adaptive responses. Within the context of this study, it refers to tourists' perception that they can adopt recommendations to respond better to how climate change may affect their tourism activities. One example is the ability to employ eco-friendly tourism behaviours, as recommended by Jopp et al. (2010).

Hypothetical shifts in destination attractiveness (the characteristics of systems)

Destination attractiveness is among the main components of tourism systems (Gunn, 1993). Tourists travel to different destinations to enjoy the attractions or natural features (Uyarra et al., 2005). Researchers have found that weather is among the attractions of tourist destinations (Hu & Ritchie, 1993). However, because of the probability of abnormal changes in weather and climate, the climatic variables of numerous destinations have been reconfigured, leading to changes in destination attractiveness (Tervo-Kankare, Hall, & Saarinen, 2013). Hence, changes in destination attractiveness can be regarded as system features that are sensitive to the climate or are responses to climate change. 'Destination attractiveness' is defined as the combination between 'the relative importance of individual benefits and the perceived ability of the destination to deliver individual benefits' (Hu & Ritchie, 1993, p. 25). 'Destination attractiveness' can be understood as tourists' perception of the ability of a destination to satisfy their holiday needs. The relative importance of different destination attributes will determine the evaluation of destination attractiveness. Based on the above definition, this study defines the hypothetical shifts in destination attractiveness as a possible alteration in a destination's environmental attributes owing to the effects of climate change; these may mitigate the destination's ability to satisfy tourists' specific destination or holiday needs.

According to the views of Thach and Axinn (1994), destination attractiveness is composed of core and augmented attributes. In previous studies related to possible alterations in the features of coastal destinations due to climate change, the core dimensions of coastal destination attractiveness include terrestrial and marine attributes such as natural settings, scenery, beaches, and water (e.g. Uyarra et al., 2005). In contrast, the augmented dimensions encompassed functional/essential attributes that can influence tourists' assessment of the core attributes, including general and niche facilities (e.g. Williams, 2011) such as the proximity of toilets and café amenities, beach access from car parks, and the availability of beach huts. However, the hypothetical shift in coastal destination attractiveness has been viewed as the collection of effects on tourism functions. Within the context of climate change, this viewpoint might be too limited because changes in destination vulnerability would cause tourists to form different levels of psychological threat according to different climatic variables (Moreno & Becken, 2009). Many researchers have also supported this idea (Scott, Hall, & Gössling, 2012), even though the functions of essential destination attributes remain crucial among certain tourists (Williams, 2011).

Since climate change involves persistent accumulation during a long period of time (U.S. EPA, 2013), past studies have commonly employed scenario analysis based on climate change's negative impacts on the environmental attributes of a destination, so as to understand changes in visitors' opinions to hypothetical shifts in destination attractiveness. Uyarra et al. (2005) presented tourists with hypothetical changes involving severe coral bleaching and mortality, and a large disappearance of beach land owing to a rise in sea level. They found that if the environmental attributes preferred by tourists (e.g. coral reefs and beaches) were affected by climate change, their willingness to revisit tropical islands would lessen drastically. Their study is similar to those that have examined adaptive behaviours using hypothetical changes in destination attributes, or attractiveness due to climate change; for example, the vulnerability of coastal sites (Moreno & Becken, 2009) or the willingness to revisit skiing destinations (Tervo-Kankare et al., 2013).

Climate change perception (the characteristics of stress/sources of information)

Climate change perception is another acknowledged determinant of tourist adaptation (Morrison & Pickering, 2013). 'Weather' refers to atmospheric conditions within a specific location or time and includes atmospheric temperature and pressure, wind speed and direction, and precipitation. Climate is the concept of average weather and is a generalized phenomenon (Gómez-Martín, 2005). Climate change refers to significant variations in the climate over the course of an extended period (U.S. EPA, 2013). Spence, Poortinga, Butler, and Pidgeon (2011) defined 'climate change perception' as people's beliefs, ideas, and impressions toward the climatic features of extreme weather events, or climatic interferences that are greater than long-term changes in the average climatic variables. In general, tourists can experience climate change via mass media (newspapers and magazines), friends and family, or in person to obtain information on significant long-term shifts in weather attributes or climatic features at a tourist destination. The formation of tourists' climate change perception involves the information processing of abnormal atmospheric temperature, rainfall, humidity, or other climate indicators at a tourist destination (Smith, 1993).

Belle and Bramwell (2005) say that tourists all hope to engage in tourism activities in 'good weather'. Also, temperature and other climatic variables are key influencing factors in tourists' decision-making; these factors affect their choice of destination (Jeuring & Peters, 2013; Woodside & Lysonski, 1989). For some tourists, identified desired climatic variables is a major reason to visit a given destination (e.g. Williams, 2011). Due to the differences in tourists' personal characteristics or views on weather or climatic variables, they might use various methods to assess climate change and hence have different climate change perceptions (Spence et al., 2011; Tervo-Kankare et al., 2013). Denstadli, Jacobsen, and Lohmann (2011) suggested that thermal, aesthetic, and physical sensations support the perception of summer weather in Scandinavia. In addition, the perception of most tourists is formed by their experiences of local weather when engaging in activities and experiences at the destination. Gómez-Martín (2005) found that different tourist activities led to varying climate needs. Interestingly, Landauer et al. (2014) believe that the people's climate change concept is a social construct; it could be developed based on potential conflicts in society or opposing methods and other forms of discourse to establish adaptive strategies or behaviours. Hence, although the difference in tourists' climate change perception is a type of subjective assessment, its presence is fairly important for climate change adaptation (Reser et al., 2012). Gonzalez (2012) and Tervo-Kankare et al. (2013) advocate that more research is needed to understand tourists' climate change perception of destinations or activities in different climate zones.

Methodology

Proposed theoretical model

Based on the theories and review of empirical studies mentioned above, we developed an integrated model using perception of climate change as the exogenous construct; tourists' adaptation intention as the endogenous construct; and hypothetical shifts in destination attractiveness, threat appraisals, and coping appraisals as the mediating constructs. Six hypotheses were proposed, in support of this model (see Figure 1).

The relationship between climate change and hypothetical shifts in destination attractiveness has been examined in many previous studies. Nyaupane and Chhetri (2009) found that rich biodiversity and natural landscapes were the main attractions for tourists who engaged in mountain tourism; the attractiveness of mountrainous regions has altered because of changes in the climatic characteristics. Moreno and Becken (2009) and Scott et al. (2012) had similar findings. Therefore, we put forward the following hypothesis:

> H1: Climate change perception has a significant positive relation with hypothetical shifts in destination attractiveness.

In the literature on climate change and tourism, evidence supporting the relationship among hypothetical shifts in destination attractiveness, threat appraisal, and coping appraisal are reported. However, certain researchers have observed that climate change could lead to changes in destination attractiveness (Nyaupane & Chhetri, 2009), and could thus be used to assess threat and coping appraisals in tourist adaptation (Schliephack & Dickinson, 2017). Based on the observations from the above studies, we inferred the following hypotheses:

> H2: Hypothetical shift in destination attractiveness has a significant positive relation with threat appraisal.
> H3: Hypothetical shift in destination attractiveness has a significant positive relation with coping appraisal.

Studies have explored the relationships among threat appraisal, fear appraisal, and tourists' adaptation intention toward climate change. However, many scholars have attempted to verify these relationships in other fields. Prentice-Dunn, Mcmath, and Cramer (2009) found that, when college women understood the severity and coping methods of threats (how skin cancer can harm or threaten their lives), they transitioned from pre-contemplation to contemplation of healthy sun behaviours. Horng et al. (2014) found that a coping appraisal was better than a threat appraisal in predicting the subsequent energy-saving behavioural intention among tourists. Additionally, when tourists are presented with the threat of climate change, as well as actual adjustments and responses to environmental impacts, they often adjust their subsequent behaviours due to mediation in their internal psychological processes (Scott et al., 2009). Prentice-Dunn et al. (2009) proposed that, in threatening situations, people's attitudes might be reinforced, causing coping strategies to become more prominent. The relationship between threat appraisals and coping appraisals has not been verified. According to the results and recommendations of empirical studies on the relationships among threat appraisals, coping appraisals, and adaptation intention in PMT, we formulated the following hypotheses.

H4: A threat appraisal has a significant positive relation with a coping appraisal.
H5: A threat appraisal has a significant positive relation with tourists' adaptation intention.
H6: A coping appraisal has a significant positive relation with tourists' adaptation intention.

Study site and subjects

The data of previous studies, which have focused on tourists' adaptations to the possible effects of climate change on a destination's attributes, have been collected mostly from Western tourists travelling in Western countries (Landauer et al., 2014; Scott et al., 2012). Less attention has been paid to the hypothetical shifts in destination attractiveness of domestic and international tourist destinations in Asian countries. This is a large gap in the tourism literature. Currently, the Asian tourism market is the world's fastest growing outbound market; it has increased by 27% from 2015 to 2016 (United Nations World Tourism Organization, 2016). A better understanding of cross-cultural differences in the field is not only important (Gössling et al., 2012) – it is imperative that scholars gradually begin to include non-Western views in their research.

The study sample was from Kenting National Park in southern Taiwan (Figure 2). From 2001 to 2016, the number of local tourists who were over 18 at the time and had spent at least one night at the destination was 5,834,294–6,341,160 person-time. It is among the most representative domestic tourist destinations in Taiwan (Kenting National Park, 2014a). The Guanshan sunset spot is among the 12 superb sunset spots around the world (CNN, 2017). In addition to its weather, the region also has coral reefs, sea-eroded and cliff terrains, and other geographical landscapes. It has a tropical climate that has nurtured its rich ecological morphology, and a large number of migratory birds arrive from the north for winter each year. Destinations with diverse tourism activities can mainly be divided into marine and terrestrial activities. Marine activities include sunbathing, recreational water activities, snorkelling and surfing. Terrestrial activities involve cultural, natural, and eco-environmental educational activities such as observing coral reefs, limestone terrains, or visiting lighthouses.

Kenting National Park has a tropical climate, with an annual average temperature of 20°C to 28°C, and annual rainfall of about 2,200 mm. The rainy season is between May and October, while the dry season is between November and April (Kenting National Park, 2014b). There were two reasons for choosing this region as our study site. First, the tourism type of this region is predominantly coastal or ecological, and scholars have found that coastal or ecological resources are more susceptible to the effects of climate

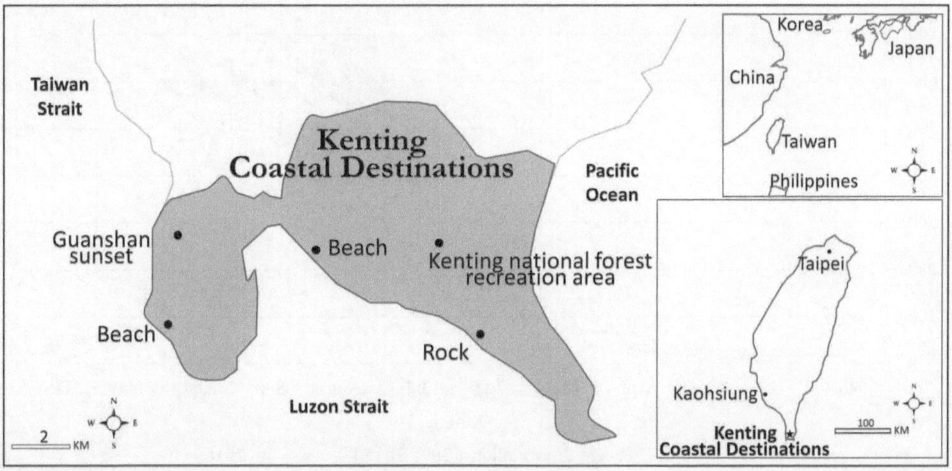

Figure 2. Study site.

change (Moreno & Becken, 2009; Schliephack & Dickinson, 2017); for example, climate change has caused coral bleaching (Dai, 2002) and loss of coastline (Ministry of the Interior of Taiwan, 2017). Second, the impression of tourists about this region is clearly a summer destination (Taiwan Tourism Bureau, 2017). According to data on temperature and rainfall collected between 2001 and 2016 from the regional climate monitoring stations (Taiwan Central Weather Bureau, 2017a), the average temperature rises by 2.5°C in summer (June–August), and the average rainfall decreases by 25% in autumn and winter (September–February). A comparison of the climate data with the distribution of monthly tourists in the same period (Figure 3) indicates that tourists to Kenting National Park advance or postpone their time of travel depending on the increase in temperature and decrease in rainfall. For coastal destination management, knowing tourists' adaptation to climate change in advance will allow them to prepare for future responses. Hence, this region is a representative study site.

Data collection

We collected data during randomly selected time periods (e.g. daytime, nighttime, weekdays and holidays) for three months, from July–October 2014. The survey period coincided with the seventh month of the traditional Taiwanese lunar calendar, known as the Ghost Month, during which tourists

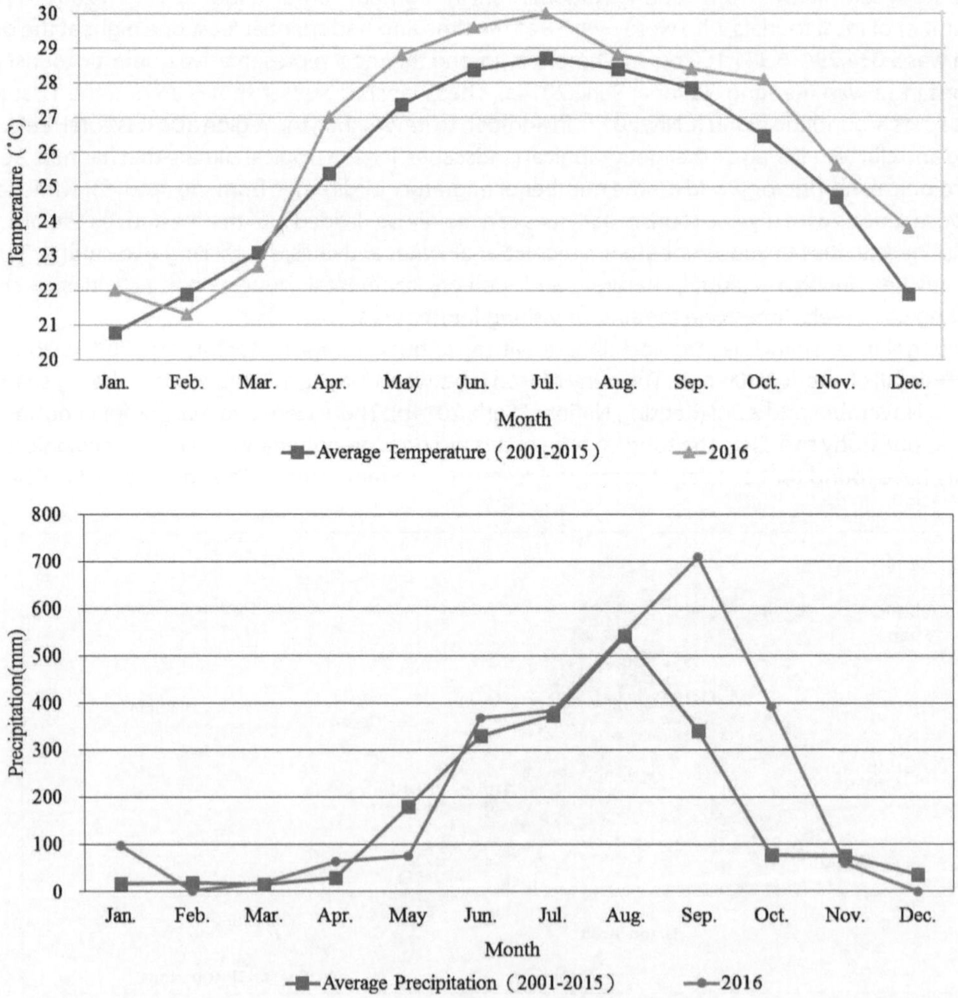

Figure 3. Average month temperatures (in centigrades) and precipitation in kenting. 2001–2016.

generally avoid water-based locations (e.g. Kenting). In addition, because of the gas explosions in Kaohsiung (a city that tourists may pass when travelling to the Kenting National Park) that led to several casualties, the intentions of tourists passing through Kaohsiung may be affected (Lee, 2014). All research interviewers wore name tags and carried their student IDs. They were located at sites where tourists engaged in water activities. The tourists closest to the interviewers were asked whether they would be willing to participate in our survey. If the participants were not willing, the interviewer approached the next participant. (Note that a limitations of the study is that the reasons for not participating were not collected.) If the participants were unclear regarding the questionnaire's content, the interviewers explained the questionnaire and guided them on how to fill it out.

The research tool was an on-site self-administered questionnaire. The questionnaire included four major parts (Wang, 2015). First, climate change perception, which sought participant's views on climate change. The items were based on the results of a study by Tiller and Schott (2013) on the contributions of tourism within a destination to global climate change. Twelve items were developed, and each item was scored on a five-point Likert scale (5 = strong agreement; 1 = strong disagreement). Second, hypothetical shifts in Kenting's destination attractiveness. It comprised 12 items and was developed in a two-stage process that combined structured and non-structured methods. Since the effects of climate change may alter various destinations differently (Bujosa et al., 2015; Wall, 1998), we reviewed relevant literature to identify the key changes in the environmental or resource features of Kenting National Park that could be induced by climate change (Dai, 2002) and its main destination attractiveness (Kenting National Park, 2014b, 2014c; Pan, 2014; Yeh, 2014). Fifteen items were collected in this first stage. In the second stage, we gathered destination attributes important to tourists, and asked 30 local tourists, 'What are the features of Kenting National Park that satisfy your tourism or activity needs?' Sixteen items were collected in this stage. In order to compile a comprehensive list of hypothetical shifts in destination attractiveness, we combined the interview results and literature review and reference was made to the scenario analysis on the negative impact of climate change on destination attractiveness (Uyarra et al., 2005). The list was then modified based on a panel discussion by three tourism researchers. They removed repeated items and revised the appropriateness of the statements to fit the context of Taiwan. Thus, we formulated twelve items. When conducting the survey, we asked the participants: 'Given the following changes in tourism characteristics of Kenting National Park due to climate change, how attractive would Kenting be to you as a destination for tourism?' This part was measurement based on prior literature (Hu & Ritchie, 1993) and scored on a five-point Likert scale (5 = strong unattractiveness, 1 = strong attractiveness).

Third, threat and coping appraisals. We asked the participants for their views on threat and coping appraisals, given the changes in Kenting National Parks' destination attractiveness (tourism features or characteristics). The items were based on studies by Plotnikoff and Higginbotham (1995); Cox, Koster, and Russell (2004); Kellstedt, Zahran, and Vedlitz (2008); and Reser et al. (2012), and included nine and seven items on threat and coping appraisals, respectively. Fourth, tourists' adaptation intention. This part was created referring to the studies by Dawson et al. (2011); Denstadli et al. (2011); and Reser et al. (2012). We compiled 10 items, which were measured on a five-point Likert scale (5 = strong agreement, 1 = strong disagreement). The final part involved collecting data on the participants' socioeconomic background (i.e. gender, age, marital status, education level, and average monthly household income) and personal experiences (i.e. main tourism activities in which they engaged and source of information on climate change). The research tool was written in Taiwan's official language – traditional Mandarin Chinese. As the scales used were obtained from English studies, two colleagues who were fluent in both Mandarin Chinese and English, and who were familiar with the relevant literature, evaluated the face validity of the scales. The items in this study reflect those items that have been back translated from Chinese into English.

These constructs have rarely been explored in past tourism research, and specific items were based on pioneering research results and references from other fields. To ensure that the survey questionnaire could be read and understood easily by the general population, a pretest was conducted with 30 local tourists in the Kenting Coastal Destination who were selected by convenience sampling. The textual descriptions of a few items were revised based on the pretest results. In addition, previous studies on climate change perception, hypothetical shifts in destination attractiveness, and tourists' adaptation intention have mostly remained a conceptual discourse and rarely mentioned the composition of its theoretical constructs. Hence, this study conducted a second questionnaire survey using convenience sampling. The survey was conducted in July 2014, including holidays and weekdays. Of the 350 questionnaires distributed, 320 valid questionnaires were returned. An exploratory factor analysis was performed to analyse and identify the latent factors of climate change perception, hypothetical shifts in destination attractiveness, and tourists' adaptation intention. The factors obtained were used as indicators to measure these constructs. Implementing this procedure helped resolve the possible issues with multicollinearity among the indicators, or correlation among the indicator error variances during structural equation modelling (Bollen, 1989). After the data analysis, two cross-factor variables were removed from climate change perception. Two dimensions were extracted, namely 'indirect perception' (six items; $\alpha = 0.82$) and 'direct perception' (four items; $\alpha = 0.75$); the total variance explained was 55.93%. The hypothetical shift in destination attractiveness comprised two dimensions: 'changes in terrestrial attractiveness (eight items; $\alpha = 0.85$) and 'changes in marine attractiveness' (four items; $\alpha = 0.87$). Tourists' adaptation intention comprised three dimensions: 'behavioural adaptation' (five items; $\alpha = 0.78$), 'physical adaptation' (three items; $\alpha = 0.74$), and 'psychological adaptation' (two items; $\alpha = 0.60$).

The theoretical model used to test the relationships proposed above was structural equation modeling (SEM), and the sample for analysis was limited to participants who: a) have perceived climate change; b) believed that if the destination attractiveness of Kenting Coastal Destination were to be altered by climate change, they would not travel there; and c) if they were to experience the hypothetical shifts in destination attractiveness (threat), they would adjust their tourism behaviours related to the Kenting Coastal Destination. Samples that had no concept of climate change, or did not meet the precondition of perceiving climate change as a tourism threat, were eliminated because adaptation is not possible in these participants.

Results

Of the 600 tourists who participated in this study, 556 agreed to complete the on-site survey, yielding a participation rate of about 93%. However, among the 556 questionnaires, 223 did not meet the qualification criteria for data analysis (e.g. did not perceive climate change, believed that the hypothetical shift in Kenting destination attractiveness would not affect tourism, no possibility of adaptation); thus, 333 questionnaires were analysed (60%). Among the valid samples, the gender distribution indicated that there were more women (58.4%) than men (41.6%); 66.4% of the sample were aged 22–40 years; marital status was almost the same between single (49.2%) and married (50.8%) participants. Most participants had university (undergraduate) education (64.8%), and their average monthly personal income was between TWD 30,000 and TWD 60,000 (47.1%). The demographic profile of the participants was consistent with the profile of domestic tourists visiting Kenting Coastal Destination (Hsu, Yeh, & Lin, 2013; Liu, Chia, & Chang, 2009; Wang, 2015). In addition, regarding the tourists' sources of information on climate change, the most commonly used channels were television (45%), the internet (26.8%), and newspapers/magazines (25.7%). Most tourism activities in Kenting Coastal Destination were enjoying natural landscapes (42.2%) and engaging in recreational water activities (32.1%).

Measurement model testing

A confirmatory analysis was performed on the measurement structure for each latent construct of climate change perception, hypothetical shifts in destination attractiveness, threat appraisal, coping appraisal, and tourists' adaptation intention, before analysing the measurement model (see Table 1). The measurement model's adequacy was evaluated, based on its fitness indicators and the reliability and validity of construct components (discriminant and aggregation validities). The fitness indicators showed that all latent constructs had acceptable data fitness: climate change perception (2 dimensions, 10 indicators), hypothetical shifts in destination attractiveness (2 dimensions, 12 indicators), threat appraisal (2 dimensions, 9 indicators), coping appraisal (2 dimensions, 7 indicators), and tourists' adaptation intention (3 dimensions, 10 indicators). The results showed that all latent constructs had a reliability above 0.7. Moreover, aggregation validity can be ensured if the factor loading of each indicator is two times higher than the standard error (Anderson & Gerbing, 1988). The target loadings in this study had significant *t*-values, ranging from 7.74 to 14.96, thus, verifying the aggregation validity of each latent construct. Moreover, the discriminant validity is established when the estimated variance among any group of variables is greater than the square of the correlation between two of the variables. The results demonstrated there was discriminant validity (Jöreskog & Sörbom, 1993).

According to the two-step procedure recommended by Anderson and Gerbing (1988), we used the LISREL 8.52 software (Jöreskog & Sörbom, 1993) to test the hypothesized relationships among the variables influencing tourists' adaptation intention. The first step required the validation of the measurement model comprising climate change perception (two composite indicators), hypothetical shifts in destination attractiveness (two composite indicators), threat appraisal (two composite indicators), coping appraisal (two composite indicators), and tourists' adaptation intention (three composite indicators). The adequacy assessment of the measurement model indicated that the model had rational data fit ($\chi2 = 70.62$, $df = 34$, $\chi2/df = 2.08$, $p < .01$, $SRMR = 0.027$, $RMSEA = 0.042$, $NNFI = 0.98$, $CFI = 0.99$). The composite reliability of climate change perception, hypothetical shifts in destination attractiveness, threat appraisal, coping appraisal, and tourists' adaptation intention were 0.72, 0.73, 0.79, 0.75, and 0.75, respectively. The values were all above 0.7, indicating an acceptable level of reliability. In terms of validity, the *t*-values for the loadings of each indicator reached statistical significance ($p < .001$) and were all above 0.5, thus, providing evidence for aggregation validity (Anderson & Gerbing, 1988). In addition, discriminant validity was tested; the results supported the discriminant validity of the five latent constructs. The measurement model has thus been validated and be utillised in subsequent structural model analysis.

Structural model testing

The second step was to conduct a path analysis of the research model. The results indicated $\chi2 = 108.81$, $df = 38$, $\chi2/df = 2.86$, $SRMR = 0.055$, $RMSEA = 0.056$, $NNFI = 0.97$, and $CFI = 0.98$. A composite evaluation of the fitness of all indicators showed that the sample data provided a good fit with the structural model; all path coefficients were significant and positive (see Figure 4). The results of the analysis supported the hypothesized relationships of climate change perception, hypothetical shift in destination attractiveness, threat appraisal, and coping appraisal with tourists' adaptation intention. Hence, the six hypotheses were accepted: climate change perception has a significant positive relation to hypothetical shifts in destination attractiveness ($\beta = 0.63$; H1); hypothetical shifts in destination attractiveness has a significant positive relation to threat appraisal ($\beta = 0.50$; H2); hypothetical shifts in destination attractiveness has a significant positive relation to coping appraisal ($\beta = 0.30$; H3); threat appraisal has a significant positive relation to coping appraisal ($\beta = 0.69$; H4); threat appraisal has a significant positive relation to tourists' adaptation intention ($\beta = 0.48$; H5); and coping appraisal has a significant positive relation to tourists' adaptation intention ($\beta = 0.32$; H6) (Table 2). Overall, the results indicated that climate change perception, hypothetical shifts in destination attractiveness, and threat and coping appraisals are important determinants of tourists' adaptation intention. Moreover, hypothetical shifts in destination attractiveness, and threat and coping appraisals were mediating variables in the relationship between perception of climate change and tourists' adaptation intention. The squared multiple correlation (R2) was 58% of our target construct, that

Table 2. Summary of structural model.

	Hypothetical shift indestination attractiveness	Threat appraisal	Coping appraisal	Tourist's adaption intention
Climate Change Perception				
Direct effect	0.63	–	–	–
Indirect effect	–	–	–	–
Total effect	0.63	0.32	0.41	0.28
Hypothetical shift in destination attractiveness				
Direct effect	–	0.50	0.30	–
Indirect effect	–	–	0.35	0.45
Total effect	–	0.50	0.65	0.45
Threat appraisal				
Direct effect	–	–	0.69	0.48
Indirect effect	–	–	–	0.22
Total effect	–	–	0.69	0.70
Coping appraisal				
Direct effect	–	–	–	0.32
Indirect effect	–	–	–	–
Total effect	–	–	–	0.32

model's purpose was to explain the adaptation intention of tourists and the interrelationships of tourists' adaptation intention with climate change perception, hypothetical shifts in destination attractiveness, and threat and coping appraisals. The results suggest that (a) there are two direct effects on tourists' adaptation intention: threat and coping appraisals; (b) threat appraisal had one indirect effect on tourists' adaptation intention: coping appraisal; (c) hypothetical shifts in destination attractiveness had two indirect effects on tourists' adaptation intention: threat and coping appraisals; (d) climate change perception had three indirect effects on tourists' adaptation intention: destination attractiveness, threat, and coping appraisals; and (e) climate change perception and hypothetical shifts in destination attractiveness were both multidimensional constructs. These results support the findings of the existing literature, emphasizing the importance of climate change perception, hypothetical shifts in destination attractiveness, and threat and coping appraisals in the formation of tourists' adaptation intention. However, these results also reveal that existing studies have neglected the complexity of tourists' adaptation intention towards coastal destinations within the context of climate change. In this study, threat appraisal and hypothetical shifts in destination attractiveness were the most important predictors of adaptation intention. Additionally, vulnerability was the most significant dimension of threat appraisal, while changes in terrestrial and marine attractiveness were the most important dimensions in the hypothetical shifts in destination attractiveness. Therefore, future understanding of the process of tourists' adaptation intention to climate change must consider the possible changes in destination attractiveness because of climate change, and the influence of probability changes on tourism behaviour and threat of inconvenience.

Discussions

Researchers have suggested that climate change perception, hypothetical shifts in destination attractiveness, and threat and coping appraisals are critical factors influencing tourists' adaptation intention (Bujosa et al., 2015; Dang et al., 2012; Simpson, Gössling, Scott, Hall, & Gladin, 2008). However, few studies have constructed an adaptation intention model of tourists' behaviours toward climate change. In this study, we present the first attempt to test and improve a model that incorporates the relationships of climate change perception, hypothetical shifts in destination attractiveness, and threat and coping appraisals with adaptation intention. The results suggest that climate change perception, hypothetical shifts in destination attractiveness, and threat and coping appraisal are important determinants of tourists' adaptation intention. Our results can contribute to tourism research on the sustainable development of coastal destinations by providing a deeper understanding of the concept, and the formation of tourists' adaptation intention (Belle & Bramwell, 2005; Moreno & Becken, 2009).

To interpret the adaptation intention of these tourists, scholars have called for more studies to discuss climate change in places in different climate zones and terrains (Gonzalez, 2012; Tervo-Kankare et al., 2013). We developed a structural model to explain tourists' adaptation intention at a specific (coastal) destination that experiences the impact of climate change. The resulting structural model has good explanatory power. Among tourists in coastal Taiwan, climate change perception and hypothetical shifts in destination attractiveness were important in tourists' adaptation intention; they were aware that 'climate change can lead to hypothetical shifts in destination attractiveness'. Moreover, threat appraisal also directly affected tourists' adaptation intention. Hence, the adaptation intention of coastal tourists emphasizes hypothetical alterations in destination attractiveness caused by climate change, and tourists perceive the vulnerability of this threat. Our results integrate the suggestions of existing studies since they include the characteristics of the system, a key component in an individual's (farmer's) climate change adaptation process (Bryant et al., 2000). Parallelly, the findings allow us to judge threat appraisal in PMT (Rogers, 1975). According to Tervo-Kankare et al. (2013), even when presented with a climate change phenomenon (stimulus), people can have different responses (adaptations). Moreno and Becken (2009) believed that varied responses (adaptations) might be attributed to differences in the system's characteristics, or individuals' contexts.

Meanwhile, the core concept of PMT depends on the cognitive process of threat appraisal undertaken by individuals when faced with threatening events. Therefore, tourists' adaptation intention to climate change stems from their encounters with possible transformations in the characteristics of tourism systems, and their perception or worry that these changes will affect the probability or inconvenience of tourism behaviours in these places. These results can be explained if we claim that tourists' consideration for coastal destinations is not only influenced by the attractiveness of water-based tourism, which is is susceptible to climatic effects, (Williams, 2011) but also a subjective experience of other activities (Schliephack & Dickinson, 2017). In Asian countries, coastal tourist activities are easily influenced by climate, perceived risks, and personal factors (Cheablam & Shrestha, 2015; Yang, Sharif, & Khoo-Lattimore, 2015). Many researchers have shown that for Asians, the weather's effect on the attractiveness (change) of water-based destinations is a crucial factor that influences their engagement in specific activities (e.g. Cheablam & Shrestha, 2015; Morachat, 2003). Existing tourism research has not included empirical studies on tourists' adaptation intention to climate change, with hypothetical shifts in destination attractiveness at its core. This study encompassed hypothetical shifts in destination attractiveness while investigating the roles of climate change perception, hypothetical shifts in destination attractiveness, and threat and coping appraisals in tourists' adaptive behaviours. Past studies have emphasized that tourists' use of climate change information is effective in the adaptation of tourist destinations (Simpson et al., 2008; Tiller & Schott, 2013).

The results support the relationships between threat appraisal and adaptation intention, and coping appraisal and adaptation intention. This suggests that threat and coping appraisals can effectively predict adaptive behaviours. These results are consistent with those of Rogers (1975), whose influential study found that when individuals perceive the impact of interfering threats and have a greater strategy to avoid them, the probability of adaptation rises. These empirical results are also consistent with the previous SEM analysis of the farmer's adaptation intention (Dang et al., 2012). Moreover, the hypothesized relationship between threat and coping appraisals was supported, showing that coping appraisal had a direct effect on threat appraisal, thus suggesting that threat appraisal can affect coping appraisal in tourists' adaptation. These results confirm the claims by Prentice-Dunn et al. (2009), who proposed that people's attitudes may be reinforced when they are faced with threats, making their coping strategy more prominent. However, this hypothesis was developed based on a small number of previous studies.

The importance of destination attributes in our understanding of climate change adaptation has been advocated by recent studies (Bujosa et al., 2015; Moreno & Becken, 2009). Similarly, we pointed out that hypothetical shifts in destination attractiveness is a determinant of threat and coping appraisals. The importance of changes in destination attractiveness is not surprising, since the tourism industry is based on enjoying attractions (Buzinde et al., 2010). Moreover, changes in destination attractiveness are gradually becoming more important in determining threat and coping appraisals. Our findings are consistent with the results of Schliephack and Dickinson (2017), who proposed that possible changes in destination attractiveness cause tourists to perceive threats and reduce coping skills. Since few studies examined the effects of threat and coping appraisals on hypothetical shifts in destination attractiveness, our study cannot be compared against other studies. Therefore, these two relationships warrant further examination. Our study found that the two dimensions of hypothetical shifts in destination attractiveness were the changes in terrestrial and marine attractiveness, which has been noted in other studies (see Uyarra et al., 2005). Scott et al. (2012) and Tervo-Kankare et al. (2013) found that destinations may lose their attractiveness because of environmental or climate changes; for example, reducing the attractiveness of terrestrial landscapes, decreasing the quality, or reducing the size of beaches. Similar to Schliephack and Dickinson (2017) and Williams (2011), we noticed differences in tourists' perceived attractiveness of coastal destinations, especially for the environmental attributes of land and ocean.

Finally, climate change perception was identified a determining factor in hypothetical shifts of a destination's attractiveness. Our result was consistent with the result of Moreno and Becken (2009), who found that perceptions of changes in climatic conditions were related to perceived threat of hypothetical shifts in destination attractiveness caused by climate change. Tourists' perception of climate change can be divided into direct and indirect perception. Evidently, in the indirect perception dimension, climate change will not appear in the climatic features of Taiwan (e.g. snow or glaciers), or immediately perceived by tourists in their surroundings (e.g. ocean currents, coastline, or sea level). Instead, they arrive at these perceptions through global or regional climate change information reported in the media. Surprisingly, although typhoons and dust storms are a part of Taiwan's climatic features, they were included in the indirect perception dimension. Based on monitoring data, a possible reason could be that the survey period (2014) was among the years with the lowest number of typhoons, on par with that of 2009 and 2012 (twice) (Taiwan Central Weather Bureau, 2017b); dust storms did not significantly impact Taiwan during the survey period (Environmental Protection Administration, Taiwan, 2014). Conversely, in the direct perception dimension, the features of climate change included temperature, rainfall, or seasons, with which participants had direct contact in their daily lives. For example, the warming of annual average temperature and reduction in rainfall in Taiwan are both obvious trends, whereas the increase in temperature during winter has been greater than in the other three seasons (Zhuo & Lu, 2014). The climate change phenomenon perceived subjectively by tourists conformed to the objective climate characteristics found in monitoring reports.

Practical implications

The results suggest that tourists tend to develop a destination behavioural intention to adapt to the impacts of climate change. Specifically, tourists perceive that climate change may alter the attractiveness of the destination, and engage in threat and coping appraisals, thus forming a tendency towards adaptation intention. Moreover, hypothetical shifts in destination attractiveness, climate change perception, and threat appraisal have an indirect effect on adaptation intention via coping appraisal. These results provide hints on sustainable development for use by DMOs. First, for tourists visiting the Kenting Coastal Destination, their climate change perception comprised direct perception (i.e. temperature, rainfall, seasonal variation) and indirect perception (i.e. ocean currents, coastline, change in sea level). Scholars have found that changes in temperature, rainfall, seasons, and sea level could threaten the public health of coastal destinations (Uyarra et al., 2005), and increase the seasonal population (Martinez et al., 2014). Hence, prolonging the operating period is not only an adaptation to climate change but could also promote social sustainability by protecting the health of tourists during the tourism period. It is also an important consideration for economic sustainability, as it responds to the market changes in tourist demands, and avoids the increased competition caused by substitution. Therefore, we suggest that the general marketing strategies of coastal DMOs provide special packages to attract tourists from May–June or October–November. Moreover, slogans must be designed to emphasize that the extended operating period is less harmful to tourists' health and has more desirable weather to engage in tourism activities. Although peak summer in Kenting will not be altered substantially, an extended operating period that matches the desirable weather for tourism and actual weather data must enable coastal DMOs to enhance tourists' adaptation and increase the probability of tourism behaviours.

Second, this study showed that hypothetical shifts in destination attractiveness of Kenting comprised changes in terrestrial and marine attractiveness. According to Uyarra et al. (2005) and Schliephack and Dickinson (2017), climate change can affect the environmental features of coastal destinations, and as tourists have preferences for environmental features, it will also affect their choice of destination. Hence, coastal DMOs must adopt strategies to lessen the impact of climate change on their destinations and to prioritize environmental features that increases the attractiveness of the destination. Therefore, the health of coral reefs and fish populations must be maintained, and beach and marine activities must be improved. In addition, scholars have proposed that when tourists perceive the threat of possible change, they employ their self-efficacy to generate the probability of adjustments in tourism behaviours (Horng et al., 2014). Thus, we recommend that DMOs implement tourism incentive programmes that encourage tourists to engage in pro-environmental behaviours in their accommodation, transportation, and reduce the environmental burden when engaging in tourism activities in the Kenting Coastal Destination. Adopting green tourism as a coping method can improve Kenting destination's attractiveness. Moreover, DMOs could specially mark out attractions (landscapes) in the destination that can be influenced easily by climate change, and identify the possible changes that could occur in these attractions if climate change persists. As climate change is a process that occurs over time, tourists can understand the severity of these possible threats, and promote pro-environmental behaviours for environmental sustainability. However, there may be other marine or terrestrial attributes not included in our questionnaire that are important to tourists.

Limitations and further research

Overall, this study provided an effective structural behavioural model to explain and predict the intention of tourists to adapt to climate change. This is only the first step in filling the gap in tourism research related to tourists' adaptation intention to climate change. There are certain limitations that must be addressed in future studies. First, we only collected data from local tourists in the Kenting Coastal Destination, so the conclusions cannot be generalized to other destinations in Taiwan. This suggests that future studies must repeat the investigations in this study on tourists' adaptation intention process in different climate zones and types of tourism. This way, we can gain a comprehensive understanding of tourists' psychological threat and adaptive behaviours caused by climate change within tourism systems. Second, as suggested by other researchers (Landauer et al., 2014; Martinez et al., 2014), tourists' adaptive strategies or behaviours related to climate change are influenced by their culture. The scope of this study is limited to Taiwan's domestic coastal tourism, and the development of tourists' adaptation intention is related to a destination adaptation model centred on the hypothetical shifts in destination attractiveness caused by climate change within a tourism system. Testing this theoretical model across different groups within the same society, or groups from diverse societies (e.g. samples from different countries) could be important. Third, this study showed that literature on changes in destination attributes/attractiveness regarding climate change previously have been descriptive, whereas empirical studies have rarely been conducted, and appropriate measurement scales have not been developed. However, scholars have emphasized that changes in destination attributes/attractiveness related to climate change is a key factor in tourists' adaptation intention (Dawson et al., 2011). Future research must construct a set of reliable and effective scales to deepen our understanding of how climate change relates to coastal destination attractiveness. Fourth, this study only included the hypothetical shifts in destination attractiveness as a threat factor for the dangers of climate change to tourisms systems. Since there are other threat factors in the tourism activities or process of tourists, subsequent studies must integrate the increased tourism expenses or landscape safety issues caused by climate change into the research model, to increase our knowledge on the threat factors of climate change that could harm tourism systems. Finally, this study collected data on maladaptive tourists. However, to reduce the possible interference factors when testing the effectiveness of the proposed theoretical framework, this study only analysed the data of tourists who perceived the hypothetical shifts in destination attractiveness (threat) and would adjust their tourism behaviours related to the Kenting Coastal Destination. Subsequent studies must compare the model differences of adaptive and maladaptive tourists using our framework, to expand our understanding of tourist behaviours and responses within the context of climate change.

Disclosure statement

No potential conflict of interest was reported by the authors.

References

Anderson, J., & Gerbing, D. (1988). Structural equation modeling in practice: A review and recommended two-step approach. *Psychological Bulletin, 103*, 411–423.

Belle, N., & Bramwell, B. (2005). Climate change and small island tourism: Policy maker and industry perspectives in Barbados. *Journal of Travel Research, 44*(1), 32–41.

Bollen, K. A. (1989). *Structural equations with latent variables*. New York: Wiley.

Bryant, C. R., Smit, B., Brklacich, M., Johnston, T. R., Smithers, J., Chjotti, Q., & Singh, B. (2000). Adaptation in Canadian agriculture to climatic variability and change. *Climatic Change, 45*(1), 181–201.

Bujosa, A., Riera, A., & Torres, C. M. (2015). Valuing tourism demand attributes to guide climate change adaptation measures efficiently: The case of the Spanish domestic travel market. *Tourism Management, 47*, 233–239.

Buzinde, C. N., Manuel-Navarrete, D., Yoo, E. E., & Morais, D. (2010). Tourists' perceptions in a climate of change: Eroding destinations. *Annals of Tourism Research, 37*(2), 333–354.

Cheablam, O., & Shrestha, R. P. (2015). Climate change trends and its impact on tourism resources in Mu Ko Surin Marine National Park, Thailand. *Asia Pacific Journal of Tourism Research, 20*(4), 435–454.

Cismaru, M., & Lavack, A. M. (2006). Marketing communications and protection motivation theory: Examining consumer decision-making. *International Review on Public and Nonprofit Marketing, 3*(2), 9–24.

CNN. (2017). *'Manhattanhenge' fail? 12 other superb sunset spots around the world*. Retrieved from http://edition.cnn.com/travel/article/12-sunsets/index.html

Cocolas, N., Walters, G., & Ruhanen, L. (2016). Behavioural adaptation to climate change among winter alpine tourists: An analysis of tourist motivations and leisure substitutability. *Journal of Sustainable Tourism, 24*(6), 846–865.

Cox, D. N., Koster, A., & Russell, C. G. (2004). Predicting intentions to consume functional foods and supplements to offset memory loss using an adaptation of protection motivation theory. *Appetite, 43*, 55–64.

Dai, C.-F. (2002). Climate change, coral bleaching, and the future of coral reefs. *Global Change Communication, Taiwan, 35*, 25–33.

Dang, H. L., Li, E., & Bruwer, J. (2012). Understanding climate change adaptive behavior of farmers: An integrated conceptual framework. *The International Journal of Climate Change: Impacts and Responses, 3*(2), 255–272.

Dawson, J., Havitz, M., & Scott, D. (2011). Behavioral adaptation of Alpine skiers to climate change: Examining activity involvement and place loyalty. *Journal of Travel & Tourism Marketing, 28*(4), 388–404.

Denstadli, J. M., Jacobsen, J. Kr. S., & Lohmann, M. (2011). Tourist perceptions of summer weather in Scandinavia. *Annals of Tourism Research, 38*(3), 920–940.

Environmental Protection Administration, Taiwan. (2014). *Sandstorm monitoring network for China*. Retrieved from http://taqm.epa.gov.tw/dust/tw/News.aspx?id=1156

Fishbein, M., & Ajzen, I. (1975). *Belief, attitude, intention and behavior: An introduction to theory and research*. Reading, MA: Addison-Wesley.

Floyd, D. L., Prentice-Dunn, S., & Rogers, R. W. (2000). A meta-analysis of research on protection motivation theory. *Journal of Applied Social Psychology, 30*(2), 407–429.

Gómez-Martín, M. B. (2005). Weather, climate and tourism: A geographical perspective. *Annals of Tourism Research, 32*(3), 571–591.

Gonzalez, J. A. A. (2012). Climate change and tourism: Comparing female Costa Rican and American perceptions. *Worldwide Hospitality and Tourism Themes, 4*(4), 356–377.

Gössling, S., Scott, D., Hall, C. M., Ceron, J.-P., & Dubois, G. (2012). Consumer behavior and demand response of tourists to climate change. *Annals of Tourism Research, 39*(1), 36–58.

Gunn, L. A. (1993). *Tourism planning: Basics, concepts, cases.* New York: Taylor & Feancis.

Hong, J.-M. (2012). *Can't wait till late winter and early spring! Cherry Blossom blooming October.* Retrieved from http://news.cts.com.tw/action/print.php?newsid=201210041111179

Hopkins, D. (2014). The sustainability of climate change adaptation strategies in New Zealand's ski industry: A range of stakeholder perceptions. *Journal of Sustainable Tourism, 22*(1), 107–126.

Horng, J.-S., Hu, M.-L., Teng, C. C. C., & Lin, L. (2014). Energy saving and carbon reduction behaviors in tourism: A perception study of asian visitors from a protection motivation theory perspective. *Asia Pacific Journal of Tourism Research, 19* (6), 721–735.

Hsu, M.-C., Yeh, M.-J., & Lin, Y.-H. (2013). The adoption and comparison of tourists behavioral intention models. *Journal of Leisure and Recreation Industry Management, Taiwan, 6*(2), 23–46.

Hu, Y., & Ritchie, J. (1993). Measuring destination attractiveness: A contextual approach. *Journal of Travel Research, 32*(2), 25–34.

Intergovernmental Panel on Climate Change. (2007). *Climate change 2007: Synthesis report for policy makers.* Retrieved from http://www.ipcc.ch/pdf/assessment-report/ar4/syr/ar4_syr.pdf

Jeuring, J. H. G., & Peters, K. B. (2013). The influence of the weather on tourist experiences: Analysing travel blog narratives. *Journal of Vacation Marketing, 19*(3), 209–219.

Jopp, R., DeLacy, T., & Mair, J. (2010). Developing a framework for regional destination adaptation to climate change. *Current Issues in Tourism, 13*(6), 591–605.

Jöreskog, K. G., & Sörbom, D. (1993). *LISREL 8: User's reference guide.* Chicago, IL: Scientific Software International.

Kellstedt, P. M., Zahran, S., & Vedlitz, A. (2008). Personal efficacy, the information environment, and attitudes toward global warming and climate change in the United States. *Risk Analysis, 28*(1), 113–126.

Kenting National Park. (2014a). *Data and statisticsata.* Retrieved from http://www.ktnp.gov.tw/News.aspx?n=228F1362E45E0B89&sms=830F4DD99E91DBB7

Kenting National Park. (2014b). *Introduction to Kenting.* Retrieved from http://www.ktnp.gov.tw/cht/aboutus.aspx

Kenting National Park. (2014c). *Tours of Kenting.* Retrieved from http://www.ktnp.gov.tw/Content_List.aspx?n=B7E0652B10F73024

Lam, S.-P. (2015). Predicting support of climate policies by using a protection motivation model. *Climate Policy, 15*(3), 321–338.

Landauer, M., Haider, W., & Pröbstl-Haider, U. (2014). The influence of culture on climate change adaptation strategies: Preferences of cross-country skiers in Austria and Finland. *Journal of Travel Research, 53*(1), 96–110.

Lee, C.-D. (2014). *The number of visits has dropped 30% after gas explosion: Kaochung City, Taiwan promote city tours.* Retrieved from http://www.epochtimes.com.tw/n102958/%E6%B0%A3%E7%88%86%E5%BE%8C%E5%9C%8B%E6%97%85%E5%B0%913%E6%88%90-%E9%AB%98%E5%B8%82%E4%BF%83%E9%8A%B7%E8%A7%80%E5%85%89.html

Lee, P.-F. (2008). The impacts fo climate change on ecology. *Science Development, Taiwan, 424,* 34–43.

Liu, C.-Y., Chia, F., & Chang, C.-C. (2009). The study of travel satisfaction of Ken-Ting area visitors. *Journal of National Huwei University of Science & Technology, Taiwan, 28*(1), 127–138.

Maddux, J. E. (1993). Social cognitive models of health and exercise behavior: An introduction and review of conceptual issues. *Journal of Applied Sport Psychology, 5*(2), 116–140.

Martinez, G., Orbach, M., Frick, F., Donargo, A., Ducklow, K., & Morison, N. (2014). The cultural context of climate change adaptation: Cases from the U.S. East Coast and the German Baltic Sea coast. In G. Martinez, P. Fröhle & H.-J. Meier (Eds.), *Social dimensions of climate change adaptation in coastal regions: Findings from transdisciplinary research* (pp. 85–100). München: Oekom Verlag.

Ministry of the Interior of Taiwan. (2017). *Coastal areas action programme.* Retrieved from fhttp://ws.ndc.gov.tw/Download.ashx?u=LzAwMS9hZG1pbmlzdHJhdG9yLzEwL3JlbGZpbGUvNTU2Ni83MDkxLzAwNjA2OTBfNS5wZGY%3D&n=NS7mtbflsrjpoJjln5%2FooYzli5XmlrnmoYgtMTAzMDUucGRm&icon=..pdf

Morachat, C. (2003). *A study of destination attractiveness through tourists' perspectives: A focus on Chiang Mai, Thailand.* Retrieved from http://ro.ecu.edu.au/theses/1504

Moreno, A., & Becken, S. (2009). A climate change vulnerability assessment methodology for coastal tourism. *Journal of Sustainable Tourism, 17*(4), 473–488.

Morrison, C., & Pickering, C. M. (2013). Perceptions of climate change impacts, adaptation and limits to adaption in the Australian Alps: The ski-tourism industry and key stakeholders. *Journal of Sustainable Tourism, 21*(2), 173–191.

Nyaupane, G. P., & Chhetri, N. (2009). Vulnerability to climate change of nature-based tourism in the Nepalese Himalayas. *Tourism Geographies, 11*(1), 95–119.

Pan, S.-C. (2014). *Visitors disappointed by Kenting's beach "shrinks".* Retrieved from http://g.udn.com/NEWS/DOMESTIC/DOM6/8962638.shtml

Plotnikoff, R. C., & Higginbotham, N. (1995). Predicting low-fat diet intentions and behaviors for the prevention of coronary heart disease: An application of protection motivation theory among an Australian population. *Psychology & Health, 10*(5), 397–408.

Prentice-Dunn, S., Mcmath, B. F., & Cramer, R. J. (2009). Protection motivation theory and stages of change in sun protective behavior. *Journal of Health Psychology, 14*(2), 297–305.

Reser, J. P., Bradley, G. L., Glendon, A I., Ellul, M. C., & Callaghan, R. (2012). *Public risk perceptions, understandings and responses to climate change in Australia and Great Britain.* Gold Coast: National Climate Change Adaptation Research Facility.

Rogers, R. W. (1975). A protection motivation theory of fear appeals and attitude change. *The Journal of Psychology, 91,* 93–114.

Schliephack, J., & Dickinson, E. (2017). Tourists' representations of coastal managed realignment as a climate change adaptation strategy. *Tourism Management, 59,* 182–192.

Scott, D., de Freitas, C. R., & Matzarakis, A. (2009). Adaptation in the tourism and recreation sector. In K. L. Ebi, I. Burton, & P. Hoeppe (Eds.), *Biometeorology for adaptation to climate variability and change* (pp. 171–194). Springer.

Scott, D., Hall, C. M., & Gössling, S. (2012). *Tourism and climate change: Impacts, adaptation and mitigation.* London: Routledge.

Simpson, M. C., Gössling, S., Scott, D., Hall, C. M., & Gladin, E. (2008). *Climate change adaptation and mitigation in the tourism sector: Frameworks, tools and practices.* Paris: UNEP, University of Oxford, UNWTO, WMO.

Smith, K. (1993). The influence of weather and climate on recreation and tourism. *Weather, 48*(12), 398–404.

Spence, A., Poortinga, W., Butler, C., & Pidgeon, N. F. (2011). Perceptions of climate change and willingness to save energy related to flood experience. *Nature Climate Change, 1,* 46–49.

Taiwan Central Weather Bureau. (2017a). *Statistics on climate.* Retrieved from http://www.cwb.gov.tw/V7/climate/dailyPrecipitation/dP.htm

Taiwan Central Weather Bureau. (2017b). *Climate analyses of typhoon.* Retrieved from http://www.cwb.gov.tw/V7/climate/climate_info/monitoring/monitoring_2.html

Taiwan Tourism Bureau. (2017). *Survey of travel by R.O.C. Citizens.* Taipei: Government Printing Office.

Tervo-Kankare, K., Hall, C. M., & Saarinen, J. (2013). Christmas tourists' perceptions to climate change in Rovaniemi, Finland. *Tourism Geographies, 15*(2), 292–317.

Thach, S. V., & Axinn, C. N. (1994). Patron assessments of amusement park attributes. *Journal of Travel Research, 32*(3), 51–60.

Tiller, T. R., & Schott, C. (2013). The critical relationship between climate change awareness and action: An origin-based perspective. *Asia Pacific Journal of Tourism Research, 18*(1–2), 21–34.

Tolman, E. C. A (1923). A behavioristic account of the emotions. *Psychological Review, 30*(3), 217–227.

United Nations World Tourism Organization. (2013). *Tourism highlights: 2013 edition.* Madrid: UN World Tourism Organization.

United Nations World Tourism Organization. (2016). *Tourism highlights: 2016 edition.* Madrid: UN World Tourism Organization.

U.S. EPA. (2013). *Climate change: Basic information.* Retrieved from http://www.epa.gov/climatechange/basics/

Uyarra, M. C., Côté, I. M., Gill, J. A., Tinch, R. R. T., Viner, D., & Watkinson, A. R. (2005). Island specific preferences of tourists for environmental features: Implications of climate change for tourism-dependent states. *Environmental Conservation, 32*(1), 11–19.

Wall, G. (1998). Climate change, tourism and the IPCC. *Tourism Recreation Research, 23*(2), 65–68.

Wang, W.-C. (2015). Advancing conceptual understanding of tourists' adaptation intention to climate change: Multi-theoretical approaches (Unpublished doctoral dissertation). Feng-Chia University, Taichung, Taiwan.

Williams, A. (2011). Definitions and typologies of coastal tourism beach destinations. In A. Jones & M. Phillips (Eds.), *Disappearing destinations: Climate changes and future challenges for coastal tourism* (pp. 47–65). Cambridge, MA: CABI.

Woodside, A. G., & Lysonski, S. (1989). A general model of traveler destination choice. *Journal of Travel Research, 27*(4), 8–14.

Yang, E. C. L., Sharif, S. P., & Khoo-Lattimore, C. (2015). Tourists' risk perception of risky destinations: The case of Sabah's eastern coast. *Tourism and Hospitality Research, 15*(3), 206–221.

Yeh, F.-D. (2014). *Kenting visitors above 7 million people: China tourists must go Maobitou Park.* Retrieved from http://news.tvbs.com.tw/entry/517361

Yeh, L.-B. (2014). *Global warming: Central Weather Bureau—Strong typhoons and high temperature days get more.* Retrieved from http://www.nownews.com/n/2013/10/05/985795

Zhuo, I.-M., & Lu, M.-M. (2014). *1951-2013 Taiwan's climate characteristics.* Retrieved from http://photino.cwb.gov.tw/conf/2014WAF/A3/A3-13.pdf

Understanding the role of local food in sustaining Chinese destinations

Qian Chen and Rong Huang

Food is a sustainable part of destination tourism. However, few studies attempt to separate and examine the role played by local foods within tourists' overall experience when claiming the benefits that local foods can bring to a destination. Using Mainland China as the case in point, this study aims to evaluate the potential for destinations to use local food as a sustainable marketing tool. This is achieved through examining how local food, as part of the tourist experience, can contribute to tourists' destination loyalty. This study proposes a structural model addressing the potential relationships between tourist food-related behaviours (tourists' food-related motivation, information search, food involvement, and food satisfaction) and destination loyalty among domestic tourists in China. Useable data was collected from 1353 respondents at eight destinations in China, this was screened and analysed with SPSS 21 and partial least squares-structural equation modelling. The results reveal the positive correlation between tourists' food satisfaction and destination loyalty, confirming the value of local food in China as a sustainable marketing tool. However, the paper highlights that tourists' food-related motivation, information search, and involvement are not in direct correlation to their destination loyalty. Possible implications for destination marketers and researchers are provided.

Introduction

Tourism is a holistic experience, in which food nourishes both the body and the soul of tourists (Hjalager & Johansen, 2013). Specifically, food is a necessity, not only satisfying the physical needs of tourists, but also assisting in the enhancement of tourists' social relationships, excitement, and learning. With food being an inextricable part of the touristic experience, it is widely acknowledged by researchers (Everett & Aitchison, 2008; Horng & Tsai, 2012) that numerous aspects of food add great value to destinations, including economy, culture, destination identity, and local sustainability. Against a background of wide appeals for sustainability within the development of the tourism industry, growing attention has been noted into the role of food as a sustainable resource (Everett & Aitchison, 2008; Everett & Slocum, 2013; Sims, 2009).

However, an analysis of academic literature indicates that the research into the role of food in destination sustainable development has yet to be fully explored. Very few studies manage to provide empirical data from the demand side, which could indicate how food can be utilised by destinations to contribute to the development of sustainable tourism. Moreover the Western world is the main focus of current literature in food and sustainable tourism (Everett & Aitchison, 2008; Hjalager &

Johansen, 2013; Sims, 2009) with a deficiency of literature focusing on countries such as China, Japan, Korea, Singapore, Thailand, and India.

This paper therefore focuses on Mainland China, an area with a diversity of food resources and a profound food culture (Du, 2012). An increasing number of environmental problems are associated with the development of the Chinese tourism industry, so the need to develop sustainable tourism is well noted (Zhang, Ji, & Zhu, 2013). Previous studies (Wan & Li, 2013; Xu & Fox, 2014) have contributed to the research of Chinese sustainable tourism, however, a review of both English and Chinese literature reveals that few studies have managed to offer insights into how to utilise food as a sustainable tourism resources and thus contribute to the sustainable development of the tourism industry.

Against the above background, this study aims to evaluate the role of food as a sustainable marketing tool in Chinese destinations. This study constructs a systematic model addressing whether, and how, food as part of tourist experience can influence tourist behaviours and contribute to their loyalty intentions to destinations. Although China-focused, the findings and the theoretical model constructed in this study will not only shed light on the marketing strategies for destination marketers in China, but also benefit other regions that would like to utilise food as a sustainable marketing tool for contributing to a destination's development.

Literature review

Food & sustainable tourism

As one of the essential elements, researchers (Boyne, Hall, & Williams, 2003; Lin, Pearson, & Cai, 2011) notice that many tourist destinations increasingly use food as a popular initiative within the development of local tourism. There are a myriad of benefits associated with local foods in a destination (Handszuh, 2000). In particular, local food holds a lot of potential for contributing to the development of local sustainable tourism (Berno, Kline, & Wolf, 2014; He, 2006; Sims, 2009). As can be seen below, a volume of research and information shows that the significance of food to sustainable tourism can be understood from different aspects (Mgonja, Backman, Backman, Moore, & Hallo, 2017; Sims, 2009).

Firstly, local foods contribute to the economic sustainability of destinations (Sims, 2009). Food is a sustainable way to increase tourist spending and extend the tourist season (Everett & Aitchison, 2008). Tourist consumption of local foods will support the local economy through promoting local food production, generating revenues, and potentially retaining tourism earnings in the region (Mgonja et al., 2017). Secondly, local foods assist in the development of the environmental sustainability of a destination. Tourists who consume local foods can generate market opportunities and support diversification of local agriculture. It can create benefits by conserving traditional farming landscapes, addressing environmental preservation, and encouraging sustainable development of agriculture (Everett & Aitchison, 2008; Sims, 2009). Lastly, local foods assist in improving the social and cultural sustainability of destinations. According to Sims (2009), local foods are defined as authentic products representing the place and culture of a destination. In this way the tourists' consumption of local foods can enhance a destination's identity within the tourist experience, improve the involvement of the local community and contribute to the social and cultural development of local sustainable tourism (Mgonja et al., 2017; Quan & Wang, 2004).

The use of local food can contribute, both directly and indirectly, to various elements of sustainability in a specific area (Du Rand & Heath, 2006; Everett & Slocum, 2013; Sims, 2009). Gössling and Hall (2013) highlight that the production and consumption of food are significant aspects in the development of sustainable tourism. Local food and drink products can improve the economic, environmental, social and cultural sustainability of both tourism, and the rural host community, through generating revenues, supporting local businesses, encouraging sustainable agricultural practices, and meeting the needs of tourists for local experience and a sense of place (Berno et al., 2014; Sims, 2009).

However, although several studies (Everett & Aitchison, 2008; Sims, 2009) claim the role of food as an important vehicle for delivering sustainable tourism to destinations, very few studies manage to provide empirical data from the demand side evaluating the role that local food plays in tourists' experience.

Moreover, many researchers (e.g. Lee, Packer, & Scott, 2015; McKercher & Chan, 2005; Smith, Pitts, & Litvin, 2012) express their concerns that the significance of food as a motivation for travel has been exaggerated and that many studies have assumed food-related activities are a valid proxy for travel motives. Therefore, it is important to understand the true value of food within the overall travel decisions of tourists and their consumer behaviours and how local food can be utilised by destinations to contribute to the development of sustainable tourism needs further investigation.

Food-related behaviour & tourist destination loyalty

Consumer behaviour is a constant decision-making process (Gilbert, 1991). According to Engel, Kollat, and Blackwell (1968), the grand model of consumer behaviour consists of five main phases, which include problem/need recognition, information search, and evaluation of alternatives, purchase, and post-purchase behaviour. To understand tourists' food-related behaviours, the grand model of consumer behaviour is adapted in the current research, with suitable factors identified throughout distinct phases.

Recognition of consumers' needs is the first step for marketers (Engel et al., 1968). To understand why customers do what they do is to basically understand their motivations (Solomon, Bamossy, Askegaard, & Hogg, 2013). According to Boyne et al. (2003), the importance of food in tourists' motivation varies among people and is likely to influence tourists' attitudes and behaviours. Therefore, tourists' food-related motivations assist in recognising the needs of the tourist market, and in predicting tourists' food-related behaviours at destinations. Tourist food-related motivation is thus identified as the first factor in evaluating the role of food in influencing tourists' motivation.

Once need has been recognised, potential tourists start searching for adequate information and solutions to meet that need (Fodness & Murray, 1997). Faced with large amounts of information, Ab Karim and Chi (2010) propose that the travellers responded to information based on the type and credibility of information sources. Therefore, it is vital for destination marketers to understand how tourists acquire knowledge and find out what, to them, is a perceived credible source of information (Ab Karim & Chi, 2010). Searching for information refers to travellers using different types of information sources to facilitate trip planning; and is thus identified as the second factor that might affect their travel behaviours and decision making (Gursoy & Chen, 2000). When searching for relevant information, tourists might evaluate alternative options before making their final choice. However, as this phase is the direct result of the information search, and a difficulty exists in identifying a suitable factor to evaluate this phase, no separate factor has been noted.

To accord with the stage of 'purchase' in the consumer behaviour model, tourist food involvement is claimed as the equivalent factor examining tourists' on-site food-related behaviours and activities. According to Havitz and Dimanche (1999, p. 123), the concept of involvement refers to an 'unobservable state of motivation, arousal, or interest toward a recreational activity or product' and has been widely employed to understand tourist attitudes and behaviours. In addition, Bell and Marshall (2003) highlight that a tourist's level of involvement with food is likely to vary across people. Therefore, tourist food involvement is significant in revealing tourist behaviours toward food (Kim, Suh, & Eves, 2010).

When the 'purchase' is finished, the next stage is about tourists' post-purchase behaviours, this is when tourists will form their evaluation of their on-site experience. According to Rust and Oliver (1994), the level of satisfaction is the outcome generated from a comparison between performance and expectations. Due to a positive correlation with the quality of experience, satisfaction is an important indicator of tourist experience in that it creates positive feelings (Lee & Chang, 2012). Therefore, tourist food satisfaction is identified as the most suitable factor for this stage.

Based on the above discussion, the four factors of 'food-related motivation', 'information search', 'food involvement', and 'food satisfaction' have been identified as underlying tourist food-related behaviour. However, for the purpose of understanding the importance of a tourist's food experience during their entire destination experience, and to predict future travel demand in relation to food, an evaluative factor needs to be identified. Based on a comprehensive review of relevant research (Kim & Brown, 2012; Lee, Huang, & Yeh, 2010; Yoon & Uysal, 2005), tourist destination loyalty is noted as the evaluative factor for assessing tourist food experience. Yoon and Uysal (2005) mention that tourists' positive experience of the service, products and other resources provided at tourism destinations can result in loyalty behaviours such as repeat visits and positive recommendations. Consequently, destination loyalty is identified as the assessment factor of tourists' food experience.

Following the identification of the above factors, it is important to examine potential relationships between the factors; about how tourist behaviours across various stages integrate and influence tourist destination loyalty. This is achieved in the following four ways:

Firstly, according to existing studies, there is a lack of research investigating the specific relationships between tourist food-related motivations and the tourists' food-related attitudes and behaviours. However, a few studies, in more general contexts, have presented causal relationships between motivation and different indicators including information search (Moorthy, Ratchford, & Talukdar, 1997), involvement (Kim, 2008; Kim, Woo, & Uysal, 2015; Prebensen, Woo, Chen, & Uysal, 2013), satisfaction (Lee & Beeler, 2009), and destination loyalty (Kim, 2008; Yoon & Uysal, 2005). Consequently, the following hypotheses are proposed:

H1: Food-related motivation has a positive influence on information search.
H2: Food-related motivation has a positive influence on food involvement.
H3: Food-related motivation has a positive influence on food satisfaction.
H4: Food-related motivation has a positive influence on destination loyalty.

Secondly, according to Fodness and Murray (1998) and Gursoy and Chen (2000), the information acquired and used by potential customers might affect their decision-making processes and travel behaviours. Previous studies have revealed the significance of tourist information search behaviour on tourist level of involvement (Xiang & Law, 2013), level of overall satisfaction (Gursoy & McCleary, 2004), and further behaviour intentions (Correia & Pimpão, 2008). Therefore, it is postulated:

H5: Information search has a positive influence on food involvement.
H6: Information search has a positive influence on food satisfaction.
H7: Information search has a positive influence on destination loyalty.

Thirdly, involvement is thus highlighted of significance in predicting and interpreting tourist experience (Havitz & Dimanche, 1999). Previous studies have noted the impacts of tourists' level of involvement on tourists' satisfaction (Kim, 2008; Kim et al., 2015; Lee & Beeler, 2009), and tourists' loyalty (Iwasaki & Havitz, 2004; Lee & Chang, 2012). Therefore, the following hypotheses are drawn:

H8: Food involvement has a positive influence on food satisfaction.
H9: Food involvement has a positive influence on destination loyalty.

Lastly, Baker and Crompton (2000) mention that measuring tourist' satisfaction offers predictions on tourists' future behavioural intentions. Researchers (Chen & Tsai, 2007; Kim, 2008) have shown their support through revealing a significant association between tourists' level of satisfaction and loyalty intentions. Consequently, it is predicted:

H10: Food satisfaction has a positive influence on destination loyalty.

To summarise, this study investigates the underlying dimensions of food-related behaviour and examines the impact of food-related behaviour on tourist destination loyalty. Based on the grand model of consumer behaviour, as well as the relationships shown among the constructs from previous research, specific hypotheses on the potential relationships among the constructs have all been identified. The proposed theoretical model is shown as Figure 1.

Methodology

Questionnaire design and measurement of construct

This study adopted the quantitative survey approach using a structured questionnaire to test the hypothesised theoretical model. The questionnaire was firstly designed in English and then translated into Chinese using a back-translation approach to ensure consistency (Brislin, 1976). The questionnaire contained six parts; measuring tourist food-related motivation, information search, food involvement, food satisfaction, destination loyalty and demographic information. The demographic profiles of respondents were examined in terms of gender, age, and educational level. As for the constructs of the theoretical model, each construct was designed with a multi-item measurement scale because the single-item scale might fail to capture all the attributes of the construct (Churchill, 1979). Respondents were asked to rate each item using a 7-point Likert scale, with 1 being 'strongly disagree' and 7 being 'strongly agree'.

In terms of measuring tourist food-related motivation, a six-item scale approach was developed from studies by Beer, Ottenbacher, and Harrington (2012), McKercher, Okumus, and Okumus (2008), and Kim et al. (2010). The construct of information search was optimally operationalised as the use of information sources suggested by Fodness and Murray (1997). A multi-attribute approach was developed in this study to measure tourists' ratings of the importance of various information sources (Ab Karim & Chi, 2010). To measure tourist food involvement, an eight-item uni-dimensional scale was adapted from the study by Sparks, 2007, which was originally modified from Zaichkowsky's (1985) *Personal Involvement Inventory* scale. As for the construct of food satisfaction, five items were derived from the previous studies (Bosque & Martin, 2008; Mason & Paggiaro, 2012; Žabkar, Brenčič, & Dmitrović, 2010), and for the construct of destination loyalty, a three-item scale was adapted from previous research (Žabkar et al., 2010).

To ensure the face validity of the questionnaire design, the questionnaire was subject to the expert reviews both academically and practically, in both Chinese and English contexts. Specifically, the questionnaire was sent to local tour operators in China, Chinese destination marketers, and both English and Chinese academics and professionals. Based on the evaluation of the questionnaire and comments made by the experts, a few changes were made to improve the questionnaire and thus form the second draft. Then, a pilot study was conducted in Yangzhou, Jiangsu Province in China on 30th March, 5th April, and 6th April 2014. A total of 109 usable responses were collected, and the collected data was analysed through SPSS version 21 to determine whether the

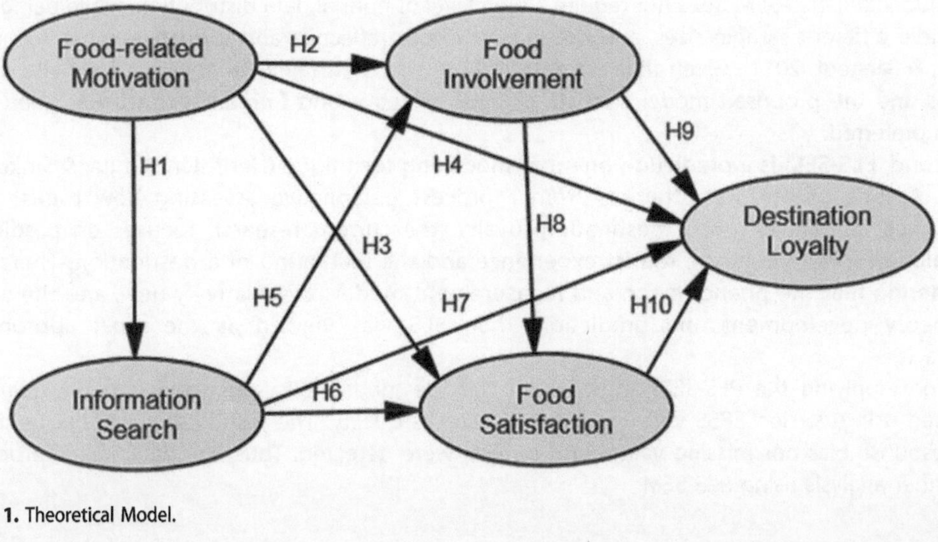

Figure 1. Theoretical Model.

measurement scales of the constructs were valid and reliable. Due to the analysis results of the constructs all being above the suggested statistical threshold (Hair, Hult, Ringle, & Sarstedt, 2014), the constructs were confirmed to be of satisfactory reliability and validity, and the questionnaire was finalised for the main survey.

Data collection

The main survey was conducted at eight chosen destinations (Changsha, Chengdu, Guangzhou, Nanjing, Hangzhou, Hefei, Qingdao, and Xiamen) in China in May and June 2014. The eight destinations were selected as they represent the 'Eight Major Schools of Cuisine' in China (Lu, Chuan, Yue, Min, Su, Zhe, Xiang, and Hui) based upon geographical regions that divide Chinese cuisine into eight main categories each with their own traditions and characteristics (Du & Li, 2010). For example, 'Chuan' is characterised by spicy and pungent flavours, 'Su' is renowned for light and sweet tastes, while 'Lu' frequently uses shallots and garlic as seasonings. The eight cuisines together show the characteristics of Chinese culinary culture.

At each survey destination, the questionnaire was conducted using an interviewer-completed approach by a few research assistants at the departure lounges of coach stations, train stations, and airports. Research assistants were instructed to approach respondents using the next-to-pass non-probability sampling technique (Tourism and Recreation Research Unit, 1983). If the approached person showed willingness to continue the questionnaire interview, screening questions were applied by the interviewers to confirm the appropriateness of each respondent. Firstly, respondents were asked whether or not they were domestic tourists in China; secondly, respondents were provided with the definition of local food, and asked whether they had a local food experience during their travels. In total 1427 questionnaires were collected of which 1353 questionnaires were useable.

Data analysis

To test the presented hypotheses in the study, Structural Equation Modelling (SEM) with a Partial Least Squares (PLS) approach was applied to test the theoretical model using the software application of SmartPLS 2.0 software (Ringle, Wende, & Will, 2005). PLS-SEM was used over other methods for two reasons.

First, PLS estimates 'all path coefficients and multiple individual item loadings simultaneously, and consequently enables researchers to avoid biased and inconsistent parameter estimates' (Loureiro & González, 2008). PLS-SEM does not require a high level of normal data distribution, while being able to handle different sample sizes, and dealing with both reflective and formative constructs (Hair, Ringle, & Sarstedt, 2011). Given that the dataset in this study is fairly large, with 1,353 valid observations and the proposed model consists of both reflective and formative constructs, then PLS-SEM is preferred.

Second, PLS-SEM is a prediction oriented modelling technique (Henseler, Ringle, & Sinkovics, 2009). As one of the first attempts from a process perspective assessing how tourist food experience influences tourist destination loyalty, the current research focuses on predicting the value of food within the tourist experience and the marketing of a destination. Therefore, considering that the phenomenon and measurement models are relatively new, and the aim is for theory development and prediction, then PLS was viewed as the most appropriate approach.

Before applying the PLS-SEM approach to test the model, it is suggested that the data was screened using version SPSS version 21.0 to ensure its quality. The results showed that neither a non-response bias nor missing values and outliers were detected. Thus, the data was appropriate for further analysis using PLS-SEM.

Results

Profile of the respondents

There was a slight higher distribution of females among respondents in terms of gender: 52.9% were female, and 47.1% male; this may be explained by the fact that females are more likely to respond to food-related surveys (Robinson & Getz, 2014). The two dominant age groups were those aged between 18 and 26 years old (30.7%) and 26 to 35 years old (37.5%). Around 86.7% of the respondents had at least a college degree, which may be a result of the recent expansion of higher education in China (Wang, Liu & Lai, 2012). The sample is characterised as being well educated. In addition, through analysis, the profile of the respondents also has certain influences on the food-related behaviours of tourists, details see Chen and Huang (2018).

Evaluation of measurement model

The theoretical model presented in this study was analysed through a two-step approach (i.e. measurement model and structural model). The measurement model is evaluated to verify both the reflective and formative constructs. Table 1 below presents the reliability and validity of constructs with reflective measurement model specifications.

The assessment of reflective construct reliability and validity was based on internal consistency reliability, indicator reliability, convergent validity, and discriminant validity. Firstly, according to Table 1, Cronbach's alpha and composite reliability values of the constructs all exceeded 0.80, which were above the minimal acceptable threshold of 0.70 (Nunnally, 1978), showing high internal consistency reliability of the constructs. Secondly, the item loadings of all the indicators were above the 0.70 thresholds, which supported the indicator reliability. Thirdly, the reflective measures have satisfactory convergent validity because the average variance extracted (AVE) values of the constructs

Table 1. Assessment of the reflective measurement model.

Constructs and Items	Loading	Cranach's Alpha	Composite reliability	AVE
Food-related Motivation				
Food plays an important role in my destination choice.	0.797	0.879	0.909	0.625
Prior to my trip, I planned food choices to experience local culture.	0.822			
I decided on destinations to visit only based on the foods I wanted to experience.	0.735			
I usually do some research about the local food or restaurants prior to my trip.	0.784			
Prior to my trip, one thing I anticipate is eating the food there.	0.846			
I like trying different styles or types of food during travel.	0.753			
Food Involvement				
Interesting	0.786	0.925	0.939	0.656
Appealing	0.821			
Valuable	0.793			
Exciting	0.847			
Stimulating	0.790			
Desirable	0.821			
Wanted	0.823			
Mean a lot	0.798			
Food Satisfaction				
The food experience gave me high satisfaction.	0.856	0.910	0.933	0.736
My choice to taste the local cuisine was a wise one.	0.866			
The food experience gave me a sense of joy.	0.887			
The food experienced met my expectations.	0.877			
The food experienced exceeded my expectation.	0.800			
Destination Loyalty				
I would like to recommend the destination to others.	0.889	0.834	0.909	0.751
I will spread positive word of mouth about the destination.	0.885			
I will visit the destination again.	0.824			

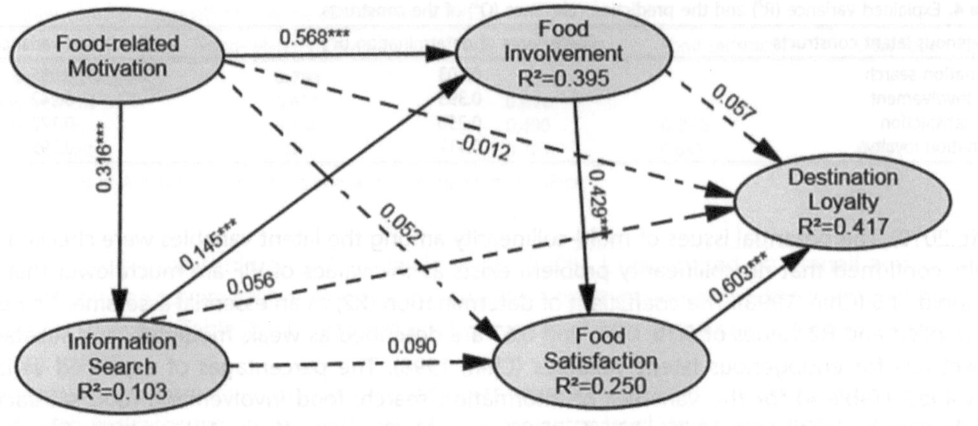

Figure 2. Results of the Structural Model Test. Note: → Indicates singificant path. → Indicates non-significant path. *Significant at $p < 0.10$. **Significant at $p < 0.05$. *** Significant at $p < 0.01$.

Table 6. Test of mediation through Bootstrapping.

		First Criteria		Second Criteria			
	Mediator variable	Direct effect (t value)	Sig. (p value)	Indirect effect (t value)	Sig. (p value)	Third Criteria (VAF%)	Mediation
MOT -> SAT	INVO	0.295 (9.221)	***(0.000)	0.292	***(0.000)	50.3	Partial mediation
MOT -> SAT	IS	0.069 (1.886)	NS				No mediation
IS->SAT	INVO	0.154 (5.190)	***(0.000)	0.062	***(0.000)	71.3	Partial mediation
IS->LOYA	SAT	0.108 (3.476)	***(0.001)	0.100	***(0.000)	51.9	Partial mediation
INVO-> LOYA	SAT	0.319 (8.900)	***(0.000)	0.259	***(0.000)	55.2	Partial mediation
MOT-> LOYA	INVO	0.015 (0.576)	NS				No mediation
MOT-> LOYA	SAT	0.019 (0.539)	NS				No mediation

Note: NS, Non-significant.
*Significant at $p < 0.10$.
**Significant at $p < 0.05$.
***Significant at $p < 0.01$.

conducting an analysis of potential mediation among constructs was suggested. Based on the analysis procedures noted by Hair et al. (2014), Table 6 (below) shows the results of the potential mediation effects analysis in the current study.

The results confirmed the partial mediating role of food involvement on the relationship between food-related motivation and food satisfaction (VAF = 50.3%), as well as the relationship between information search and food satisfaction (VAF = 71.3%). In addition, the partial mediating effects of food satisfaction on the relationship between information search and destination loyalty (VAF = 51.9%), and the relationship between food involvement and destination loyalty were verified (VAF = 55.2%).

Discussions

This study integrated key variables of tourist food-related behaviour and destination loyalty into a research model focused on domestic tourists in China. The model evaluated whether and how food, as part of tourist experience, contributes to their loyalty intentions towards destinations in China.

The positive correlation between tourist food satisfaction and destination loyalty acknowledged the value of developing food as a marketing tool in China. Many studies (Chen & Tsai, 2007; Kim, 2008) acknowledged the positive relationship between tourists' overall satisfaction and their loyalty intentions. According to Weiermair and Fuchs (1999), the overall satisfaction that tourists have towards a destination is positively linked with each dimension of that destination. It is thus understandable that tourists' food satisfaction, as one dimension of their overall satisfaction, has an influence on their loyalty intentions towards the destination. In addition, Kivela and Crotts (2006) noticed that the positive beliefs that tourists have toward their local cuisine experience could prompt them to visit the destination and experience the cuisine again (Kivela & Crotts, 2006). Therefore, the value of developing food as a marketing tool was supported.

However, according to the results, no positive direct relationship was shown between destination loyalty with tourist food-related motivation, information search, and food involvement, respectively. The results implied that tourists who have higher motivation in food, higher engagement with information search, and higher involvement with food would not necessarily have higher loyalty intentions towards a particular destination (Zhang, 2012). Nevertheless, these results seems to disagree with a previous study by Kim et al. (2010), which showed that tourists who perceive higher importance of food in their travel motivation are more likely to be satisfied with their food festival experience, and more likely to return to the food festival. In addition, Kim et al. (2010) noted a significant positive relationship between tourist food involvement and tourist loyalty and thus it contrasted with the results of this study. It is suggested that the contrast might be explained by the different settings of the two studies with their study set in the context of a food festival in South Korea, while this study is set at a destination level in China. Therefore, our results indicated that destinations in China that would like to use food as a marketing tool should pay equal attention to individuals with both low and high levels of food-related motivation, information search, and food involvement, as every domestic tourist could be a loyal and repeat visitor to that destination.

Because tourist destination loyalty was only in direct correlation to tourist food satisfaction, the focus for destination marketers should be to maximise the tourists' satisfaction level with their food experience (Wan & Chan, 2013; Zhang, 2012). According to the results, tourist food satisfaction was directly influenced by tourist food involvement, which was positively related to tourist food-related motivation and information search. In addition to the direct impacts, the mediation analysis above also highlighted the mediating role of food involvement, and food satisfaction, respectively, in the model. Specifically, the mediating role of food involvement further enhanced the importance of food involvement in monitoring the satisfaction level of tourist with their food experience regardless of their food-related motivation and information search behaviours. It was explained that involvement is a precondition to some types of behavioural consequences and satisfaction (Lee & Beeler, 2009; Prebensen et al., 2013). The level of involvement affects the level of participation, which positively influences the creation of experience satisfaction (Kim, 2008; Prebensen et al., 2013). Therefore, in line with the study by Kim (2008) that identified the mediating role of involvement between motivation and satisfaction to predict the travel demand within the university student market, tourist food involvement might be useful in predicting tourists' travel demand within the food tourism market.

Moreover, the study also highlighted the mediating effects of food satisfaction. Although the mediating role of satisfaction has not been well addressed in the food tourism context; in the similar context of wine tourism, Lee and Chang (2012) highlighted the role of satisfaction in mediating the relationship between activity involvement and wine tourists' loyalty intentions. In addition, the result of the study was in line with several studies (Castro, Armario, & Ruiz, 2007; Lee, 2009) that have acknowledged the mediating role of satisfaction in the tourist behavioural model. Consequently, maximising tourists' satisfaction level with their local food experience is of foremost significance if that destination wishes to retain the loyalty intentions of those tourists.

Conclusions and implications

This study is of an exploratory nature to understand how, from a tourist behaviour perspective, a destination could utilise food for marketing purposes. The results confirmed the value of local food to a destination, and supported the development of local food as a marketing tool for a destination. With a thorough investigation from the perspective of tourist experience, the specific relationships are supposed to provide suggestions to destination marketers about how to utilise effectively local food to retain and attract tourists. A range of practical implications is listed below:

Firstly, destinations should create positive, unique images of destination food through effective positioning and differentiation to attract tourists' attention (Chen & Huang, 2016; Lin et al., 2011; Okumus, Okumus, & McKercher, 2007). Based on the positive relationship shown between tourist food-related motivation and information search, tourists who perceive higher importance of food in their travel motivation tend to do more information search on food prior to travel. Therefore destinations that are already promoting themselves to the public using sophisticated materials (e.g good images and relevant information) related to local foods are likely to attract this potential group of tourists (Okumus et al., 2007). Thus it is suggested that destinations should appreciate their local food resources and pay detailed attention to how those resources are presented and promoted to the public.

Secondly, based on the positive relationship shown between information search and tourist food involvement, destinations should utilise a range of different channels of information sources to distribute the information (Ab Karim & Chi, 2010). Destination marketers should aim to create positive 'word of mouth' publicity of that destination's food so as to increase people's willingness to visit destinations for food purposes, and also actively seek regular feedback from tourists who had recently visited that destinations in order to monitor, or possibly avert, any negative 'word-of-mouth' publicity of the food at that destination (Cheng, Lam, & Hsu, 2006). Furthermore, destinations should use a multi-channel approach towards the distribution of their information; for example publishing brochures, magazines and booklets on local food culture and restaurants, presenting destination food-related documentaries or TV shows, sharing information on websites and social media, and developing good relationship with travel agencies (Ab Karim & Chi, 2010; Chen & Huang, 2016).

Thirdly, the significance of food involvement has been highlighted in the model on influencing tourists' food satisfaction. Consequently, destination marketers should develop different types of food-related activities to attract and encourage tourists to get involved with destination food, such as identifying distinctive or traditional food products, holding food shows or food festivals, providing cooking demonstrations or cookery classes, and organising distinctive food-themed tours (Chen & Huang, 2016; Huang, 2009). In addition to the development of different food-related activities, destination marketers should pay attention to ensuring and improving the performance and quality of the food-related activities offered by the destination (Croce & Perri, 2010). Having tourists' expectations of the destination food involvement met and further exceeded will contribute to tourists' satisfaction with their food experience, and lead to strengthening their loyalty intentions.

This study is one of the first attempts to construct a structural model examining the interrelationships between tourist food-related behaviours and their destination loyalty. Through the employment of a new generation of SEM (PLS-SEM), it offers insights to destination marketers on how to effectively assess the value of local food to the destination marketing. Additionally, as an aspect of cultural tourism, this study might also benefit similar research areas such as wine tourism, and other forms of cultural tourism. Specifically, the model might be tested and applied in different contexts of cultural tourism revealing how to achieve successful destination marketing through understanding tourist experience (Charters & Ali-Knight, 2002; Horng & Tsai, 2012).

This study examined tourists' food-related behaviours across different stages, and provided insights for destination marketers to delve deeper into the food-related behaviours of tourists prior to conducting food-related marketing activities. Kim, Eves, and Scarles (2013) argued that research into tourists' local food consumption and food experience is still in its infancy. Compared to the existing dominating research perspective from the supply side in the field, this study contributed to the body of existing research on tourists' food experience, and provided empirical evidence on the food experience of Chinese tourists when they travelled in China.

Limitations and future research

Despite the research contributions, several limitations of this study need to be considered to provide direction for future research. Firstly, in terms of the proposed theoretical model, due to the exploratory nature, the current model was limited in the number of variables used. Additional factors such as food-related personal traits, previous experience, perceived value/quality might also have impacts on tourists' food experience and possibly influence tourists' destination loyalty. Future research should focus on testing the theoretical model with the inclusion of additional factors and proposing a more comprehensive model.

Secondly, limitation was noted on the selection of data collection locations. Concerns might be raised that the destinations with rich food resources and specialities might have impacts on the role of food influencing tourists' experience and loyalty. Respondents might be argued to be more likely to have a higher rating of the importance of food at those destinations. However, in the special context of China, due to every destination possessing its own food resources, speciality dish and variations of Chinese cuisine, the limitations are inevitable.

Thirdly, this study was conducted at eight different destinations at a city level in China and there are concerns on the generalisation of the results to represent the whole of China. Future study could be replicated in a destination of a smaller size in China. Additionally, the sample used in the study was limited to Chinese domestic tourists. It may be of concern that the sample could fail to verify the wider generalizability of the results. The theoretical model in the study could be tested with different samples from different destinations or regions in the future.

Disclosure statement

No potential conflict of interest was reported by the authors.

References

Ab Karim, S., & Chi, C. G. Q. (2010). Culinary tourism as a destination attraction: An empirical examination of destinations' food image. *Journal of Hospitality Marketing & Management, 19*(6), 531–555. doi:10.1080/19368623.2010.493064

Baker, D. A., & Crompton, J. L. (2000). Quality, satisfaction and behavioral intentions. *Annals of Tourism Research, 27*(3), 785–804.

Beer, C. L., Ottenbacher, M. C., & Harrington, R. J. (2012). Food tourism implementation in the Black Forest destination. *Journal of Culinary Science & Technology, 10*(2), 106–128. doi:10.1080/15428052.2012.677601

Bell, R., & Marshall, D. W. (2003). The construct of food involvement in behavioral research: Scale development and validation⋆. *Appetite, 40*(3), 235–244. doi:10.1016/S0195-6663(03)00009-6

Berno, T., Kline, C., & Wolf, E. (2014). Food tourism in academia. In E. Wolf, J. Bussell, C. Campbell, W. Lange-Faria, & K. McAree (Eds.), *Have fork will travel: A practical handbook for food and drink tourism professionals* (pp. 309–321). Portland: World Food Travel Association.

Bosque, I. R. D., & Martin, H. S. (2008). Tourist satisfaction a cognitive-affective model. *Annals of Tourism Research, 35*, 551–573. doi:10.1016/j.annals.2008.02.006

Boyne, S., Hall, D., & Williams, F. (2003). Policy, support and promotion for food-related tourism initiatives: A marketing approach to regional development. In C. M. Hall (Ed.), *Wine, food and tourism marketing* (pp. 131–153). Birmingham: The Harworth Press.

Brislin, R. W. (1976). Comparative research methodology: Cross-cultural studies. *International Journal of Psychology, 11*(3), 215–229. doi:10.1080/00207597608247 35

Castro, C. B., Armario, M. E., & Ruiz, D. M. (2007). The influence of market heterogeneity on the relationship between a destination's image and tourists' future behaviour. *Tourism Management, 28*(1), 175–187. doi:10.1016/j.tourman.2005.11.013

Charters, S., & Ali-Knight, J. (2002). Who is the wine tourist? *Tourism Management, 23*(3), 311–319.

Chen, Q., & Huang, R. (2018). Local food in China: A viable destination attraction. *British Food Journal, 120*(1), 146–157.

Chen, Q., & Huang, R. (2016). Understanding the importance of food tourism to Chongqing, China. *Journal of Vacation Marketing, 22*(1), 42–54. doi:10.1177/1356766715589427

Chen, C. F., & Tsai, D. (2007). How destination image and evaluative factors affect behavioral intentions? *Tourism Management, 28*(4), 1115–1122.

Cheng, S., Lam, T., & Hsu, C. H. C. (2006). Negative word-of-mouth communication intention: An application of the theory of planned behavior. *Journal of Hospitality and Tourism Research, 30*(1), 95–116. doi:10.1177/1096348005284269

Chin, W. W. (1998). The partial least squares approach for structural equation modelling. In G. A. Marcoulides (Ed.), *Modern methods for business research* (pp. 295–336). Mahwah, NJ: Lawrence Erlbaum Associates.

Churchill, G. A. (1979). A paradigm for developing better measures of marketing constructs. *Journal of Marketing Research, 16*(1), 64–73. doi:10.2307/3150876

Correia, A., & Pimpão, A. (2008). Decision-making processes of Portuguese tourist travelling to South America and africa. *International Journal of Culture, Tourism and Hospitality Research, 2*(4), 330–373.

Croce, E., & Perri, G. (2010). *Food and wine tourism: Integrating food, travel and territory*. Wallingford, Oxfordshire: CABI.

Diamantopoulos, A., & Winklhofer, H. M. (2001). Index construction with formative indicators: An alternative to scale development. *Journal of Marketing Research, 38*(2), 269–277. Retrieved from http://www.jstor.org/stable/1558630

Du, F. X., & Li, X. Q. (2010). *Chinese culinary culture*. Beijing: China Travel & Tourism Press.

Du, Y. (2012). 'A bite of China' brings along the heat of food tourism. *Hangzhou Daily*. [Online]. Retrieved from http://fashion.ifeng.com/travel/roll/detail_2012_06/08/15152455_0.shtml

Du Rand, G. E., & Heath, E. (2006). Towards a framework for food tourism as element of destination marketing. *Current Issues in Tourism, 9*(3), 206–234. doi:10.1300/J073v14n03_06

Engel, J. F., Kollat, D. J., & Blackwell, R. D. (1968). *Consumer behaviour*. New York: Holt, Rinehart, and Winston.

Everett, S., & Aitchison, C. (2008). The role of food tourism in sustaining regional identity: A case study of Cornwall, South West England. *Journal of Sustainable Tourism, 16*(2), 150–167. doi:10.2167/jost696.0

Everett, S., & Slocum, S. L. (2013). Food and tourism: An effective partnership? A UK-based review. *Journal of Sustainable Tourism, 21*(6), 789–809. doi:10.1080/09669582.2012.741601

Fodness, D., & Murray, B. (1998). A typology of tourist information search strategies. *Journal of Travel Research, 37*, 108–119.

Fodness, D., & Murray, B. (1997). Tourist information search. *Annals of Tourism Research, 24*(3), 503–523. doi:10.1016/S0160-7383(97)00009-1

Fornell, C., & Larcker, D. F. (1981). Evaluating structural equation models with unobservable variables and measurement error. *Journal of Marketing Research, 18*(2), 161–188. doi:10.2307/3151312

Geisser, S. (1975). The predictive sample reuse method with applications. *Journal of the American Statistical Association, 70* (350), 320–328.

Gilbert, D. C. (1991). Consumer behaviour in tourism. In C. P. Cooper (Ed.), *Progress in tourism, recreation and hospitality management* (pp. 78–105). Lymington, Hants: Belhaven Press.

Gössling, S., & Hall, M. (2013). Sustainable culinary systems: An introduction. In M. Hall, & S. Gössling (Eds.), *Sustainable culinary systems: Local foods, innovation, tourism and hospitality* (pp. 3–44). Abingdon: Routledge.

Götz, O., Liehr-Gobbers, K., & Krafft, M. (2010). Evaluation of structural equation models using the partial least squares (PLS) approach. In V. E. Vinzi, W. W. Chin, & J.W Henseler (Eds.), *Handbook of Partial Least Squares: Concepts, Methods and Applications*. (pp. 691–711). Heidelberg: Springer.

Gursoy, D., & Chen, J. S. (2000). Competitive analysis of cross cultural information search behavior. *Tourism Management, 21*(6), 583–590. doi:10.1016/S0261-5177(00)00005-4

Gursoy, D., & McCleary, K. W. (2004). An integrative model of tourists' information search behavior. *Annals of Tourism Research, 31*(2), 353–373.

Hair, J. F., Hult, G. T., Ringle, C. M., & Sarstedt, M. (2014). *A primer on partial least squares structural equation modelling (PLS-SEM)*. Thousand Oaks, CA: Sage.

Hair, J. F., Ringle, C. M., & Sarstedt, M. (2011). PLS-SEM: Indeed a silver bullet. *Journal of Marketing Theory and Practice, 19* (2), 139–152. doi:10.2753/MTP1069-6679190202

Handszuh, H. (2000). *Local food in tourism policies*. Paper presented at the international conference on local food and tourism, Larnaka, Cyprus.

Havitz, M., & Dimanche, F. (1999). Leisure involvement revisited: Drive properties and paradoxes. *Journal of Leisure Research, 31*(2), 122–149.

He, L. P. (2006). Discussing the food tourism resources in changsha. *Journal of Changsha University, 20* (3), 10–12.

Henseler, J., Ringle, C. M., & Sinkovics, R. R. (2009). The use of partial least squares path modelling in international marketing. *Advances in International Marketing, 20*(1), 277–320. doi:10.1108/S1474-7979

Hjalager, A. M., & Johansen, P. H. (2013). Food tourism in protected areas – sustainability for producers, the environment and tourism? *Journal of Sustainable Tourism, 21*(3), 417–433. doi:10.1080/09669582.2012.708041

Horng, J. S., & Tsai, C. Y. (2012). Constructing indicators of culinary tourism strategy: An application of resource-based theory. *Journal of Travel & Tourism Marketing, 29*(8), 796–816. doi:10.1080/10548408.2012.730945

Huang, R. (2009). The role of food in promoting Chinese regions on the web. *International Journal of Tourism Policy, 2*(4), 289–305. doi:10.1504/IJTP.2009.028715

Iwasaki, Y., & Havitz, M. E. (2004). Examining relationships between leisure involvement, psychological commitment and loyalty to a recreation agency. *Journal of Leisure Research, 36*, 45–72.

Kim, A. K., & Brown, G. (2012). Understanding the relationships between perceived travel experiences, overall satisfaction, and destination loyalty. *Anatolia: An International Journal of Tourism and Hospitality Research*, *23*(3), 328–347. doi:10.1080/13032917.2012.696272

Kim, Y. G., Eves, A., & Scarles, C. (2013). Empirical verification of a conceptual model of local food consumption at a tourist destination. *International Journal of Hospitality Management*, *33*(1), 484–489. doi:10.1016/j.ijhm.2012.06.005

Kim, K. (2008). Analysis of structural equation model for the student pleasure travel market: Motivation, involvement, satisfaction, and destination loyalty. *Journal of Travel & Tourism Marketing*, *24*(4), 297–313. doi:10.1080/10548400802156802

Kim, Y. G., Suh, B. W., & Eves, A. (2010). The relationships between food-related personality traits, satisfaction, and loyalty among visitors attending food events and festivals. *International Journal of Hospitality Management*, *29*(2), 216–226. doi:10.1016/j.ijhm.2009.10.015

Kim, H. L., Woo, E. J., & Uysal, M. (2015). Tourism experience and quality of life among elderly tourists. *Tourism Management*, *46*, 465–476.

Kivela, J., & Crotts, J. C. (2006). Tourism and gastronomy: Gastronomy's influence on how tourists experience a destination. *Journal of Hospitality & Tourism Research*, *30*(3), 354–377. doi:10.1177/1096348006286797

Lee, J., & Beeler, C. (2009). An investigation of predictors of satisfaction and future intention: Links to motivation, involvement, and service quality in a local festival. *Event Management*, *13*(1), 17–29. doi:10.3727/152599509789130584

Lee, T. H., & Chang, Y. S. (2012). The influence of experiential marketing and activity involvement on the loyalty intentions of wine tourists in Taiwan. *Leisure Studies*, *31*(1), 103–121.

Lee, C. F., Huang, H. I., & Yeh, H. R. (2010). Developing an evaluation model for destination attractiveness: Sustainable forest recreation tourism in Taiwan. *Journal of Sustainable Tourism*, *18*(6), 811–828. doi:10.1080/09669581003690478

Lee, K. H., Packer, J., & Scott, N. (2015). Travel lifestyle preferences and destination activity choices of slow food members and non-members. *Tourism Management*, *46*(0), 1–10.

Lee, T. H. (2009). A structural model to examine how destination image, attitude, and motivation affect the future behaviour of tourists. *Leisure Sciences: An Interdisciplinary Journal*, *31*(3), 215–236.

Lin, Y., Pearson, T., & Cai, L. (2011). Food as a form of destination identity: A tourism destination brand perspective. *Tourism and Hospitality Research*, *11*, 30–48. doi:10.1057/thr.2010.22

Loureiro, S. M., & González, F. J. (2008). The importance of quality, satisfaction, trust, and image in relation to rural tourist loyalty. *Journal of Travel and Tourism Marketing*, *25*, 117–136.

Mason, M. C., & Paggiaro, A. (2012). Investigating the role of festivalscape in culinary tourism: The case of food and wine events. *Tourism Management*, *33*(6), 1329–1336. doi:10.1016/j.tourman.2011.12.016

McKercher, B., & Chan, A. (2005). How special is special interest tourism? *Journal of Travel Research*, *44*(1), 21–31.

McKercher, B. F., Okumus, F., & Okumus, B. (2008). Food tourism as a viable market segment: It's all how you cook the numbers!. *Journal of Travel &Tourism Marketing*, *25*(2), 137–148.

Mgonja, J. T., Backman, K. F., Backman, S. J., Moore, D. D., & Hallo, J. C. (2017). A structural model to assess international visitors' perceptions about local foods in Tanzania. *Journal of Sustainable Tourism*, *25*, 796–816. *Published Online*. doi:10.1080/09669582.2016.1250768

Moorthy, S., Ratchford, B. R., & Talukdar, D. (1997). Consumer information search revisited: Theory and empirical analysis. *Journal of Consumer Research*, *23*(4), 263–277.

Nunkoo, R., Ramkissoon, H., & Gursoy, D. (2013). Use of structural equation modelling in tourism research: Past, present and future. *Journal of Travel Research*, *52*(6), 759–771. doi:10.1177/0047287513478503

Nunnally, J. C. (1978). *Psychometric theory* (2nd ed.). New York: McGraw-Hill.

Okumus, B., Okumus, F., & McKercher, B. F. (2007). Incorporating local and international cuisines in the marketing of tourism destinations: The cases of Hong Kong and turkey. *Tourism Management*, *28*(1), 253–261. doi:10.1016/j.tourman.2005.12.020

Prebensen, N., Woo, E., Chen, J., & Uysal, M. (2013). Motivation and involvement as antecedents of the perceived value of the destination experience. *Journal of Travel Research*, *52*(2), 253–264. doi:10.1177/0047287512461181

Quan, S., & Wang, N. (2004). Towards a structural model of the tourist experience: An illustration from food experiences in tourism. *Tourism Management*, *25*(3), 297–305. doi:10.1016/S0261-5177(03)00130-4

Ringle, C. M., Wende, S., & Will, A.. (2005). SmartPLS 2.0 (Beta). SmartPLS, Hamburg, Germany. Available at: http://www.smartpls.de.

Robinson, R. N. S., & Getz, D. (2014). Profiling potential food tourists: An Australian study. *British Food Journal*, *116*(4), 690–706.

Rust, R., & Oliver, R. L. (1994). *Service quality: New directions in theory and practice*. London: Sage Publications.

Sims, R. (2009). Food, place and authenticity: Local food and the sustainable tourism experience. *Journal of Sustainable Tourism*, *17*(3), 321–336. doi:10.1080/09669580802359293

Smith, W. W., Pitts, R. E., & Litvin, S. W. (2012). Travel and leisure activity participation. *Annals of Tourism Research*, *39*(4), 2207–2210.

Solomon, M., Bamossy, G., Askegaard, S., & Hogg, M. (2013). *Consumer behaviour: A European perspective*. (5th ed.). Harlow: Pearson Higher Education.

Sparks, B. (2007). Planning a wine tourism vacation? Factors that help to predict tourist behavioural intentions. *Tourism Management*, *28*(5), 1180–1192.

Tourism and Recreation Research Unit. (1983). *Recreation site survey manual - methods and techniques for conducting visitor surveys*. London: E. & F. N. Spon.

Wan, Y. K., & Chan, S. H. (2013). Factors that affect the levels of tourists' satisfaction and loyalty towards food festivals: A case study of macau. *International Journal of Tourism Research, 15*(3), 226–240.

Wan, Y. K., & Li, X. P. (2013). Sustainability of tourism development in Macao, China. *International Journal of Tourism Research, 15*(1), 52–65. doi:10.1002/jtr.873

Wang, D., Liu, D., & Lai, C. (2012). Expansion of higher education and the employment crisis: Policy innovations in China. *On the Horizon, 20*, 336–344.

Weiermair, K., & Fuchs, M. (1999). Measuring tourist judgment on service quality. *Annals of Tourism Research, 26*, 1004–1021. doi:10.1016/S0160-7383(99)00037-7

Xiang, Z., & Law, R. (2013). Online competitive information space for hotels: An information search perspective. *Journal of Hospitality Marketing & Management, 22*(5), 530–546.

Xu, F. F., & Fox, D. (2014). Modelling attitudes to nature, tourism and sustainable development in national parks: A survey of visitors in China and the UK. *Tourism Management, 45*, 142–158. doi:10.1016/j.tourman.2014.03.005

Yoon, Y., & Uysal, M. (2005). An examination of the effects of motivation and satisfaction on destination loyalty: A structural model. *Tourism Management, 26*(1), 45–56. doi:10.1016/j.tourman.2003.08.016

Žabkar, V., Brenčič, M. M., & Dmitrović, T. (2010). Modelling perceived quality, visitor satisfaction and behavioural intentions at the destination level. *Tourism Management, 31*(4), 537–546. doi:10.1016/j.tourman.2009.06.005

Zaichkowsky, J. L. (1985). Measuring the involvement construct. *Journal of Consumer Research, 12*, 341–352.

Zhang, J. K., Ji, M., & Zhu, P. X. (2013). Research on Chinese sustainable tourism in the 21st century. *Economic Management Journal, 1*, 121–129. Retrieved from http://kreader.cnki.net/Kreader/CatalogViewPage.aspx?dbCode=CJFQ&filename=JJGU201301015&tablename=CJFDLAST2015&compose=&first=1&uid=

Zhang, T. (2012). A study on the effect of food tourism motivation on tourist satisfaction and behaviour intention. *Tourism Tribune, 27*(10), 78–84.

Appendix. Survey Questionnaire

A1. Food-related motivation

1. Please indicate the overall role of food in affecting your motivation for travel in Chinese regions (1 means strongly disagree, 7 means strongly agree).

	1	2	3	4	5	6	7
Food plays an important role in my destination choice.	O	O	O	O	O	O	O
Prior to my trip, I planned food choices to experience local culture.	O	O	O	O	O	O	O
I decided on destinations to visit only based on the foods I wanted to experience.	O	O	O	O	O	O	O
I usually do some research about the local food or restaurants prior to my trip.	O	O	O	O	O	O	O
Prior to my trip, one of the things I anticipate is eating the food there.	O	O	O	O	O	O	O
I like trying different styles or types of food during travel.	O	O	O	O	O	O	O

A2. Information search

2. Please indicate your rating of the importance of the listed information sources (1 means not important, 7 means very important)

	1	2	3	4	5	6	7
Overall importance of communication sources	O	O	O	O	O	O	O
Newspaper/magazines ads	O	O	O	O	O	O	O
Brochures	O	O	O	O	O	O	O
TV, Radio commercials	O	O	O	O	O	O	O
Internet advertising	O	O	O	O	O	O	O
Travel agency ads	O	O	O	O	O	O	O
Outdoor media ads	O	O	O	O	O	O	O
Print media columns/travel news	O	O	O	O	O	O	O
TV/Movies/Travel channels /Radio programmes	O	O	O	O	O	O	O
Websites information/travel news	O	O	O	O	O	O	O
Social media (forums, blogs, twitter)	O	O	O	O	O	O	O
Friends or families recommendation	O	O	O	O	O	O	O
Word of mouth from other tourists	O	O	O	O	O	O	O
Past experience/general knowledge	O	O	O	O	O	O	O

A3. Involvement of local food

3. Please indicate your rating with the following statements. To you, tasting local food and getting involved with food-related activities, such as visiting good-rated restaurants; reading articles or watching TV shows about local cuisine; and buying locally produced food products are . . . . (1 means strongly disagree, 7 means strongly agree)

	1	2	3	4	5	6	7	
Are not interesting	O	O	O	O	O	O	O	Are interesting
Are not appealing	O	O	O	O	O	O	O	Are appealing
Are not valuable	O	O	O	O	O	O	O	Are valuable
Are not exciting	O	O	O	O	O	O	O	Are exciting
Are not stimulating	O	O	O	O	O	O	O	Are stimulating
Are not desirable	O	O	O	O	O	O	O	Are desirable
Are not wanted	O	O	O	O	O	O	O	Are wanted
Do not mean a lot	O	O	O	O	O	O	O	Mean a lot

A4. Food satisfaction

4. Please rate your agreement with the following statements based on your satisfaction of your overall food experience (1 means strongly disagree, 7 means strongly agree).

	1	2	3	4	5	6	7
The food experience gave me high satisfaction.	O	O	O	O	O	O	O
My choice to taste the local cuisine was a wise one.	O	O	O	O	O	O	O
The food experience gave me a sense of joy.	O	O	O	O	O	O	O
The food experienced met my expectations.	O	O	O	O	O	O	O
The food experienced exceeded my expectation.	O	O	O	O	O	O	O

A5. Destination loyalty

5. Please rate your agreement with the following statements describing your destination loyalty (1 means strongly disagree, 7 means strongly agree).

	1	2	3	4	5	6	7
I would like to recommend the destination to others.	O	O	O	O	O	O	O
I will spread positive word of mouth about the destination.	O	O	O	O	O	O	O
I will visit the destination again.	O	O	O	O	O	O	O

A6. Basic information

6. Gender:

A. Male	B. Female

7. Your age:

A. below 26	B. 26–35	C.36–45
D. 46–55	E. 56–65	F. Above 65

8. Your education level:

A. Junior school or below	B. High School
C. College/University-diploma/degree	D. Postgraduate degree

How to improve tourism energy efficiency to achieve sustainable tourism: evidence from China

Lamei He, Jianping Zha and Hui Ann Loo

ABSTRACT

This study constructs the biennial Malmquist index, including undesirable outputs, to evaluate tourism energy efficiency and decompose its changes using a sample of 30 provinces in China over the period of 9 years from 2005 to 2013. The changes in tourism energy efficiency can be attributed to five factors, including total factor productivity, capital-energy ratio, labour-energy ratio, energy supply composition and output composition. The study results show that the energy efficiency of the tourism industry is much lower than that of the industry as a whole and not greener. Total factor productivity is a key positive driving force to boost tourism energy efficiency. Capital-labour ratio gradually overtook the total factor productivity as the tourism energy efficiency change's largest contribution since 2010 and thus became the biggest inhibitor for increasing tourism energy efficiency. The rational utilization of capital and new capital investment should accompany energy-saving technology and is essential for sustainable tourism.

1. Introduction

The rapid development of industrial civilization and urban economies has exacerbated pressure on the natural environment, resulting in a sharp reduction in resources, environmental pollution and ecological damage in the world, seriously restricting the survival of mankind and long-term economic development (Boggia & Cortina, 2010; Cornillie & Fankhauser, 2004). With the promulgation of *Our Common Future* by the World Commission on Environment and Development in 1987 (Brundtland, 1987), the concept of sustainable development has been promoted and popularized. More and more countries are beginning to pay attention to resource consumption and environmental problems in the process of economic development and attach importance to the balance between economic growth and environmental protection. In this context, sustainable tourism development remains one of the most popular concepts in tourism research with the aim of minimizing environmental and socio-cultural impacts while maximizing economic benefits for tourist destinations (Godde, Price, & Zimmermann, 1999; Lu & Nepal, 2009).

The production, distribution, and consumption of energy determine all basic functions of our existence, including work and production, housing, education, communication, transport, and even tourism and recreation (Brucher, 2001). Energy in the form of basic infrastructure, electricity, fuels consumed in travelling, heating, cooling, catering, and other services are the basic prerequisites of tourism. The supply and price of energy (especially in the form of travel costs) directly affect the development or stagnation of tourism and related economies (Becken, 2011; Lennox, 2012; Pentelow & Scott, 2010). Energy is also the leading cause of environmental problems. Atmospheric pollution, such as greenhouse gases, acid rain and dust, is directly related to fossil energy combustion. Therefore, researching energy-related issues is very necessary and important for sustainable development (Bilgen & Sarıkaya, 2015; Fath, 2015; Idahosa, Marwa, & Akotey, 2017; Olabi, 2012). Since the reform and opening, China has maintained rapid economic growth for over 30 years and has become a 'China miracle', highly regarded by the international academic community and many developing countries (Lin, Cai, & Li, 2003). Behind the sustained high economic growth and along with the drastic increase of energy consumption, China has now become the largest energy producer and consumer in the world (Liu, Lei, Xu, & Yuan, 2018). How to ensure that the energy supply supports economic growth, meets the increasing demand for energy, and addresses the climate change caused by energy consumption have become major challenges.

The extant research shows that improving energy efficiency in all policy toolkits is the vital means of addressing these challenges because the change of energy structure is unlikely to be achieved at this moment (Feng & Wang, 2017; Li & Lin, 2018; Mao, Du, Cai, & Jiang, 2010; Sun, Xu, & Yin, 2016). In general, energy efficiency is used to measure the extent to which the actual output capacity reaches the maximum output with fixed energy input, or the minimum energy input can be achieved under fixed output (Balachandra, Ravindranath, & Ravindranath, 2010; Fisher-Vanden, Jefferson, Liu & Tao, 2004; Hu & Wang, 2006; Streimikiene, Ciegis, & Grundey, 2008). Energy efficiency is calculated by comparing the output value of the tourism industry in a certain period of time with the corresponding energy consumption. Improving tourism energy efficiency means using a smaller amount of energy to achieve the same output, which helps promote the green technology revolution, serve the multiple goals of energy conservation and emissions reduction, and achieve climate change mitigation and sustainable development. Moreover, energy efficiency reflects the quality of economic development and is an important symbol of technological progress and transformation of the economic development model. It is also the key to pushing forward supply-side reform, optimizing resource allocation and raising total factor productivity. As such, it is of great significance to discuss how to improve energy efficiency in the tourism industry.

2. Literature review

Since the 1990s, the eco-environmental problems induced by tourism have gained worldwide attention. In 1995, the 21st Convention of the Tourism and Travel Industry noted that resource management and energy consumption are the core fields in the tourism industry (Meng, Xu, Hu, Zhou, & Wang, 2016). Most studies have focused on energy demands and carbon dioxide emissions in the tourism industry covering New Zealand, Sweden, the Maldives and other countries (Becken & Patterson, 2006; Bernard, Khelil, Pichon, & Tissot, 2010; Dwyer, Forsyth, Spurr, & Hoque, 2010; Konan & Chan, 2010; Moutinho, Costa, & Bento, 2015; Perch-Nielsen, Sesartic, & Stucki, 2010), the influencing factors of tourism energy consumption (Becken, Simmons, & Frampton, 2003; Katircioglu, 2014; Zaman, Shahbaz, Loganathan, & Raza, 2016) and the strategies of energy savings and emissions reduction in tourism (Becken et al., 2003; Buckley, 2011; Scott, Hall, & Gössling, 2016). Few studies have been conducted on the issue of tourism energy efficiency. Hence, the literature review section is divided into three subsections.

2.1. Energy efficiency and tourism

Since the 1970s, energy efficiency and conservation have become a key component in addressing energy security. Patterson (1996), an earlier scholar, noted that energy efficiency is a generic issue and that there is no quantitative measure. Subsequently, the measurement of the energy efficiency becomes popular among manufacturing industries (Blomberg, Henriksson, & Lundmark, 2012) or high energy consuming industries (Spoladore, Borelli, Devia, Mora, & Schenone, 2016). Few studies have looked at the energy efficiency of the service industry. Fang, Hu, and Lou (2013) assessed the total factor energy efficiency of Taiwan's service sectors and suggested that finance, insurance and real estate had higher energy efficiency than the tourism industry (lodging and catering). Becken et al. (2003) noted the difference in energy efficiency of tourism transport, accommodations, and attractions for international tourists to New Zealand. Tourism accommodation thus far has been analysed as the usual cases in the field of tourism energy efficiency. Khemiri and Hassairi (2005) discussed the development of the Tunisian hotel sector's energy efficiency improvement. Önüt and Soner (2006) evaluated the Antalya Region hotels' energy efficiency and identified the relatively best performing hotels through data envelopment analysis methodology. Pace (2016) studied different patterns of environmental innovations and the adoption of energy efficiency measures in hotels in Malta by a qualitative study. Tang et al. (2016) measured the energy performance of hotel buildings in Lijiang, China by defining annual energy use divided by the gross floor area of a hotel. Bianco, Righi, Scarpa, and Tagliafico (2017) modelled the maximum potential energy savings and assessed the implementation of a realistic energy efficiency scenario. In terms of energy efficiency of the tourism industry, Tang, Zhong, and Ng (2017) took the Wulingyuan Scenic and Historic Interest Area in China as a case to assess its energy efficiency combined with the stage of the tourism life cycle. To our knowledge, tourism energy efficiency is in its infancy, and the existing body of research focuses on the assessment of energy efficiency of hotels and the implementation of energy savings improvements.

2.2. Energy consumption and CO_2 emissions from the tourism industry

Tourism-related carbon emissions attracted considerable attention of scholars at the turn of this century, and Gössling (2000) first proposed the evaluation method for carbon emissions of tourism. This topic has since been taken seriously in academic circles under different spatial scales, mainly measuring carbon emissions from tourism transportation (Becken, 2002; Becken et al., 2003; Gössling et al., 2005; Howitt, Revol, Smith, & Rodger, 2010; Huang, Cao, Jin, Yu, & Huang, 2016; Luo, Becken, & Zhong, 2018; Peeters & Schouten, 2006), tourism accommodations (Becken, Frampton, & Simmons, 2001; Filimonau, Dickinson, Robbins, & Huijbregts, 2011; Lai, 2015; Rosselló-Batle, Moià, Cladera, & Martínez, 2010; Teng, Horng, Hu, Chien, & Shen, 2012; Tsai, Lin, Hwang, & Huang, 2014), and regional or national tourism industries (Bagliani, Galli, Niccolucci, & Marchettini, 2008; Becken & Patterson, 2006; Dubois & Ceron, 2006; Gössling, 2002; Katircioglu, Feridun, & Kilinc, 2014; Kelly & Williams, 2007; Kuo & Chen, 2009; Paramati, Alam, & Chen, 2017; Perch-Nielsen et al., 2010; Pereira, Ribeiro, & Filimonau, 2017; Robaina-Alves, Moutinho, & Costa, 2016; Sun, 2014; Tang, Shang, Shi, Liu, & Bi, 2014). Although it has been difficult to measure the carbon emissions of tourism for many years, research on energy consumption and carbon emissions of the tourism industry has still been widely carried out and discussed. Various methods have been developed and introduced to measure the energy consumption and carbon emissions from the tourism industry, including top-down or bottom-up methodologies (Becken & Patterson, 2006; Gössling, 2013), input-output models (Cadarso, Gómez, López, & Tobarra, 2016; Surugiu, Surugiu, Breda, & Dinca, 2012), production and expenditure approaches (Dwyer et al., 2010; Sharp, Grundius, & Heinonen, 2016), and environmentally extended Tourism Satellite Accounts (Munday, Turner, & Jones, 2013). These methods have their own advantages, disadvantages and applicability (Munday et al., 2013). Considering the situations of China's tourism industry, this paper uses both the bottom-up approach and the top–bottom approach to evaluate the energy consumption and CO_2 emissions from the tourism industry. Specifically, sampling data on the composition of tourist consumption is used as well as annual statistical data, input and output of tourism-related industries. Based on this data, carbon emissions from the tourism industry are calculated using the tourism consumption stripping coefficient proposed by Li and Li (1999), stripping out the amount of tourism consumption from commodity trade, catering, transport, postal industries, and related services (including accommodation, entertainment, and other services).

2.3. Determinants of energy efficiency

The main determinants of energy efficiency at the macro-economic level are technology effects, capital investment, structure effects, and labour. Phylipsen, Blok, and Worrell (1997) identified structural differences in energy intensive industries and suggested that structural differences should be taken into account in comparisons of energy efficiency. Efficient technologies can promote energy savings (Khemiri & Hassairi, 2005). Capital-labour investment should be accompanied with energy-saving technology to improve energy efficiency in the service sectors (Fang et al., 2013). Subrahmanya (2006) also claimed that labour positively affects energy efficiency. A widely used method is to decompose the energy efficiency change into factors based on the Malmqusit index. The Malmquist index is a nonparametric method introduced by Caves, Christensen, and Diewert (1982). Wang (2011) explicated how to decompose energy efficiency change into factors by series of formulas through the Malmquist index. The key argument posed against the use of this decomposition is that it may result in linear programming infeasibilities. Pastor, Asmild, and Lovell (2011) proposed a biennial Malmquist index to avoid infeasibilities.

In the context of tourism studies, Becken et al. (2003) combined different patterns of tourism subsectors to adjust the energy efficiency. However, few published papers have identified the determinants that affect tourism energy efficiency, especially at the industry level. Seeking to overcome the above-mentioned limitations, this study first estimates tourism-related energy consumption by adopting the tourism consumption stripping coefficient, then evaluates tourism energy efficiency using the ratio of tourism income to energy consumption. The vital contribution of this study is to analyse the factors that influence the tourism energy efficiency by proposing a new biennial Malmquist index under the nonparametric decomposition framework. Thus, this study discovers the factors hidden behind the improvement of tourism energy efficiency so that it can be used as a reference in the development of sustainable tourism.

3. Model specification and data

3.1. The measurement model of energy consumption from the tourism industry

The tourism industry is comprised of several sectors, which should all be considered if accurate energy consumption from tourism in specific areas is to be obtained. However, energy data of the tourism industry is lacking in China since tourism is neither one of the 90 industries in the Chinese Industrial Classification for National Economic Activities nor a separate industry in the input-output table of the Chinese economy. In this context, the tourism consumption stripping coefficient is adapted to calculate tourism energy consumption. Thus, the tourism consumption stripping coefficients must be identified. They are calculated by dividing the added value contributed by tourism by the overall added value of an industry. The added value contributed by tourism is derived by multiplying the gross value of tourism by the added value ratio. Once the tourism consumption stripping coefficient is identified, tourism energy consumption can be separated from the overall energy consumption of an industry (Wu, Han, & Tian, 2015). The formula is as follows:

$$P_i = T_i/V_i, \tag{1}$$

where P_i is the tourism consumption stripping coefficient of i industry, referring to transport, storage, postal, wholesale, retail sales and catering industries, V_i is the added value of i industry, and T_i is the added value of i industry contributed by tourism. The added value ratio is calculated by dividing the added value by the total value of the industry.

Energy consumption from the tourism industry can be estimated based on the following formula (1):

$$E_{ij} = E'_{ij} \times P_i, \tag{2}$$

where E_{ij} refers to consumption of j energy caused by tourism in i industry involving 15 types of energy: raw coal, cleaned coal, coke, coke oven gas, other gas, crude oil, gasoline, kerosene, diesel oil, fuel oil, liquefied petroleum gas (LPG), refinery gas, natural gas, other petroleum products and other coking products. E'_{ij} means the consumption of j energy of i industry.

It should be noted that the industry classification of the national economic census is basically consistent with that of the China Energy Yearbook. In the national economic census, industries related to tourism include three sectors: transport, storage and postal industries; wholesale and retail industries; and accommodation and catering industries. The regional energy balance of the China Energy Yearbook includes transport, storage and postal, wholesale, retail, accommodation and catering industries. To calculate the tourism consumption stripping coefficients, the tourism consumption types in the investigation are mapped to the industry classification of the national economic census and the China Energy Yearbook. As a result, long-haul transport, city transport, postal and communication, and four other industries are grouped into the transport, storage and postal industries; tourist site visit, shopping and entertainment are grouped into the wholesale and retail industries; and catering and accommodation correspond to the catering and accommodation industries. Because of different compositions, the consumption of domestic and foreign tourists is calculated separately and then added together. Considering the data availability of energy consumption at industrial level, wholesale, retail, catering and accommodation industries are further grouped in this study. The purpose is to improve the accuracy of the energy consumption. Therefore, the tourism consumption stripping coefficients of two industry clusters are calculated in this paper for transport, storage and post industries, as well as wholesale, retail, accommodation and catering industries.

3.2. Definition of tourism energy efficiency

The energy efficiency of the tourism industry could be measured as the ratio of the revenue of the tourism industry to its total energy consumption. The total tourism energy consumption equals the sum of E_{ij} by converting various types of energy consumption into an average low calorific value referred to in Table 1.

$$EP = \frac{Y}{E} = \frac{\text{Tourism revenue}}{\sum_{i,j} E_{ij}}, \tag{3}$$

where EP denotes tourism energy efficiency, Y represents the revenue of tourism industry, and E is total tourism energy consumption. The tourism energy efficiency of specific areas in a specific year can be acquired through formula (3).

3.3. Decomposition of tourism energy efficiency change

While the undesirable output is not taken into account in the biennial Malmquist index proposed by Pastor et al. (2011), the negative impact of tourism development is ignored. Therefore, this study proposes a new biennial Malmquist index for the decomposition of tourism energy efficiency change. The process is stated as follows.

3.3.1. The biennial Malmquist index

Consider that in the time period t, the tourism industry of specific area k is using inputs $x^t \in R^n_+$ meanwhile to produce desirable outputs $y^t \in R^m_+$ and undesirable outputs $b^t \in R^s_+$. Define the biennial production technology of period t in terms of the input requirement set, which is:

$$T^B(x) = \Big\{(y^t, b^t) \Big| \sum_{i=1}^{K} z_i^t y_i^t + \sum_{i=1}^{K} z_i^{t+1} y_i^{t+1} \geq y_{i'}^t;\ \sum_{i=1}^{K} z_i^t b_i^t + \sum_{i=1}^{K} z_i^{t+1} b_i^{t+1} = b_{i'}^t;\ \sum_{i=1}^{K} z_i^t x_i^t + \sum_{i=1}^{K} z_i^{t+1} x_i^{t+1} \leq x_{i'}^t;\ z_i \geq 0, i = 1, \cdots, K \Big\}, \tag{4}$$

where $T^B(x)$ is non-empty, closed, convex, bounded and satisfies strong disposability of inputs and outputs. Then, y_i, b_i,x_i are the known desirable outputs, undesirable outputs and inputs of the k area, $y_{i'}$, $b_{i'}$, $x_{i'}$ are values of y_i, b_i,x_i that maximize the objective function in the fractional programming problem, and z_i are the variable weights to be determined by the solution of this problem.

Table 1. Average low calorific values and conversion factors of different energy types.

Energy types	Unit	Average low calorific values (kJ)	Conversion factors from physical unit to coal equivalent (f_j)	Energy types	Unit	Average low calorific values (kJ)	Conversion factors from physical unit to coal equivalent (f_j)
Raw coal	10^4 tons	20,908	0.7143 kgce	Kerosene	10^4 tons	43,070	1.4714 kgce/kg
Cleaned coal	10^4 tons	26,344	0.9000 kgce	Diesel oil	10^4 tons	42,652	1.4571 kgce/kg
Coke	10^4 tons	28,435	0.9714 kgce	Fuel oil	10^4 tons	41,816	1.4286 kgce/kg
Coke oven gas	10^8 cu.m	17,354	0.5929 kgce/m^3	LPG	10^4 tons	50,179	1.7143 kgce/kg
Other gas	10^8 cu.m	16,970	0.5798 kgce/m^3	Refinery gas	10^4 tons	46,055	1.5714 kgce/kg
Other coking products	10^4 tons	34,289	0.6449 kgce/kg	Other petroleum products	10^4 tons	37,634	0.5857 kgce/kg
Crude oil	10^4 tons	41,816	1.4286 kgce/kg	Natural gas	10^8 cu.m	38,931	1.3300 kgce/m^3
Gasoline	10^4 tons	43,070	1.4714 kgce/kg				

To add the undesirable outputs into the biennial Malmquist index proposed by Pastor et al. (2011), we introduced the directional distance function proposed by Chung, Färe, and Grosskopf (1997):

$$\vec{D}^{B}(x, y, b; g) = \sup\{\beta:(y, b) + \beta g \in T^{B}(x)\}, \tag{5}$$

where g is the vector of directions in which outputs are scaled. Seeking to increase the desirable outputs while simultaneously decreasing the undesirable outputs, $g = (y, -b)$. β is the maximum proportion of the increase in desirable output or decrease in undesirable output.

Based on the directional distance function, the new biennial Malmquist index is constructed as follows:

$$\mathrm{BML}_t^{t+1} = \frac{1 + \vec{D}^{B}(x^t, y^t, b^t; y^t, -b^t)}{1 + \vec{D}^{B}(x^{t+1}, y^{t+1}, b^{t+1}; y^{t+1}, -b^{t+1})} \tag{6}$$

3.3.2. The decomposition of tourism energy efficiency change

Capital, labour and energy are regarded as the common inputs of production. Revenue from the tourism industry is taken as a desirable output, and CO_2 emissions caused by tourism development are undesirable output.

Using production technology in time period *t*as a reference, the tourism energy efficiency changes between time period *t* can be decomposed as:

$$\begin{aligned} EP_t^{t+1} &= \frac{Y^{t+1}/E^{t+1}}{Y^t/E^t} \\ &= \left[\frac{1 + \vec{D}^{t}(L^t, K^t, E^t, Y^t, C^t; Y^t, C^t)}{1 + \vec{D}^{t+1}(L^{t+1}, K^{t+1}, E^{t+1}, Y^{t+1}, C^{t+1}; Y^{t+1}, C^{t+1})}\right] \\ &\times \left[\frac{1 + \vec{D}^{B}(L^t, K^t, E^t, Y^t, C^t; Y^t, C^t)}{1 + \vec{D}^{t}(L^t, K^t, E^t, Y^t, C^t; Y^t, C^t)} \times \frac{1 + \vec{D}^{t+1}(L^{t+1}, K^{t+1}, E^{t+1}, Y^{t+1}, C^{t+1}; Y^{t+1}, C^{t+1})}{1 + \vec{D}^{B}(L^{t+1}, K^{t+1}, E^{t+1}, Y^{t+1}, C^{t+1}; Y^{t+1}, C^{t+1})}\right] \\ &\times \left[\frac{(Y^{t+1}/E^{t+1})/(1 + \vec{D}^{B}(L^t, K^t, E^t, Y^t, C^t; Y^t, C^t))}{(Y^t/E^t)/(1 + \vec{D}^{B}(L^{t+1}, K^{t+1}, E^{t+1}, Y^{t+1}, C^{t+1}; Y^{t+1}, C^{t+1}))}\right], \end{aligned} \tag{7}$$

where the variables are labour L, capital K, energy E, revenue Y, and CO_2 emissions C of the tourism industry. Through Wang's (2011) guideline for decomposition, formula (7) can be finally written as:

$$\frac{Y^{t+1}/E^{t+1}}{Y^t/E^t} = \Bigg[\underbrace{\frac{1+\vec{D}^t(L^t,K^t,E^t,Y^t,C^t;Y^t,C^t)}{1+\vec{D}^{t+1}(L^{t+1},K^{t+1},E^{t+1},Y^{t+1},C^{t+1};Y^{t+1},C^{t+1})}}_{\text{EC}}\Bigg]$$

$$\times\Bigg[\underbrace{\frac{1+\vec{D}^B(L^t,K^t,E^t,Y^t,C^t;Y^t,C^t)}{1+\vec{D}^t(L^t,K^t,E^t,Y^t,C^t;Y^t,C^t)}\times\frac{1+\vec{D}^{t+1}(L^{t+1},K^{t+1},E^{t+1},Y^{t+1},C^{t+1};Y^{t+1},C^{t+1})}{1+\vec{D}^B(L^{t+1},K^{t+1},E^{t+1},Y^{t+1},C^{t+1};Y^{t+1},C^{t+1})}}_{\text{TC}}\Bigg]$$

$$\times\Bigg[\underbrace{\frac{f^B(l^{t+1},k^{t+1},1,y^{t+1},c^{t+1};Y^{t+1},C^{t+1})}{f^B(l^t,k^{t+1},1,y^{t+1},c^{t+1};Y^{t+1},C^{t+1})}}_{\text{LE}}\Bigg]\times\Bigg[\underbrace{\frac{f^B(l^t,k^{t+1},1,y^{t+1},c^{t+1};Y^{t+1},C^{t+1})}{f^B(l^t,k^t,1,y^{t+1},c^{t+1};Y^{t+1},C^{t+1})}}_{\text{KE}}\Bigg]$$

$$\times\Bigg[\underbrace{\frac{f^B(l^t,k^t,1,y^{t+1},c^{t+1};Y^{t+1},C^{t+1})}{f^B(l^t,k^t,1,y^{t+1},c^t;Y^{t+1},C^{t+1})}}_{\text{CE}}\Bigg]\times\Bigg[\underbrace{\frac{f^B(l^t,k^t,1,y^{t+1},c^t;Y^{t+1},C^{t+1})}{f^B(l^t,k^t,1,y^{t+1},c^t;Y^t,C^t)}\times\frac{f^B(l^t,k^t,1,y^{t+1},c^t;Y^t,C^t)}{f^B(l^t,k^t,1,y^t,c^t;Y^t,C^t)}}_{\text{OSE}}\Bigg], \tag{8}$$

where $l = L/E$, $k = K/E$, $y = Y/E$, $c = C/E$ and the tourism energy efficiency change can be decomposed into six factors. The first measures technical efficiency change (EC) where values greater than one signify improvements in technical efficiency; the second factor denotes technological change (TC), which is the shift of the production frontier between two periods. The first two factors in formula (8) are the total factor productivity (TP), which is widely studied in the area of productivity and efficiency analysis (Färe, Grosskopf, Norris, & Zhang, 1994). The last four factors measure effects on tourism energy efficiency change from changes in the labour-energy ratio (LE), capital-energy ratio (KE), energy supply composition (CE), and output composition (OSE).

3.4. Data sources

Based on the above method, the data of the input–output variables of the tourism industry in specific areas in China is required to calculate and decompose tourism energy efficiency. Thirty provinces are considered in this empirical study due to data availability; the observations do not contain Tibet, Hong Kong, Macao and Taiwan. First, energy consumption of the tourism industry should be calculated through formula (1) and formula (2). This involves the added value of the industry, sample surveys of tourism consumption structures and energy consumption of the industry in the 30 provinces. Table 2 provides the corresponding data sources. Second, the input-output variables are needed for the decomposition of tourism energy efficiency, including capital, labour, tourism energy consumption, revenue, and CO_2 emissions. CO_2 emissions of the tourism industry can be calculated by multiplying the tourism energy consumption by the energy-emission ratio, involving the coefficient used to convert it into the consumption of standard coal (refer to (f_j) in Table 1) and CO_2 emissions from a unit of standard coal. Descriptive statistics for the data are presented in Table 3.

4. Empirical results and discussion

This section focuses on the results of tourism energy efficiency scores and their influencing factors, which are estimated by MATLAB 2014b software based on the above formulas.

Table 2. Data sources.

Variable	Representatives of variables	Data resources
Capital (*K*)	Fixed capital of tourism industry	China Tourism Statistics Yearbook (2005–2014)
Labour (*L*)	Number of employees in tourism industry	China Tourism Statistics Yearbook (2005–2014)
Tourism energy consumption (*E*)	Added value of industry	China Statistics Yearbook (2005–2014)
	Sample surveys of tourism consumption structures	Tourism Statistic Communiques, Summaries and Fact Sheets
	Energy consumption of industry	China Energy Statistical Yearbook (2005–2014)
Revenue (*Y*)	Overall income of tourism industry	Statistical Bulletin on National Economy and Social Development
CO_2 emissions (*C*)	$\sum_{ij} E_{ij} \times f_j \times \beta$	China Energy Statistical Yearbook (2005–2014)

Note: The overall income of the tourism industry contains domestic tourism income and inbound tourism income, and combines the current exchange rate with inbound tourism income. β is CO_2 emissions from a unit of standard coal, which is set as 2.45 through previous studies (Wu et al., 2015). The monetary value of data of capital and revenue are measured in 10^8 yuan and held at 2000 constant prices to eliminate the effect of currency fluctuations.

Table 3. Descriptive statistics.

Variable	Mean	St.dev	Min	Max
Fixed capital of tourism industry (10^8 yuan)	1141.074	1038.781	9.606	1336.621
Number of employees (people)	79,843.748	63,445.005	6526.000	491,829.000
Energy consumption (GJ)	127,438.147	111,821.345	3562.981	689,946.738
Overall income of tourism industry (10^8 yuan)	1141.074	1038.781	16.591	6058.240
CO_2 emissions (10^4 tons)	974.937	871.859	28.696	6309.034

4.1. Characteristics of tourism energy efficiency

Figure 1 presents that energy efficiency of the industrial industry is much higher than the tourism industry. Therefore, compared with the industrial industry, tourism energy efficiency is at a low level and is not greener than the industry as a whole in terms of energy efficiency. Tourism energy efficiency shows a fluctuating downward trend from 0.0088 in 2005–0.0085 in 2013, with a peak in 2010. What drives this fluctuating change? As shown in Table 4, tourism energy efficiency declined in the periods of 2007~2008, 2011~2012, and 2012~2013. From 2007 to 2008, all five factors decreased and their contribution from the largest to the smallest is capital-energy ratio (33.51%), energy supply composition (28.45%), labour-energy ratio (15.27%), total factor productivity (15.23%) and output composition (10.01%). Therefore, the tourism energy efficiency decrease is mainly attributed to the capital-energy ratio, which means that the substitution effect of capital is low. In the period of 2011~2012, the decrease in tourism energy efficiency was mainly caused by the capital-energy ratio (74.85%), while output composition and total factor productivity inhibit this reduction with the contribution rates −21.35% and −17.43%. From 2012 to 2013, the capital-energy ratio (96.95%) is still the dominant reason for tourism energy efficiency reduction, while output composition (−20.57%) and total factor productivity (−5.81%) suppress this reduction.

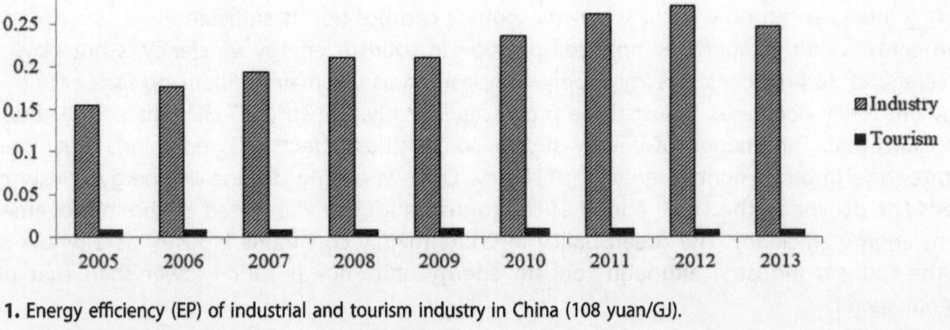

Figure 1. Energy efficiency (EP) of industrial and tourism industry in China (108 yuan/GJ).

Table 4. Average tourism energy efficiency changes and its factors of 30 provinces from 2005 to 2013.

Year	TP	KE	LE	CE	OSE	EP_t^{t+1}
2005–2006	1.010	1.007	1.000	0.999	0.996	1.012
2006–2007	1.003	1.004	0.995	1.002	1.010	1.014
2007–2008	0.990	0.979	0.990	0.982	0.994	0.938
2008–2009	1.063	1.037	1.020	1.030	1.000	1.157
2009–2010	1.094	0.954	1.003	1.005	1.002	1.054
2010–2011	1.010	1.020	1.017	1.012	0.990	1.050
2011–2012	1.011	0.953	0.978	0.981	1.013	0.937
2012–2013	1.002	0.967	0.994	0.996	1.007	0.966
Average	1.022	0.990	1.000	1.001	1.001	1.014

4.2. Decomposition factors for tourism energy efficiency changes

Table 4 reports that total factor productivity (164.79%) is the greatest contribution to tourism energy efficiency changes in the entire period, followed by capital-energy ratio (−76.61%), output composition (10.62%), energy supply composition (5.45%) and labour-energy ratio (−2.69%).

In the entire period, total factor productivity showed a decrease in the year of 2008 in most provinces, except that Tianjin, Jiangsu and Shaanxi always kept growing. The corresponding contribution rate of total factor productivity declined in 2008 at the average of 30 provinces. Total factor productivity usually plays a positive impact on the improvement of tourism energy efficiency, which is the same as the function of total factor productivity in Industry. Liaoning, Jiangsu, Zhejiang, Fujian, Guangdong, Jilin, and Sichuan have greater than a 100% contribution of total factor productivity.

A higher capital-energy ratio can improve tourism energy efficiency, while capital-energy ratio has a negative impact throughout the period due to its low value. Thus, there is a huge potential for optimization of tourism capital allocation and reduction of energy consumption and waste, meanwhile reducing capital investment per unit of output.

Output composition is the mapping result of the input structure of the elements and indirectly reflects the green level of the inputs. The ratio of tourism revenue to CO_2 emissions further measures the output structure. That is, the higher the amount of pollutants per unit of added value, the more serious the pollution of the tourism industry and the lack of environmental friendliness of the output structure, whereas pollution is lighter, and the output structure is good. As presented in Table 4, output composition influenced tourism energy efficiency in a positive way in the years 2007, 2010, 2012 and 2013; the rest of the years were negatively impacted. It is evidenced that output structure has improved since 2012. The ratio of tourism revenue to CO_2 emissions is larger due to the decrease of pollution under the green policy in China. Only 16 provinces' output composition had a positive contribution to tourism energy efficiency improvement, which means that the output composition is not good enough to boost tourism energy efficiency despite the fact that it has been improved in the last three years. In theory, energy efficiency can be improved by simply increasing tourism revenue or higher ratio value of output composition; that is, the growth rate of tourism revenue is higher than that of CO_2 emissions, or the decline in CO_2 emissions is greater than tourism revenue. While tourism revenue has been showing a growth trend, a positive contribution of output composition to tourism energy efficiency increase can only occur when the output composition is sufficient.

The contribution of energy supply composition in tourism energy efficiency is not obvious in the average of 30 provinces. Energy supply composition as the main influencing factor for tourism energy efficiency decreases in just three provinces, namely, Gansu (92.71%), Shanxi (46.46%), and Hainan (32.00%). The change of energy supply composition affects CO_2 emissions, but does not guarantee the improvement of energy efficiency. Only when the decline in energy consumption exceeds the decline in the value added of the tourism industry will it lead to the improvement of tourism energy efficiency. The green policy in China mainly constrains industry as a whole rather than the tourism industry, although tourism energy efficiency is much lower than that of the industrial sector.

Labour-energy ratio has a negative impact on tourism energy efficiency. The tourism industry is regarded as a labour-intensive industry; hence the increase in energy price cannot result in energy being replaced by labour. However, low education or a shortage of high education tourism service talents makes a great potential improvement of the labour-energy ratio's positive contribution.

Figure 2 shows the five factors contributing to tourism energy efficiency changes as well as demonstrating the main contribution factor in each period. From the dynamic perspective, capital-labour ratio overtook the total factor productivity as the tourism energy efficiency change's largest contribution since 2010. The averages of capital-energy ratio and labour-energy ratio are 0.990 and 1.000, which shows that the substitution effect of capital and labour is at a low level, and the impact of capital investment and labour on energy-saving is negative, suggesting that new capital investment should be accompanied with energy-saving technology in the tourism industry.

4.3. Analysis of tourism energy efficiency differences among provinces

To identify production technology heterogeneity among areas, 30 provinces are divided into three regions: eastern, central, and western referring to the work of Wang, Zhao, Zhou, and Zhou (2013), Zhang, Kong, and Yu (2015), and Feng, Wang, Liu, and Huang (2017). The eastern region consists of 11 provinces, i.e. Beijing, Tianjin, Hebei, Liaoning, Shanghai, Jiangsu, Zhejiang, Fujian, Shandong, Guangdong, Hainan; the central region consists of 10 provinces, i.e. Shanxi, Inner Mongolia, Jilin, Heilongjiang, Anhui, Jiangxi, Henan, Hubei, Hunan, Guangxi; and the western region consists of 9 provinces, i.e. Sichuan, Chongqing, Guizhou, Yunnan, Shaanxi, Gansu, Qinghai, Ningxia, Xinjiang. In the entire period, the average of tourism energy efficiency in the eastern region is highest, followed by the central region, and finally the western region. The factors responsible for the disparity in tourism energy efficiency are different among the three regions. Specifically, for the eastern region, total factor productivity (39.01%) is the dominant factor behind the increase in tourism energy efficiency. Meanwhile, capital-energy ratio (−4.80%) and output composition (−1.06%) have a slightly negative impact on tourism energy efficiency. Compared to the eastern region, the inhibitor of low tourism energy efficiency in the western region is the capital-energy ratio (−147.04%) and labour-energy ratio (−74.15%). The capital-energy ratio has the highest correlation coefficient (0.958) with the differences between the eastern and western regions, followed by the

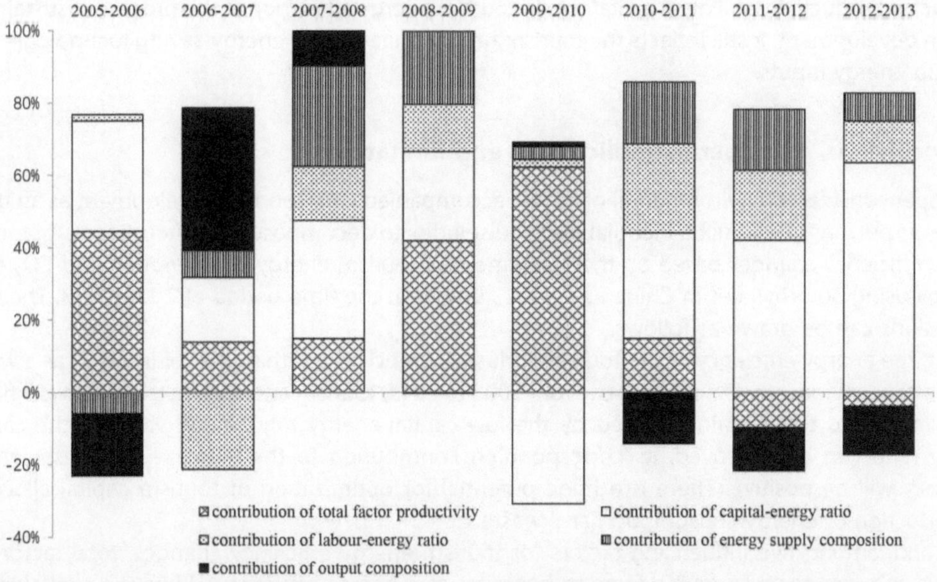

Figure 2. Contribution rate of the five factors to the changes of tourism energy efficiency.

labour-energy ratio with a correlation coefficient of 0.746. The evidence is shown that redundant inputs and misallocation of resources are serious in the western region, which might be due to the scarcity of talent.

Table 5 reports the tourism energy efficiency change and its influencing factors in 30 provinces. Tourism energy efficiency of 13 provinces, including Tianjin, Hebei, Liaoning, Jiangsu, Shandong, Guangdong, Jilin, Hunan, Guangxi, Sichuan, Shaanxi, Ningxia, and Xinjiang is on the rise. The highest average growth rate of tourism energy efficiency is in Tianjin with a rate of 12.37%, and the main driving factor is labour-energy ratio with a 78.33% contribution. Higher labour productivity makes the energy dependence of unit labour very low; that is, unit labour can use less energy to produce higher output, which will increase tourism energy efficiency. The labour-energy ratio also represents the highest contribution (28.63%) to the tourism energy efficiency increase in Hunan; the second contribution factor is total factor productivity (28.07%). Total factor productivity is an important driver for most provinces' tourism energy efficiency increase except for Tianjin, Hunan and Xinjiang. The dominant driver of Xinjiang is output composition.

The remaining 17 provinces experienced a tourism energy efficiency decline. These provinces include Beijing, Shanghai, Zhejiang, Fujian, Hainan, Shanxi, Inner Mongolia, Heilongjiang, Anhui, Jiangxi, Henan, Hubei, Chongqing, Guizhou, Yunnan, Gansu, and Qinghai. Anhui has the highest average decline of 8.00%; its key inhibitor is capital-energy ratio with a contribution rate of 56.16%. The lack of capital investment results in greater energy consumption per unit capital. The low capital-energy ratio prevents the increase of tourism energy efficiency in Inner Mongolia, Heilongjiang, Jiangxi, Henan, Chongqing, Yunnan. More energy could be saved to produce the same amount of output if capital stock is higher. Because capital accumulates and production technology tends to save labour (Wang, 2013), as shown in Table 5, the low labour-energy ratio is usually the second influencing factor after low capital-energy ratio for tourism energy efficiency decline in those provinces. The labour-energy ratio overtakes capital-energy ratio as the dominant factor for the tourism energy efficiency decrease in Beijing and Shanghai. The tourism industry does not yet have the attributes of intensive technology, and the intensive use of the labour force is not high. In addition, the soft constraints of energy and environmental protection make the labour input of the tourism industry become a resistance rather than an impetus for energy efficiency improvement.

Capital investment and labour obviously affect energy efficiency (Comin & Hobijn, 2004), while total factor productivity still plays a vital role in tourism energy efficiency. In the process of sustainable tourism development, it still reflects the tourism growth that ignores energy-saving technologies and relies on energy inputs.

5. Conclusions, managerial implications and limitations

This paper considered environmental pollution accompanied with tourism development as an undesirable output, and then built a biennial Malmquist index to decompose the determinants of tourism energy efficiency changes based on the assessment of tourism energy consumption and CO_2 emissions by using 30 provinces in China as a case study from the time period of 2005–2013. The main conclusions can be drawn as follows.

First, the energy efficiency of the tourism industry is much lower than that of industry as a whole, and the tourism industry is not greener. From 2005 to 2013, tourism energy efficiency shows a fluctuating downward trend mainly caused by the low capital-energy ratio. If the value of the capital-energy ratio can be improved, its corresponding contribution to the increase in tourism energy efficiency will be positive. There is a huge potential for optimization of tourism capital allocation and reduction of energy consumption and waste.

Second, among five influencing factors for tourism energy efficiency changes, total factor productivity is a key positive driving force to boost tourism energy efficiency. While the capital-labour ratio gradually overtook the total factor productivity as the tourism energy efficiency change's

Table 5. Priority ordering of five factors contribution to tourism energy efficiency changes.

Province	5 star	4 star	3 star	2 star	1 star	Province	5 star	4 star	3 star	2 star	1 star
Beijing↓	LE (+)	OSE (–)	CE (+)	KE (+)	TP (–)	Anhui↓	KE (+)	CE (+)	LE (+)	OSE (–)	TP (+)
Tianjin↑	LE (+)	TP (+)	KE (–)	CE (+)	OSE (–)	Jiangxi↓	KE (+)	CE (+)	TP (–)	LE (+)	OSE (–)
Hebei↑	TP (+)	KE (+)	LE (+)	CE (+)	OSE (+)	Henan↓	KE (+)	LE (+)	CE (+)	OSE (–)	TP (+)
Liaoning↑	TP (+)	OSE (–)	KE (–)	CE (+)	LE (+)	Hubei↓	TP (–)	KE (+)	LE (+)	CE (+)	OSE (+)
Shanghai↓	LE (+)	OSE (–)	TP (–)	CE (+)	KE (+)	Hunan↑	LE (+)	TP (+)	KE (+)	CE (+)	OSE (–)
Jiangsu↑	TP (+)	KE (–)	CE (+)	OSE (+)	LE (–)	Guangxi↑	TP (+)	LE (+)	CE (+)	OSE (–)	KE (–)
Zhejiang↓	TP (–)	KE (+)	OSE (+)	CE (+)	LE (+)	Chongqing↓	KE (+)	CE (+)	LE (+)	OSE (+)	TP (–)
Fujian↓	OSE (+)	KE (–)	TP (–)	CE (+)	LE (+)	Sichuan↑	TP (+)	KE (–)	OSE (–)	CE (+)	LE (+)
Shandong↑	TP (+)	CE (+)	LE (+)	OSE (–)	KE (+)	Guizhou↓	TP (–)	KE (+)	LE (+)	OSE (–)	CE (–)
Guangdong↑	TP (+)	KE (–)	CE (+)	LE (+)	OSE (v)	Yunnan↓	KE (+)	CE (+)	LE (+)	OSE (+)	TP (–)
Hainan↓	CE (+)	KE (+)	LE (+)	TP (+)	OSE (–)	Shaanxi↑	TP (+)	LE (+)	CE (+)	KE (+)	OSE (+)
Shanxi↓	CE (+)	KE (+)	LE (+)	TP (–)	OSE (+)	Gansu↓	CE (+)	KE (+)	LE (+)	OSE (–)	TP (+)
Inner Mongolia↓	KE (+)	LE (+)	OSE (+)	CE (+)	TP (+)	Qinghai↓	TP (+)	KE (+)	LE (+)	OSE (+)	CE (+)
Jilin↑	TP (+)	KE (–)	CE (+)	LE (–)	OSE (–)	Ningxia↑	TP (+)	KE (+)	LE (+)	CE (+)	OSE (+)
Heilongjiang↓	KE (+)	LE (+)	CE (+)	TP (–)	OSE (–)	Xinjiang↑	OSE (+)	CE (–)	LE (+)	KE (+)	TP (–)

Notes: ↓ means that tourism energy efficiency declines in the whole period, ↑on the contrary. (+) is the positive contribution, (–) is the negative contribution. The higher star, the greater contribution.

largest contribution since 2010, improving the capital-energy ratio and labour-energy ratio is a vital strategy to avoid the decline of tourism energy efficiency.

Third, after identifying production technology heterogeneity among provinces, the average of tourism energy efficiency in the eastern region is higher than that of the western region. The most relevant reason for this discrepancy is the low capital-energy ratio in the western region. The evidence shows that redundant inputs and misallocation of resources are serious in the western region, which might be due to the scarcity of talent.

From a theoretical perspective, the biennial decomposition method considers the undesirable output when analysing the factors for tourism energy efficiency change and fills in the shortcomings of the decomposition method used. Meanwhile, it is applicable to other cases. From a practical perspective, an empirical analysis of tourism energy efficiency changes in 30 provinces of China provides managerial implications for governments. Tourism energy efficiency is different in specific areas; therefore, energy and environmental policies, as well as different strategies on the improvement of tourism energy efficiency should be formulated and implemented. In the process of sustainable tourism development, the application of environmental regulations is ignored because the tourism industry is usually regarded as a green industry. Incentives should be made for lodging and catering businesses to offer green hotels or restaurants. To accomplish energy-saving benefits in the accommodation sector, green criteria could be added into the hotel star-rating system. Offering incentives to use solar water heaters and green furniture in the tourism service sectors is yet another strategy. New capital investment should be accompanied with energy-saving technology in the tourism industry. In addition, the resource allocation efficiency of various elements of the tourism industry is not high, and there is wasted input or output. Insufficient phenomenon, the energy effect of capital and labour did not come into play. Giving full play to high-quality capital and high-end talents can not only reduce capital, labour, and energy waste but also increase output levels and reduce pollutant emissions.

Regarding the limitation of this study, the main drawback lies in the fact that the secondary statistical data we used to calculate the tourism energy consumption might be a systematic underestimation. Fifteen energy types are accessed rather than full energy types leading to the underestimation. Thus it results in overestimating tourism energy efficiency. Future research is encouraged to examine the comparison study of the energy efficiency of tourism industry and other industries to gap above limitation.

Acknowledgements

The authors gratefully acknowledge the anonymous reviewers providing critical comments and suggestions to the original version of this paper. Meanwhile, the authors also gratefully acknowledge the financial support from National Natural Science Foundation of China (Grant number: 71704125), China's Major University Research Project Funds from Sichuan University (Grant number: SKQY201772), and the Innovation Spark Project Funds from Sichuan University (Grant number: 2018hhf-38).

Disclosure statement

No potential conflict of interest was reported by the authors.

ORCID

Lamei He http://orcid.org/0000-0002-0373-0530

References

Bagliani, M., Galli, A., Niccolucci, V., & Marchettini, N. (2008). Ecological footprint analysis applied to a sub-national area: The case of the Province of Siena (Italy). *Journal of Environmental Management*, *86*(2), 354–364.

Balachandra, P., Ravindranath, D., & Ravindranath, N. H. (2010). Energy efficiency in India: Assessing the policy regimes and their impacts. *Energy Policy*, *38*(11), 6428–6438.

Becken, S. (2002). Analysing international tourist flows to estimate energy use associated with air travel. *Journal of Sustainable Tourism, 10*(2), 114–131.

Becken, S. (2011). A critical review of tourism and oil. *Annals of Tourism Research, 38*(2), 359–379.

Becken, S., Frampton, C., & Simmons, D. (2001). Energy consumption patterns in the accommodation sector – the New Zealand case. *Ecological Economics, 39*(3), 371–386.

Becken, S., & Patterson, M. (2006). Measuring national carbon dioxide emissions from tourism as a key step towards achieving sustainable tourism. *Journal of Sustainable Tourism, 14*(4), 323–338.

Becken, S., Simmons, D. G., & Frampton, C. (2003). Energy use associated with different travel choices. *Tourism Management, 24*(3), 267–277.

Bernard, F., Khelil, T. B., Pichon, V., & Tissot, L. (2010). *The Maldives' 2009 carbon audit*. Paris: BeCitzen.

Bianco, V., Righi, D., Scarpa, F., & Tagliafico, L. A. (2017). Modeling energy consumption and efficiency measures in the Italian hotel sector. *Energy and Buildings, 149*, 329–338.

Bilgen, S., & Sarıkaya, İ. (2015). Exergy for environment, ecology and sustainable development. *Renewable and Sustainable Energy Reviews, 51*, 1115–1131.

Blomberg, J., Henriksson, E., & Lundmark, R. (2012). Energy efficiency and policy in Swedish pulp and paper mills: A data envelopment analysis approach. *Energy Policy, 42*, 569–579.

Boggia, A., & Cortina, C. (2010). Measuring sustainable development using a multi-criteria model: A case study. *Journal of Environmental Management, 91*(11), 2301–2306.

Brucher, W. (2001). Geography of energy. In N. J. Smelser & P. B. Baltes (Eds.), *International encyclopedia of the social & behavioural science* (pp. 4520–4523). Detroit, MI: Gale-Cengage.

Brundtland, G. H. (1987). Report of the World Commission on Environment and Development: "our common future." United Nations.

Buckley, R. (2011). Tourism under climate change: Will slow travel supersede short breaks? *AMBIO: A Journal of the Human Environment, 40*(3), 328–331.

Cadarso, MÁ, Gómez, N., López, L. A., & Tobarra, MÁ. (2016). Calculating tourism's carbon footprint: Measuring the impact of investments. *Journal of Cleaner Production, 111*, 529–537.

Caves, D. W., Christensen, L. R., & Diewert, W. E. (1982). The economic theory of index numbers and the measurement of input, output, and productivity. *Journal of the Econometric Society, 50*(6), 1393–1414.

Chung, Y. H., Färe, R., & Grosskopf, S. (1997). Productivity and undesirable outputs: A directional distance function approach. *Journal of Environmental Management, 51*(3), 229–240.

Comin, D., & Hobijn, B. (2004). Cross-country technology adoption: Making the theories face the facts. *Journal of Monetary Economics, 51*(1), 39–83.

Cornillie, J., & Fankhauser, S. (2004). The energy intensity of transition countries. *Energy Economics, 26*(3), 283–295.

Dubois, G., & Ceron, J. P. (2006). Tourism/leisure greenhouse gas emissions forecasts for 2050: Factors for change in France. *Journal of Sustainable Tourism, 14*(2), 172–191.

Dwyer, L., Forsyth, P., Spurr, R., & Hoque, S. (2010). Estimating the carbon footprint of Australian tourism. *Journal of Sustainable Tourism, 18*(3), 355–376.

Fang, C. Y., Hu, J. L., & Lou, T. K. (2013). Environment-adjusted total-factor energy efficiency of Taiwan's service sectors. *Energy Policy, 63*, 1160–1168.

Färe, R., Grosskopf, S., Norris, M., & Zhang, Z. (1994). Productivity growth, technical progress, and efficiency change in industrialized countries. *American Economic Review, 84*, 66–83.

Fath, B. D. (2015). Quantifying economic and ecological sustainability. *Ocean & Coastal Management, 108*, 13–19.

Feng, C., & Wang, M. (2017). Analysis of energy efficiency and energy savings potential in China's provincial industrial sectors. *Journal of Cleaner Production, 164*, 1531–1541.

Feng, C., Wang, M., Liu, G. C., & Huang, J. B. (2017). Sources of economic growth in China from 2000–2013 and its further sustainable growth path: A three-hierarchy meta-frontier data envelopment analysis. *Economic Modelling, 64*, 334–348.

Filimonau, V., Dickinson, J., Robbins, D., & Huijbregts, M. A. (2011). Reviewing the carbon footprint analysis of hotels: Life cycle energy analysis (LCEA) as a holistic method for carbon impact appraisal of tourist accommodation. *Journal of Cleaner Production, 19*(17–18), 1917–1930.

Fisher-Vanden, K., Jefferson, G. H., Liu, H., & Tao, Q. (2004). What is driving China's decline in energy intensity? *Resource and Energy Economics, 26*(1), 77–97.

Godde, P. M., Price, M. F., & Zimmermann, F. M. (1999). *Tourism and development in mountain regions*. New York, NY: CABI Publishing.

Gössling, S. (2000). Sustainable tourism development in developing countries: Some aspects of energy use. *Journal of Sustainable Tourism, 8*(5), 410–425.

Gössling, S. (2002). Global environmental consequences of tourism. *Global Environmental Change, 12*(4), 283–302.

Gössling, S. (2013). National emissions from tourism: An overlooked policy challenge? *Energy Policy, 59*, 433–442.

Gössling, S., Peeters, P., Ceron, J. P., Dubois, G., Patterson, T., & Richardson, R. B. (2005). The eco-efficiency of tourism. *Ecological Economics, 54*(4), 417–434.

Howitt, O. J., Revol, V. G., Smith, I. J., & Rodger, C. J. (2010). Carbon emissions from international cruise ship passengers' travel to and from New Zealand. *Energy Policy, 38*(5), 2552–2560.

Hu, J. L., & Wang, S. C. (2006). Total-factor energy efficiency of regions in China. *Energy Policy, 34*(17), 3206–3217.

Huang, Z. F., Cao, F. D., Jin, C., Yu, Z. Y., & Huang, R. (2016). Carbon emission flow from self-driving tours and its spatial relationship with scenic spots – a traffic-related big data method. *Journal of Cleaner Production, 142*, 946–955.

Idahosa, L. O., Marwa, N. N., & Akotey, J. O. (2017). Energy (electricity) consumption in South African hotels: A panel data analysis. *Energy and Buildings, 156*, 207–217.

Katircioglu, S. T. (2014). International tourism, energy consumption, and environmental pollution: The case of Turkey. *Renewable and Sustainable Energy Reviews, 36*, 180–187.

Katircioglu, S. T., Feridun, M., & Kilinc, C. (2014). Estimating tourism-induced energy consumption and CO_2 emissions: The case of Cyprus. *Renewable and Sustainable Energy Reviews, 29*, 634–640.

Kelly, J., & Williams, P. W. (2007). Modelling tourism destination energy consumption and greenhouse gas emissions: Whistler, British Columbia, Canada. *Journal of Sustainable Tourism, 15*(1), 67–90.

Khemiri, A., & Hassairi, M. (2005). Development of energy efficiency improvement in the Tunisian hotel sector: A case study. *Renewable Energy, 30*(6), 903–911.

Konan, D. E., & Chan, H. L. (2010). Greenhouse gas emissions in Hawaii: Household and visitor expenditure analysis. *Energy Economics, 32*(1), 210–219.

Kuo, N. W., & Chen, P. H. (2009). Quantifying energy use, carbon dioxide emission, and other environmental loads from island tourism based on a life cycle assessment approach. *Journal of Cleaner Production, 17*(15), 1324–1330.

Lai, J. H. (2015). Carbon footprints of hotels: Analysis of three archetypes in Hong Kong. *Sustainable Cities and Society, 14*, 334–341.

Lennox, J. (2012). Impacts of high oil prices on tourism in New Zealand. *Tourism Economics, 18*(4), 781–800.

Li, J. F., & Li, M. Y. (1999). On the calculation of tourism industry and tourist adding value. *Tourism Tribune, 14*(5), 16–19.

Li, K., & Lin, B. (2018). How to promote energy efficiency through technological progress in China? *Energy, 143*, 812–821.

Lin, J. Y., Cai, F., & Li, Z. (2003). *The China miracle: Development strategy and economic reform* (Revised ed.). Hongkong: Chinese University Press.

Liu, Q., Lei, Q., Xu, H., & Yuan, J. (2018). China's energy revolution strategy into 2030. *Resources, Conservation and Recycling, 128*, 78–89.

Lu, J. Y., & Nepal, S. K. (2009). Sustainable tourism research: An analysis of papers published in the journal of sustainable tourism. *Journal of Sustainable Tourism, 17*(1), 5–16.

Luo, F., Becken, S., & Zhong, Y. (2018). Changing travel patterns in China and 'carbon footprint' implications for a domestic tourist destination. *Tourism Management, 65*, 1–13.

Mao, J., Du, Y., Cai, H., & Jiang, P. (2010). Energy efficiencies of industrial sectors for China's major cities. *Procedia Environmental Sciences, 2*, 781–791.

Meng, W., Xu, L., Hu, B., Zhou, J., & Wang, Z. (2016). Quantifying direct and indirect carbon dioxide emissions of the Chinese tourism industry. *Journal of Cleaner Production, 126*, 586–594.

Moutinho, V., Costa, C., & Bento, J. P. C. (2015). The impact of energy efficiency and economic productivity on CO2 emission intensity in Portuguese tourism industries. *Tourism Management Perspectives, 16*, 217–227.

Munday, M., Turner, K., & Jones, C. (2013). Accounting for the carbon associated with regional tourism consumption. *Tourism Management, 36*, 35–44.

Olabi, A. G. (2012). Developments in sustainable energy and environmental protection. *Energy, 39*(1), 2–5.

Önüt, S., & Soner, S. (2006). Energy efficiency assessment for the Antalya region hotels in Turkey. *Energy and Buildings, 38* (8), 964–971.

Pace, L. A. (2016). How do tourism firms innovate for sustainable energy consumption? A capabilities perspective on the adoption of energy efficiency in tourism accommodation establishments. *Journal of Cleaner Production, 111*, 409–420.

Paramati, S. R., Alam, M. S., & Chen, C. F. (2017). The effects of tourism on economic growth and CO_2 emissions: A comparison between developed and developing economies. *Journal of Travel Research, 56*(6), 712–724.

Pastor, J. T., Asmild, M., & Lovell, C. K. (2011). The biennial Malmquist productivity change index. *Socio-Economic Planning Sciences, 45*(1), 10–15.

Patterson, M. G. (1996). What is energy efficiency? Concepts, indicators and methodological issues. *Energy Policy, 24*(5), 377–390.

Peeters, P., & Schouten, F. (2006). Reducing the ecological footprint of inbound tourism and transport to Amsterdam. *Journal of Sustainable Tourism, 14*(2), 157–171.

Pentelow, L., & Scott, D. (2010). The implications of climate change mitigation policy and oil price volatility for tourism arrivals to the Caribbean. *Tourism and Hospitality Planning & Development, 7*(3), 301–315.

Perch-Nielsen, S., Sesartic, A., & Stucki, M. (2010). The greenhouse gas intensity of the tourism sector: The case of Switzerland. *Environmental Science & Policy, 13*(2), 131–140.

Pereira, R. P. T., Ribeiro, G. M., & Filimonau, V. (2017). The carbon footprint appraisal of local visitor travel in Brazil: A case of the Rio de Janeiro-São Paulo itinerary. *Journal of Cleaner Production, 141*, 256–266.

Phylipsen, G. J. M., Blok, K., & Worrell, E. (1997). International comparisons of energy efficiency-methodologies for the manufacturing industry. *Energy Policy*, *25*(7–9), 715–725.

Robaina-Alves, M., Moutinho, V., & Costa, R. (2016). Change in energy-related CO_2 (carbon dioxide) emissions in Portuguese tourism: A decomposition analysis from 2000 to 2008. *Journal of Cleaner Production*, *111*, 520–528.

Rosselló-Batle, B., Moià, A., Cladera, A., & Martínez, V. (2010). Energy use, CO_2 emissions and waste throughout the life cycle of a sample of hotels in the Balearic Islands. *Energy and Buildings*, *42*(4), 547–558.

Scott, D., Hall, C. M., & Gössling, S. (2016). A report on the Paris climate change Agreement and its implications for tourism: Why we will always have Paris. *Journal of Sustainable Tourism*, *24*(7), 933–948.

Sharp, H., Grundius, J., & Heinonen, J. (2016). Carbon footprint of inbound tourism to Iceland: A consumption-based life-cycle assessment including direct and indirect emissions. *Sustainability*, *8*(11), 1147.

Spoladore, A., Borelli, D., Devia, F., Mora, F., & Schenone, C. (2016). Model for forecasting residential heat demand based on natural gas consumption and energy performance indicators. *Applied Energy*, *182*, 488–499.

Streimikiene, D., Ciegis, R., & Grundey, D. (2008). Promotion of energy efficiency in Lithuania. *Renewable and Sustainable Energy Reviews*, *12*(3), 772–789.

Subrahmanya, M. B. (2006). Labour productivity, energy intensity and economic performance in small enterprises: A study of brick enterprises cluster in India. *Energy Conversion and Management*, *47*(6), 763–777.

Sun, Y. Y. (2014). A framework to account for the tourism carbon footprint at island destinations. *Tourism Management*, *45*, 16–27.

Sun, Q., Xu, L., & Yin, H. (2016). Energy pricing reform and energy efficiency in China: Evidence from the automobile market. *Resource and Energy Economics*, *44*, 39–51.

Surugiu, C., Surugiu, M. R., Breda, Z., & Dinca, A. I. (2012). An input-output approach of CO_2 emissions in tourism sector in post-communist Romania. *Procedia Economics and Finance*, *3*, 987–992.

Tang, M., Fu, X., Cao, H., Shen, Y., Deng, H., & Wu, G. (2016). Energy performance of hotel buildings in Lijiang. *China. Sustainability*, *8*(8), 780.

Tang, Z., Shang, J., Shi, C., Liu, Z., & Bi, K. (2014). Decoupling indicators of CO_2 emissions from the tourism industry in China: 1990–2012. *Ecological Indicators*, *46*, 390–397.

Tang, C., Zhong, L., & Ng, P. (2017). Factors that influence the tourism industry's carbon emissions: A tourism area life cycle model perspective. *Energy Policy*, *109*, 704–718.

Teng, C. C., Horng, J. S., Hu, M. L. M., Chien, L. H., & Shen, Y. C. (2012). Developing energy conservation and carbon reduction indicators for the hotel industry in Taiwan. *International Journal of Hospitality Management*, *31*(1), 199–208.

Tsai, K. T., Lin, T. P., Hwang, R. L., & Huang, Y. J. (2014). Carbon dioxide emissions generated by energy consumption of hotels and homestay facilities in Taiwan. *Tourism Management*, *42*, 13–21.

Wang, C. (2011). Sources of energy productivity growth and its distribution dynamics in China. *Resource and Energy Economics*, *33*(1), 279–292.

Wang, C. (2013). Changing energy intensity of economies in the world and its decomposition. *Energy Economics*, *40*, 637–644.

Wang, Q., Zhao, Z., Zhou, P., & Zhou, D. (2013). Energy efficiency and production technology heterogeneity in China: A meta-frontier DEA approach. *Economic Modelling*, *35*, 283–289.

Wu, P., Han, Y., & Tian, M. (2015). The measurement and comparative study of carbon dioxide emissions from tourism in typical provinces in China. *Acta Ecologica Sinica*, *35*(6), 184–190.

Zaman, K., Shahbaz, M., Loganathan, N., & Raza, S. A. (2016). Tourism development, energy consumption and environmental Kuznets Curve: Trivariate analysis in the panel of developed and developing countries. *Tourism Management*, *54*, 275–283.

Zhang, N., Kong, F., & Yu, Y. (2015). Measuring ecological total-factor energy efficiency incorporating regional heterogeneities in China. *Ecological Indicators*, *51*, 165–172.

The interaction effect of tourism and foreign direct investment on urban–rural income disparity in China: a comparison between autonomous regions and other provinces

Joon Ho Kim and Kyung Ho Kang

ABSTRACT

Considering the prevalence of income distribution between urban and rural areas especially in autonomous regions with specific ethnic features and relatively inferior economic conditions, and the probable role of tourism and FDI in reducing urban–rural income disparity, it is academically and practically critical to examine a combined role of tourism and FDI on urban–rural income disparity in autonomous regions in comparison with those effects in other provinces. Thus, this study aims to compare the interaction effect of tourism and FDI on alleviating urban–rural income disparity in autonomous regions with other provinces in China. This study used fixed effects regression models, employing 423 province-year observations composed of 30 provinces including five autonomous regions from 1993 to 2014. To accomplish research purposes, this study employs tourism income, FDI, and autonomous region dummy as main independent variables. And, the ratio of urban to rural consumption expenditures is included as a dependent variable. The result of this study shows that the interaction effect between tourism and FDI on reducing urban–rural income disparity in autonomous regions is significantly greater than other provinces. Based on this result, autonomous regions can optimize the effect of tourism on mitigating urban–rural income disparity through the adequate utilization of FDI.

1. Introduction

Supporting Kuznets' (1955) suggestion, Williamson (1965) proposed that regional income disparity decreases in the mature growth stage, while it increases during the early stage of economic development. Particularly, urban–rural income disparity, the gap between urban and rural standards of living in terms of monetary condition, has been increasingly important especially in developing countries as a critical dimension of income disparities.

Meanwhile, it is widely acknowledged that tourism can mitigate income disparity between urban and rural areas due to increased tourism income in rural areas by virtue of abundant natural amenities and other exotic tourism attributes that attract foreign and local tourists (Hwang & Lee, 2015). In addition, tourism in urban areas generates spillover effects such as tourism flows to adjacent rural areas, which may serve as a critical driver for less-developed areas' economic growth (Li, Chen, Li, & Goh, 2016). In this context, rural areas' precious tourism resources may accelerate tourism spillover effects by providing chances to promote tourism attractions in rural areas for subsequent tourists.

Especially, with unique culture and authenticity in rural areas, tourism in China may contribute to mitigating urban–rural income disparity within a province by benefitting rural residents with additional profits.

Another factor that may influence urban–rural income disparity is foreign direct investment (FDI). Especially, in developing countries, inward FDI generates several positive effects on economic growth by enhancing productivity efficiency, promoting technology transfers, and introducing new management processes (Alfaro, Chanda, Kalemli-Ozcan, & Sayek, 2004). However, FDI may aggravate income disparity through the composition effect which implies that direct investments are concentrated on skill-intensive sectors (Feenstra & Hanson, 1997). In this regard, FDI may have a significant effect on urban–rural income disparity in developing countries, such as China, given that unskilled labour intensive industries are mainly located in rural areas.

Whereas tourism is one possible factor that adjusts the effect of FDI on income disparity, reversely, it is also probable that a mitigating effect of tourism for income disparity is dependent on the level of FDI. That is, tourism development may require corresponding investments on infrastructure, such as transportation, hotel, recreational facilities, etc. Thus, an interaction with FDI should be considered when investigating the impact of tourism on income distribution, especially in rural areas with relatively little accumulation of capital for implementing tourism infrastructure.

China has 31 provinces, including four centrally administered cities (i.e. Beijing, Tianjin, Shanghai, Chongqing) and five autonomous regions(i.e. Inner Mongolia, Ningxia, Xinjiang, Tibet, Guangxi) (Ma, 2005). Particularly, regarding autonomous regions, in accordance with growing demands from minorities' representatives, Chinese government established relevant legislations to provide economic and cultural self-determinations of autonomous regions (Heberer, 2017). Each province is divided into urban and rural areas and especially in autonomous regions most ethnic minorities live in rural areas where inhabitants face higher poverty risks, accounting for a large proportion of rural population (Gustafsson & Sai, 2009). In general, urban–rural disparity has been especially severe in autonomous regions due to deteriorated economic capabilities, natural resource, and market conditions, compared to other provinces, which is reinforced by the central government's policy emphasizing faster development of the coastal regions in which most other provinces located (Zhang & Zou, 2012).

Meanwhile, unique features of autonomous regions, including exotic appearances, affluent natural beauties, traditional fashion with a distinctive style, and indigenous cuisine make those regions attractive ethnic tourism destinations (Gustafsson & Shi, 2003). Thus, the mitigating effect of tourism on urban–rural income disparity may be greater in autonomous regions than other provinces, given that many ethnic minorities are domiciled in rural areas. On the other hand, autonomous regions are especially lacking in relevant infrastructures and facilities required to maintain tourists induced by those attractive features for the sustainable development of tourism. Moreover, since the main industries in autonomous regions are farming, animal husbandry, mining, etc., rural labour force's competency is restricted to those traditional sectors. Based on the statistics of national bureau of China, the contribution of primary industries to regional gross domestic product in autonomous regions is over 10%, which is a lot higher than other provinces. In this context, FDI may play a more critical role for an alleviating effect of tourism on urban–rural income disparity in autonomous regions than other provinces, by providing infrastructure to facilitate tourism activities and creating employment opportunity in second and tertiary industries, including the tourism industry.

Considering the prevalence and increasing importance of income distribution between urban and rural areas especially in developing countries, it is academically and practically critical to empirically examine the impacts of tourism and FDI on urban–rural income disparity in autonomous regions containing specific ethnic features and insufficient capital resources in comparison with those effects in other provinces. However, most studies on the aforementioned subject in the current literature are case studies focusing on individual economic effect of tourism or FDI in a particular autonomous region (Zhao, 2009). Consequently, to fill a void in the literature, this study aims to examine the difference in the effects of tourism and FDI, respectively, on urban–rural income disparity between ethnic minorities' autonomous regions and other provinces in China. Moreover, considering a combined role of tourism and FDI, this study attempts to compare the interaction effect of tourism and FDI on alleviating urban–rural income disparity in autonomous regions with other provinces. To accomplish research purposes, this study employs the ratio of urban to rural consumption expenditures per capita as a proxy of urban–rural income disparity, and utilizes tourism income per capita and actually utilized FDI per capita as main independent variables to operationalize tourism and FDI. These two main independent variables along with autonomous region dummy constitute interaction terms used in this study. This study expects to provide unique insights for comprehensive understanding of the role of tourism and FDI on urban–rural income disparity and practical guidelines for tourism policy makers involving especially in ethnic minorities' autonomous regions to effectively utilize FDI for enhancing economic effect of tourism. Following theoretical backgrounds and hypotheses development, the study explains research methods in the next section. Subsequently, results and discussion are provided, and limitations and suggestions for future research finalize the study.

2. Theoretical backgrounds and hypotheses

2.1. The effects of tourism and FDI on urban–rural income distribution

Although, to some extent, variation exists in the definition of tourism due to different structures of related industries in diverse countries, previous studies, in common, appear to include accommodation, travel agents and tour operators, transport, shopping, and production of handicrafts associated with tourism activities (Lickorish & Jenkins, 2007). Reflecting multi-faceted features of tourism to facilitate economic, social, and cultural resources, tourism is considered a significant contributor to the nation's economy. Specifically, the tourism industry may narrow regional income gaps by distributing developmental achievements from economic centres to less-developed areas. More specifically, tourism spreads economic achievements from urban areas to rural areas, thereby enhancing job opportunities for unemployed and unskilled labours in rural regions. Besides, rural tourism plays a role of non-farming sources of income for rural households, based on affluent natural amenities and idiosyncratic culture to attract tourists.

Due to diversified landscapes and cultures across the country, China has plentiful tourism resources, such as over 50 UNESCO natural and cultural heritage sites, agricultural villages offering various vegetation and activities, and large-scaled resorts. Especially, China has a great potential for developing specialized tourism opportunities, focusing on a variety of ethnic groups and colourful cultures (Lew, Chon, Guangrui, Yu, & Ap, 2003). In China, tourism has emerged as an important strategy avenue since open-door policy in the late 1970s. In accordance with fast economic growth, the tourism industry in China has unprecedentedly developed in recent years, accounting for 6.6% of total GDP as of 2017 according to national statistics of China. Besides, tourism is regarded as an efficient solution to mitigate economic polarization in communities, such as inland provinces, rural areas, and ethnic minorities' regions (Zeng & Ryan, 2012). Especially, rural tourism promotion in China has contributed consistently to uplifting rural areas' standard of living by upgrading the industrial structure of rural areas, improving the quality of agricultural products, and retaining viable distribution channels for the sales of products, all of which boost income in rural areas (Park, 2014).

Population in China officially comprises 56 ethnic groups. Among them, Han nationality is the largest ethnicity covering over 90% of the whole population, and other 55 ethnic minorities account for the rest, varying greatly in terms of population size from three thousand Lhoba nationality to more than 15 million Zhuang nationality (Guo, 2017). China is subdivided into 31 provinces, including four municipalities and five autonomous regions as the first-level administration divisions. Especially, each autonomous region is dominated by a major minority group, that is, Mongolian in Inner Mongolia, Hui in Ningxia, Uyghur in Xinjiang, Tibetans in Tibet, and Zhuang in Guangxi (Gustafsson & Shi, 2003). Ethnic minority group makes up a considerably higher proportion of the population in each autonomous region than other provinces as a whole in China (Sun, 2019). Moreover, ethnic minority group in each autonomous region has specific attributes of its own. For example,

Hui ethnicity is the Muslim who rejects the consumption of pork that is one of the most popular food ingredients in China, also following Islamic tradition to wear headscarves and white caps (Gustafsson & Sai, 2014). And Uyghur is conservative to socialize with Han nationality, and hesitates to intermarry with Han Chinese because of different cultural practices between two groups (Smith, 2002). In order to efficiently unify various ethnic groups, Chinese central government granted autonomous regions' governments the authority to implement regulations corresponding to local minority culture as long as they keep the core state policy (Yang, Wall, & Smith, 2008). In autonomous regions, ethnic minorities can preserve their own traditions and cultures of, for example, officially writing public signs in both Chinese language (i.e. Mandarin Chinese) and the minority's language and appointing minority government officials for carrying out administrative procedures for the sake of minorities. Five autonomous regions located in the western regions are isolated by Chinese government's regional development policies that especially favour the eastern and coastal regions. Moreover, about 80% of ethnic minorities in China reside in rural areas, almost distributed in the border areas between provinces. Therefore, autonomous regions generally have lower urbanization rate with larger rural population than other provinces, which leads to greater income disparity between urban and rural areas (Sicular, Ximing, Gustafsson, & Shi, 2007). Given that the countryside of autonomous regions has been attractive tourism destinations for urban tourists and urban residents spend tourism expenditure about 5 times more than rural residents, tourism can be a critical source of economic dynamics in autonomous regions (Li, Zhang, Zhang, & Abrahams, 2019). However, autonomous regions have represented the lowest level of infrastructure, which restricts sustainable development of tourism (Lee, 2001) (Figure 1).

Nevertheless, compared to other provinces, rural areas in autonomous regions have greater amounts of idiosyncratic heritages and ethnic tourism resources, such as ethnic villages, indigenous festivals, traditional performances, ethnic souvenirs, from which tourists can experience exotic culture. Moreover, in autonomous regions, the participation in tourism by minorities who mostly reside in rural areas has been growing (Toops, 1993). Thus, local governments in autonomous regions make active use of tourism as a solution for developing rural areas. Consequently, the effects of tourism on developing rural areas and tourism spillover effects are expected to be greater in autonomous regions by intensifying the linkage of rural and ethnic tourism than other provinces (Chio, 2014).

Thus, this study hypothesizes that:

> H1a: The effect of tourism on mitigating urban–rural income disparity is greater in autonomous regions than other provinces.

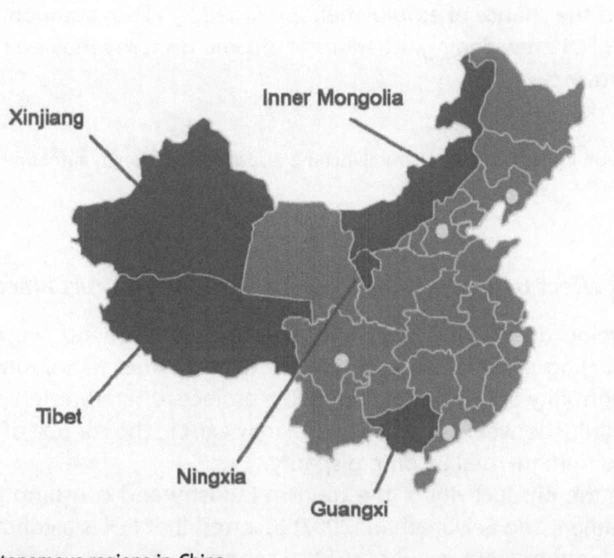

Figure 1. The location of autonomous regions in China.

Based on the definition of national bureau of statistics of China, foreign direct investment refers to foreign investment into China by establishing enterprises, developing economic resources with domestic partners, and opening a branch of foreign enterprises, in forms of cash, physical investments, technical knowledge, etc. Neoclassical models of growth and endogenous growth theory regard FDI as a reliable source of capital, which improves the efficiency of physical investments, introduces new technologies, enhances managerial skills, and facilitates innovation (Adhikary, 2011). Furthermore, FDI plays a significant role of spreading economic growth by encouraging firms to employ low-income labour according to fast industrial development, thereby mitigating income disparity (Choi, 2006).

Contrarily, a line of scholars argue that FDI aggravates income disparity by causing biased industrial structure and labour market. That is, FDI may increase skill-biased technological change ensuring improved productivity in both foreign and domestic firms, which causes only skill-intensive sectors to expand, thereby making positions and wages of skilled workers concentrating in urban areas higher than those of unskilled workers remaining in rural areas. Moreover, enhanced economic openness through FDI mainly benefits urban labourers by primarily promoting the development of manufacturing, finance, and service industries, which are accumulated in urban areas (Lu & Chen, 2006).

In China, FDI has been considered an important catalyst of economic growth since the late 1970s, serving as a critical source of constructing infrastructure and improving industrial technologies. As of 2017, the value of FDI in China reaches to over 130 billion USD, of which over 25% is invested into manufacturing, 12% is real estate projects including tourism facilities, and 5% in social infrastructure. However, FDI in agricultural industry accounts for about 1% of the total amount in China (Lv, Wen, & Xiong, 2010). Although FDI pushed China as a whole toward fast economic growth, the productivity enhanced by FDI varies from urban areas to rural ones. More specifically, FDI generally accelerates urban growth by allocating more resources into cities, thereby causing structural change of industries, which benefits urban habitants to a greater extent (Zhang, 2002).

In general, foreign firms provide much better compensations to skilled labours with a high educational level. However, there is a huge education gap between upper class and lower one in China (Rong, 2009). Given that upper class people mainly reside in urban areas and FDI usually provide employment opportunity for more skilled, intelligent urban residents, FDI may expedite urban–rural disparity. And, such effect may be magnified especially in autonomous regions where affluent ethnic minorities reside in rural areas. Consequently, considering severer regional disparities of economic advantages from FDI between urban and rural areas and a lower educational level of minorities, which limits the chance of employment provided by FDI in autonomous regions, the magnitude of the effect of FDI on widening urban–rural income disparity may be higher in autonomous regions than other provinces.

Thus, this study hypothesizes that:

H1b: The effect of FDI on intensifying urban–rural income disparity is greater in autonomous regions than other provinces.

2.2. The interaction effect between tourism and FDI on urban–rural income disparity

To appropriately develop the tourism industry, large-scale investments are required to construct infrastructure for attracting and maintaining tourists. On the other hand, tourism provides foreign investors with an opportunity of investing in profitable projects. Therefore, it is plausible that an interaction or complementarity between tourism and FDI may exist in the context of examining the effects of those two factors on urban–rural income disparity.

FDI can invigorate the productivity of the tourism industry and consumption in tourism-related sectors. Tang, Selvanathan, and Selvanathan (2007) asserted that FDI is a critical factor in expanding the tourism industry by improving venues and transportation for new tourists, and entertainment facilities to attract additional tourists. On the other hand, inward FDI into the host countries may increase, attracted by tourism demands. For example, international hotel chains expand their brands to various regions of the world, seeking viable investment opportunities (Craigwell & Moore, 2008). Chen (2017) suggested that tourism and FDI generate a beneficial interaction in combination, and thus the policies of tourism and FDI are complementary.

A rural community with substantial natural or cultural tourism resources would not be successful as a tourism destination without relevant infrastructure and high-quality tourist services (Wilson, Fesenmaier, Fesenmaier, & Van Es, 2001). In this regard, FDI not only provides underdeveloped rural areas with required resources for establishing the foundations of tourism, but also offers employment opportunities for rural residents, magnifying the tourism's effect of mitigating urban–rural income disparity (Tosun, Timothy, & Öztürk, 2003).

China is not an exception in that demand for FDI increases for the sustainable tourism development. FDI may play a role of cultivating the tourism industry especially in rural areas of autonomous regions with lower accumulation of capital but more abundant natural and cultural tourism resources than rural areas of other provinces. For example, in Tibet, foreign investments' target areas are increasingly expanding to tourism service, hotel, and F&B (Ruixue, Yang, & Ningyu, 2015). Generally, FDI in the more developed eastern regions has been utilized mainly for enhancing urban growth. Contrarily, FDI in the western regions, in which autonomous regions are located, is concentrated on social infrastructure and local specialty products (Blachford & Qi, 2009). Therefore, FDI improves tourists' accessibility to autonomous regions and facilitates travel to rural areas within autonomous regions. Consequently, FDI may augment the tourism's beneficial effects on economic improvements of less-developed rural areas in autonomous regions, thereby significantly influencing the reduction of urban–rural income disparity in those regions. And, compared to other provinces, such an interaction effect between tourism and FDI is greater in autonomous regions, in which FDI investments are more essential for utilizing affluent tourism resource in rural areas.

Thus, this study hypothesizes that:

H2: The interaction effect of tourism and FDI on mitigating urban–rural income disparity is greater in autonomous regions than other provinces.

3. Methodology

3.1. Data

Based on the classification of administrative areas in China, this study samples 30 provinces including five autonomous regions and three municipalities. Tianjin is excluded due to the inaccessibility of domestic tourism income data. Data for tourism, FDI, and urban–rural income disparity are collected from annual China Statistical Yearbook and each province's Statistical Yearbook.

The sample period spans from 1993 to 2014 to comprehensively include relevant data from data sources. After collecting all existing tourism and FDI data of each province, along with other necessary data, the final sample consists of 30 provinces with 423 province-year observations over the sample period.

3.2. Models

To examine research questions and accomplish study's purposes, this study presents two multiple regression models following the methodology of previous studies (Jaccard, Jaccard, & Turrisi, 2003; Park, Song, & Lee, 2017). The first regression model examines whether the magnitudes of the effects of tourism and FDI on urban–rural income disparity are different between autonomous regions and other provinces. The second regression model investigates whether the magnitudes of the interaction effect between tourism and FDI on urban–rural income disparity are different between autonomous regions and other provinces. The second model includes all interaction terms among all main independent variables because excluding any constitutive term (i.e. element that constitutes the interaction term) leads to inferential errors (Brambor, Clark, & Golder, 2006).

The represented model for examining the comparison of the effects of tourism and FDI on urban–rural income disparity between autonomous regions and other provinces (H1a and H1b) is:

$$\mathrm{ID} = \alpha_0 + \alpha_1\mathrm{TI} + \alpha_2\mathrm{FDI} + \alpha_3\mathrm{AR} + \alpha_4\mathrm{TI}{*}\mathrm{AR} + \alpha_5\mathrm{FDI}{*}\mathrm{AR} + \alpha_6\mathrm{FI} + \alpha_7\mathrm{URB} + \alpha_8\mathrm{GDP} + \alpha_9\mathrm{CPI} + \varepsilon,$$

where, ID represents income disparity between urban and rural areas, measured by the ratio of urban to rural consumption expenditures per capita; TI represents the value of tourism income, measured by the log of domestic and international tourism income; FDI represents the value of foreign direct investment, measured by the log of actually utilized FDI; AR represents a dummy variable for autonomous regions (autonomous regions = 1, other provinces = 0); TI*AR represents the interaction term between TI and AR; FDI*AR represents the interaction term between FDI and AR; FI represents fixed assets investment, measured by the ratio of fixed assets investment to gross domestic product; URB represents urbanization, measured by the ratio of urban to total populations; GDP represents gross domestic product, measured by the log of GDP per capita; CPI represents consumer price index, measured by the ratio of rural CPI to urban one.

The represented model to examine the comparison of the interaction effect between tourism and FDI on urban–rural income disparity between autonomous regions and other provinces (H2) is:

$$\begin{aligned}\mathrm{ID} = {} & \alpha_0 + \alpha_1\mathrm{TI} + \alpha_2\mathrm{FDI} + \alpha_3\mathrm{AR} + \alpha_4\mathrm{TI}{*}\mathrm{AR} + \alpha_5\mathrm{FDI}{*}\mathrm{AR} + \alpha_6\mathrm{TI}{*}\mathrm{FDI} \\ & + \alpha_7\mathrm{TI}{*}\mathrm{FDI}{*}\mathrm{AR} + \alpha_8\mathrm{FI} + \alpha_9\mathrm{URB} + \alpha_{10}\mathrm{GDP} + \alpha_{11}\mathrm{CPI} + \varepsilon,\end{aligned}$$

where, TI*FDI represents the interaction term between TI and FDI; TI*FDI*AR represents the interaction term among TI, FDI, and AR.

For continuous values constituting interaction terms (i.e. TI and FDI), mean-centring is conducted to mitigate a collinearity problem when they compose interaction terms (Tabachnick & Fidell, 1996). A natural logarithmic transformation is conducted for continuous values (i.e. TI, FDI, and GDP) to enhance normality by correcting positive skewness. Regarding the coefficient estimation of all models, this study employs the fixed effects methods to mitigate omitted variable bias by taking into account unobservable province- and time-specific heterogeneities across different provinces and years in a panel data set (Wooldridge, 2005). Prior to deciding to use the fixed effects model, this study conducted the Hausman test to determine whether to use fixed or random effects model for all models this study employs. Because the test results indicated the significant difference between coefficient estimations from the fixed effects model and ones from the random effects model (p-values across all models are less than 0.05), this study decided to use the fixed effects method for all models (Torres-Reyna, 2007). Additionally, for estimations of standard errors, this study adopts clustered standard errors to mitigate the bias caused by unspecified correlations between observations of the same province in different years and of different provinces in the same year, while analyzing a panel data set (Petersen, 2009).

3.3. Dependent variable

As a dependent variable, this study employs urban–rural income disparity. This study measures urban–rural income disparity by the ratio of urban to rural consumption expenditures per capita in each province. Consumption expenditure is an appropriate proxy for urban and rural incomes in the context of this study, because it reflects the level of the real standard of living in urban and rural areas (Xue, 1997), and when using the actual rural income that includes the agricultural capital expense, the underestimation of urban–rural disparity may occur (Lu, 2002). Besides, expenditure is considered to be the preferred indicator of economic welfare in the literature when conducting research on poverty and inequality (Ravallion, 2001).

3.4. Interest variables

As constituents of the main independent variables, this study employs tourism income per capita (TI), actually utilized FDI per capita (FDI), and an autonomous region dummy (AR). The first model includes two interaction terms between TI and AR and between FDI and AR as interest variables. This study uses actually utilized FDI volume reflecting the amount of FDI implemented each year, instead of contracted FDI value. Contracted FDI represents total amounts signed by foreign investors for investment projects including future years' projects, approved by Chinese government, while utilized FDI stands for actually paid-up foreign capital in a particular year (Hornstein, 2011). While contracted FDI reflects the demand of host economy, utilized FDI is a more suitable indicator to determine its actual impacts on host economy (Wang, 2016). In the data source used in this study, FDI statistics only focused on tourism do not exist and industry-specific data for FDI are not available in every year. However, the literature suggests that FDI can contribute not only to improving tourism-related infrastructure and facilities, but also to increasing business tourists in the host country, thereby positively affecting tourism development (Sokhanvar, 2019). And, many previous studies used aggregate FDI data to investigate the link between tourism and FDI (Sanford & Dong, 2000; Tang et al., 2007). Thus, this study employs aggregate-level FDI data to operationalize FDI variable in the models.

By examining the sign and the value of the each interaction term's coefficient, this study investigates whether and how much the effects of TI and FDI on urban–rural income disparity are greater in autonomous regions than other provinces. For example, the coefficient of the interaction term composed of TI (FDI) and AR represents the magnitude of the effect of TI (FDI) on urban–rural income disparity, conditional upon the level of AR.

For the second model, this study includes an interaction term among TI, FDI, and AR as a main interest variable to investigate whether the interaction effect of tourism and FDI on urban–rural income disparity is greater in autonomous regions than other provinces. By examining the coefficient of an interaction term composed of TI, FDI, and AR, this study compares the magnitudes of the interaction effects of tourism and FDI on urban–rural income disparity between autonomous regions and other provinces.

3.5. Control variables

This study adopts four control variables in each of two models: fixed assets investment (FI), urbanization (URB), gross domestic product (GDP), and consumer price index consisting of urban and rural areas' consumer price indexes (CPI).

FI, measured by the ratio of fixed assets investment to gross domestic product, controls for the effect of the discrepancy of developmental stage between urban and rural areas. As investing in fixed assets is often considered to promote the development of infrastructure, such as urban transport facilities, favouring urban economic growth (Chen & Wu, 2017), FI may positively influence urban–rural income disparity.

URB, measured by the ratio of urban to total populations, controls for the effect of the composition of urban–rural populations on urban–rural income disparity. Urbanization may mitigate urban–rural income disparity due to reduced wages of urban workers caused by labour surplus.

GDP, measured by the log of gross domestic product per capita, controls for economic developmental level of each province. Regional heterogeneity may affect urban–rural income disparity. For example, emphasizing biased development policies towards central cities and urban areas as growth-poles in regions with low GDP may aggravate the income gap between urban and rural areas (He, Bayrak, & Lin, 2017). Finally, CPI, measured by the ratio of rural consumer price index to urban one, controls for the deflating (inflating) effect of consumption when an increase (decrease) of CPI which represents annual change from 100 as a reference point in 1988 occurs (Khan & Riskin, 1998). As urban and rural CPIs affect the real values of consumption expenditures in both two areas, respectively, higher urban CPI and lower rural CPI lead to lower urban–rural income disparity.

4. Results

4.1. Descriptive statistics

Table 1 summarizes the descriptive statistics of the sampled data. ID shows the mean value of 3.093 with the standard deviation of 0.719. The results indicate that urban income is higher than rural one in all provinces and years. TI shows the mean value of 7.124 and the standard deviation of 1.570. FDI has the mean value of 5.445 and the standard deviation of 1.881. FI shows the mean value of 0.561 and the standard deviation of 0.213. URB has the mean value of 0.470 with the standard deviation of 0.153. GDP has the mean value of 9.769 with the standard deviation of 0.884. CPI has the mean value of 1.003 with the standard deviation of 0.009. All variables appear to have sufficient levels of variations for valid regression analysis.

Table 2 presents results of Pearson's correlation analysis for bivariate correlations among all continuous variables. ID negatively and significantly correlates with TI and FDI at the 1% significance level: with $r = -0.473$ and $r = -0.575$, respectively. On the other hand, ID positively and significantly correlates with interaction term between TI and FDI at the 5% significance level ($r = 0.108$).

4.2. Main analysis

Table 3 summarizes the results of main analyses. As presented in Table 3, all models present appropriate levels of goodness-of-fit; for model (1) and (2), *F*-values are 62.79 and 78.77, respectively, which are significant at the 1% level of significance.

The result of model (1) shows a positive and significant effect of TI*AR on urban–rural income disparity. The result implies that an increase in tourism leads to an increase in urban–rural income disparity to a greater extent in autonomous regions than other provinces. On the other hand, there is no significant effect of FDI*AR on urban–rural income disparity. Therefore, the results fail to support hypotheses 1a and 1b. All control variables show no significant effect on urban–rural income disparity.

The result of model (2) shows a negative and significant effect of TI*FDI*AR on urban–rural income disparity. That is, the magnitude of the interaction effect between tourism and FDI on urban–rural

Table 1. Summary of descriptive statistics[a].

Variable	N	Mean	S.D.	Minimum	Maximum
ID	423	3.09363	0.71910	1.6604	5.2095
TI	423	7.12415	1.57093	1.4501	9.8981
FDI	423	5.44503	1.88133	−2.4104	8.4338
FI	423	0.56157	0.21379	0.2329	1.2422
URB	423	0.47081	0.15357	0.1823	0.8961
GDP	423	9.76938	0.88450	7.118	11.5129
CPI	423	1.00371	0.00939	0.9706	1.0341

[a]ID: urban–rural income disparity; TI: tourism income; FDI: foreign direct investment; FI: fixed assets investment; URB: urbanization rate; GDP: gross domestic product; CPI: consumer price index.

Table 2. Summary of Pearson's correlations[a].

Variable	ID	TI	FDI	TI*FDI	FI	URB	GDP	CPI
ID	1							
TI	−0.473**	1						
FDI	−0.575**	0.836**	1					
TI*FDI	0.108*	−0.543**	−0.582**	1				
FI	−0.061	0.366**	0.078	−0.091	1			
URB	−0.490**	0.533**	0.570**	0.073	−0.070	1		
GDP	−0.543**	0.773**	0.609**	−0.038	0.481**	0.593**	1	
CPI	0.152**	−0.138**	−0.099*	0.079	−0.046	−0.077	−0.016	1

[a] * and ** denote the 5% and 1% significance level, respectively.

Table 3. Summary of the results from main analysis[a].

Model	(1)	(2)
Dependent variable	ID	
TI	−0.1690*	0.0933
	(0.0811)	(0.0999)
FDI	0.1054*	0.1058**
	(0.0430)	(0.0398)
AR	0.3281	2.4079**
	(0.6836)	(0.7844)
TI*AR	0.1795**	0.2270**
	(0.0642)	(0.0796)
FDI*AR	0.0598	0.0736
	(0.0831)	(0.0745)
TI*FDI		0.0897**
		(0.0154)
TI*FDI*AR		−0.0440*
		(0.0184)
FI	0.1470	0.3098
	(0.1964)	(0.1887)
URB	−0.3559	−0.9782**
	(0.3589)	(0.3540)
GDP	−0.1583	−0.4882*
	(0.2152)	(0.2105)
CPI	0.8180	1.7078
	(2.2542)	(2.1244)
Constant	3.2032	4.4144
	(2.8671)	(2.7145)
Observations	423	423
F-value	62.79**	78.77**
R-squared	0.8319	0.8516

[a]Standard errors are provided in the parentheses. * and **denote the 5% and 1% significance level, respectively.

income disparity is lower in autonomous regions than other provinces. In other words, the interaction effect between tourism and FDI on mitigating urban–rural income disparity is greater in autonomous regions than other provinces. Therefore, the result supports hypothesis 2. In model (2), both URB and GDP show negative and significant effects on urban–rural income disparity at the 1% and 5% significance level, respectively. However, both FI and CPI have no significant effect on urban–rural income disparity.

5. Discussion

The purpose of this study is to examine whether the effect of tourism (FDI) on urban–rural income disparity is smaller (larger) in autonomous regions than other provinces in China. Moreover, this study attempts to investigate whether the interaction effect between tourism and FDI on urban–rural income disparity is smaller in autonomous regions than other provinces.

The result shows that the effect of tourism on urban–rural income disparity is larger in autonomous regions than other provinces. Despite the tourism's possible role of mitigating urban–rural income disparity, autonomous regions with abundant ethnic tourism resources located in rural areas show a larger effect of tourism on urban–rural income disparity than other provinces. This is probably because of the connection between urban and rural areas in terms of tourism and conditions of industrial and public infrastructure to affect tourism in autonomous regions. More specifically, as tourism has rapidly developed to be a critical pillar industry in autonomous regions, the overall economy of those regions, regardless of urban and rural areas, becomes highly dependent on tourism. That is, especially in autonomous regions, tourism may contribute to not only rural residents' earnings but also urban residents' ones, failing to narrow the gap between two groups. Additionally, rural tourism in autonomous regions is lacking substantially in the systematic integration of relevant tourism-related components to effectively promote tourism activities. In this kind of situation, tourism flow is not appropriately connected to rural residents' profit earnings by engaging in sales of indigenous products or idiosyncratic cultural patterns.

Meanwhile, this study found that there is no significant difference in terms of the effect of FDI on urban–rural income disparity between autonomous regions and other provinces. A probable explanation for this result is that investment amounts of FDI are concentrated on manufacturing and exporting sectors in the coastal and eastern regions. Thus, especially in other provinces in those regions, urban real wages tremendously increase caused by the wage increment in MNCs established by FDI, which consequently intrigues the improvement of wage in domestic firms located in urban areas (Ge, 2006). Therefore, this enlarged income gap between urban and rural labour forces in other provinces leads to no significant difference in terms of urban–rural income disparity from autonomous regions with poorer rural educational environments that restrict lucrative employment opportunity from FDI in rural areas.

On the other hand, the result from the examination of model (2) shows that the degree of the interaction effect between tourism and FDI on reducing urban–rural income disparity in autonomous regions is significantly greater than other provinces. A possible explanation for this result is that in autonomous regions with relatively poor infrastructure which limits accessibility to rural tourism destinations, FDI plays a critical role of developing tourism sites and recreational facilities, encouraged by local governments' policy of actively seeking the participation of foreign investors (UNCTAD, 2014). For example, foreign lodging operators, such as Super 8 Hotel and Starwood, have opened more sites in Xinjiang and Tibet, supported by the Chinese government's policies to alleviate economic disparity and ethnic-religious instability (Warikoo, 2016). Moreover, the establishment and quality of infrastructure are not sufficient to attract more tourists towards remote rural areas due to the limitation of local governmental resources in autonomous regions. In this context, FDI offers relevant capital for constructing facilities and infrastructure for the development of rural tourism. Therefore, autonomous regions can optimize the effect of tourism on reducing the income gap between urban and rural areas, supported by FDI's investments that enhance the accessibility to rural tourism destinations and the feasibility of advanced tourism services in rural areas. And, such synergetic effect between tourism and FDI is greater in autonomous regions than other provinces because of much more tourism resources and, at the same time, less-developed infrastructure to support rural tourism in autonomous regions.

Findings of this study contribute to the body of the tourism literature by adding a unique insight regarding the role of tourism and FDI on urban–rural income disparity. Particularly, considering a growing interest of harmonious society and balanced development in China, this study fills a gap in tourism research by providing comprehensive empirical evidences for greater synergy between tourism and FDI on urban–rural income disparity in autonomous regions. Based on this study's findings, some practical implications can be derived. Given that large regional variations in the quality of road infrastructure in autonomous regions (Fan & Chan-Kang, 2008), BTL (Build-Transfer-and-Lease) or BOT (Build-Operate-and-Transfer) approach of FDI is seriously considered, which may lead to the expansion of constructing well-organized infrastructure network in countryside, thereby substantially improving rural tourism in autonomous regions. Moreover, local governments in autonomous regions should devise action plans for encouraging the involvement of indigenous local residents in rural tourism business from the planning stage, such as developing more profitable regional specialties, systemizing private rental home service, producing local representative handicrafts, etc.

6. Limitations and suggestions for future studies

There are some limitations of this study, which suggest avenues for succeeding future research. First, the findings of this study may be limited to the context of urban and rural areas in China, thus should be cautiously applied to regional income disparity outside China. In particular, considering the results of this study highlighting the role of tourism and FDI in China, future research is encouraged to conduct regional comparison study between China and other multi-ethnic countries to enhance external validity. Second, this study examines the effects of tourism and FDI on urban–rural income disparity, using secondary data. Thus, the results of this study imply the overall, on average, relationships across all provinces in China. The application of the results to a particular province should carefully consider each region's unique conditions, such as industrial structure, tourism resources, and other provincial-specific attributes. Third, characteristics of five autonomous regions are all different in terms of cultural traditions, ethnic languages, life-styles, etc. Therefore, future study that compares each minority's tourism resources reflecting inherent ethnical features may be a viable research that comprehensively reinforces practical implications for implementing tourism policies.

Disclosure statement

No potential conflict of interest was reported by the authors.

Funding

This work was supported by Changshin University Research Fund under Grant Changshin 2019-41.

ORCID

Joon Ho Kim http://orcid.org/0000-0002-9115-128X

References

Adhikary, B. K. (2011). FDI, trade openness, capital formation, and economic growth in Bangladesh: A linkage analysis. *International Journal of Business and Management*, *6*(1), 16–28.

Alfaro, L., Chanda, A., Kalemli-Ozcan, S., & Sayek, S. (2004). FDI and economic growth: The role of local financial markets. *Journal of International Economics*, *64*(1), 89–112.

Blachford, D., & Qi, J. (2009). FDI and economic development in China's minority areas: Challenges in policy and practice. *Transnational Corporations Review*, *1*(1), 36–55.

Brambor, T., Clark, W. R., & Golder, M. (2006). Understanding interaction models: Improving empirical analyses. *Political Analysis*, *14*(1), 63–82.

Chen, Y. (2017). China's tourism-led foreign direct investment inflows: An empirical study. *Modern Economy*, *8*, 39–50.

Chen, C., & Wu, Y. (2017). Impact of foreign direct investment and export on urbanization: Evidence from China. *China & World Economy*, *25*(1), 71–89.

Chio, J. T. (2014). *A landscape of travel: The work of tourism in rural ethnic China*. Seattle, WA, London: University of Washington Press.

Choi, C. (2006). Does foreign direct investment affect domestic income inequality? *Applied Economics Letters*, *13*(12), 811–814.

Craigwell, R., & Moore, W. (2008). Foreign direct investment and tourism in SIDS: Evidence from panel causality tests. *Tourism Analysis*, *13*(4), 427–432.

Fan, S., & Chan-Kang, C. (2008). Regional road development, rural and urban poverty: Evidence from China. *Transport Policy*, *15*(5), 305–314.

Feenstra, R. C., & Hanson, G. H. (1997). Foreign direct investment and relative wages: Evidence from Mexico's maquiladoras. *Journal of International Economics*, *42*(3), 371–393.

Ge, Y. (2006). The effect of foreign direct investment on the urban wage in China: An empirical examination. *Urban Studies*, *43*(9), 1439–1450.

Guo, R. (2017). *China ethnic statistical yearbook 2016*. London: Palgrave Macmillan.

Gustafsson, B., & Sai, D. (2009). Temporary and persistent poverty among ethnic minorities and the majority in rural China. *Review of Income and Wealth*, *55*, 588–606.

Gustafsson, B., & Sai, D. (2014). Why is there no income gap between the Hui Muslim minority and the Han majority in rural Ningxia, China? *The China Quarterly*, *220*, 968–987.

Gustafsson, B., & Shi, L. (2003). The ethnic minority-majority income gap in rural China during transition. *Economic Development and Cultural Change*, *51*(4), 805–822.

He, S., Bayrak, M. M., & Lin, H. (2017). A comparative analysis of multi-scalar regional inequality in China. *Geoforum; Journal of Physical, Human, and Regional Geosciences, 78*, 1–11.

Heberer, T. (2017). *China and its national minorities: Autonomy or assimilation*. London: Routledge.

Hornstein, A. S. (2011). Where a contract is signed determines its value: Chinese provincial variation in utilized vs. contracted FDI flows. *Journal of Comparative Economics, 39*(1), 92–107.

Hwang, J., & Lee, S. (2015). The effect of the rural tourism policy on non-farm income in South Korea. *Tourism Management, 46*, 501–513.

Jaccard, J., & Turrisi, R. (2003). *Interaction effects in multiple regression (No. 72)*. Newbury Park, CA: Sage.

Khan, A. R., & Riskin, C. (1998). Income and inequality in China: Composition, distribution and growth of household income, 1988 to 1995. *The China Quarterly, 154*, 221–253.

Kuznets, S. (1955). Economic growth and income inequality. *The American Economic Review, 45*(1), 1–28.

Lee, M. B. (2001). *Ethnicity, education and empowerment: How minority students in Southwest China construct identities.* Aldershot: Ashgate.

Lew, A. A., Chon, K. S., Guangrui, Z., Yu, L., & Ap, J. (2003). *Tourism in China*. London: Routledge.

Li, H., Chen, J. L., Li, G., & Goh, C. (2016). Tourism and regional income inequality: Evidence from China. *Annals of Tourism Research, 58*, 81–99.

Li, Y., Zhang, H., Zhang, D., & Abrahams, R. (2019). Mediating urban transition through rural tourism. *Annals of Tourism Research, 75*, 152–164.

Lickorish, L. J., & Jenkins, C. L. (2007). *Introduction to tourism*. London: Routledge.

Lu, D. (2002). Rural-urban income disparity: Impact of growth, allocative efficiency, and local growth welfare. *China Economic Review, 13*(4), 419–429.

Lu, M., & Chen, Z. (2006). Urbanization, urban-biased policies, and urban-rural inequality in China, 1987–2001. *The Chinese Economy, 39*(3), 42–63.

Lv, L., Wen, S., & Xiong, Q. (2010). Determinants and performance index of foreign direct investment in China's agriculture. *China Agricultural Economic Review, 2*(1), 36–48.

Ma, L. J. (2005). Urban administrative restructuring, changing scale relations and local economic development in China. *Political Geography, 24*(4), 477–497.

Park, C. H. (2014). Nongjiale tourism and contested space in rural China. *Modern China, 40*(5), 519–548.

Park, S., Song, S., & Lee, S. (2017). Corporate social responsibility and systematic risk of restaurant firms: The moderating role of geographical diversification. *Tourism Management, 59*, 610–620.

Petersen, M. A. (2009). Estimating standard errors in finance panel data sets: Comparing approaches. *Review of Financial Studies, 22*(1), 435–480.

Ravallion, M. (2001). Growth, inequality and poverty: Looking beyond averages. *World Development, 29*(11), 1803–1815.

Rong, M. A. (2009). The development of minority education and the practice of bilingual education in Xinjiang Uyghur autonomous region. *Frontiers of Education in China, 4*(2), 188–251.

Ruixue, Z., Yang, X., & Ningyu, X. (2015). Research on the potentials and countermeasures of the utilization of foreign direct investment in Tibet of China. *International Journal of Econometrics and Financial Management, 3*(3), 110–114.

Sanford Jr., D. M., & Dong, H. (2000). Investment in familiar territory: Tourism and new foreign direct investment. *Tourism Economics, 6*(3), 205–219.

Sicular, T., Ximing, Y., Gustafsson, B., & Shi, L. (2007). The urban-rural income gap and inequality in China. *Review of Income and Wealth, 53*(1), 93–126.

Smith, J. N. (2002). 'Making culture matter': Symbolic, spatial and social boundaries between Uyghurs and Han Chinese. *Asian Ethnicity, 3*(2), 153–174.

Sokhanvar, A. (2019). Does foreign direct investment accelerate tourism and economic growth within Europe? *Tourism Management Perspectives, 29*, 86–96.

Sun, Y. (2019). Debating ethnic governance in China. *Journal of Contemporary China, 28*(115), 118–132.

Tabachnick, B. G., & Fidell, L. S. (1996). *Using multivariate statistics*. New York, NY: Harper Collins.

Tang, S., Selvanathan, E. A., & Selvanathan, S. (2007). The relationship between foreign direct investment and tourism: Empirical evidence from China. *Tourism Economics, 13*(1), 25–39.

Toops, S. (1993). Xinjiang's handicraft industry. *Annals of Tourism Research, 20*(1), 88–106.

Torres-Reyna, O. (2007). *Panel data analysis fixed and random effects using Stata (v. 4.2)*. Princeton, NJ: Data & Statistical Services, Princeton University.

Tosun, C., Timothy, D. J., & Öztürk, Y. (2003). Tourism growth, national development and regional inequality in Turkey. *Journal of Sustainable Tourism, 11*(2-3), 133–161.

UNCTAD. (2014). *Investment guide to the silk road*. Geneva: United Nations.

Wang, L. (2016). *Foreign direct investment and urban growth in China*. London: Routledge.

Warikoo, K. (2016). *Xinjiang-China's Northwest frontier*. London: Routledge.

Williamson, J. G. (1965). Regional inequality and the process of national development: A description of the patterns. *Economic Development and Cultural Change, 13*(4), 1–84.

Wilson, S., Fesenmaier, D. R., Fesenmaier, J., & Van Es, J. C. (2001). Factors for success in rural tourism development. *Journal of Travel Research, 40*(2), 132–138.

Wooldridge, J. M. (2005). Fixed-effects and related estimators for correlated random-coefficient and treatment-effect panel data models. *Review of Economics and Statistics, 87*(2), 385–390.
Xue, J. (1997). Urban-rural income disparity and its significance in China. *Hitotsubashi Journal of Economics, 38*(1), 45–59.
Yang, L., Wall, G., & Smith, S. L. (2008). Ethnic tourism development: Chinese government perspectives. *Annals of Tourism Research, 35*(3), 751–771.
Zeng, B., & Ryan, C. (2012). Assisting the poor in China through tourism development: A review of research. *Tourism Management, 33*(2), 239–248.
Zhang, K. H. (2002). What explains China's rising urbanization in the reform era? *Urban Studies, 39*(12), 2301–2315.
Zhang, Q., & Zou, H. F. (2012). Regional inequality in contemporary China. *Annals of Economics & Finance, 13*(1), 113–137.
Zhao, W. (2009). The nature and roles of small tourism businesses in poverty alleviation: Evidence from Guangxi, China. *Asia Pacific Journal of Tourism Research, 14*(2), 169–182.

Asian Tourism Research Methods

Towards an understanding of Chinese tourist photography: evidence from the UK

Mohan Li, Richard Sharpley and Sean Gammon

ABSTRACT

The relationship between tourism and photography has long been subject to academic scrutiny. However, the extent research has adopted a primarily Western perspective, exploring the photographic practices of Western tourists. Conversely, more limited academic attention has been paid to the photographic practices of non-Western tourists. In particular, although China is now the largest source of international tourists, few attempts have been made to date to how and why Chinese tourists engage in photography during their international travels. This paper begins to address this gap in the literature, seeking to identify the extent to which the photographic practices of Chinese tourists are determined by what may be referred to as a 'Chinese tourist gaze'. Drawing on research undertaken amongst Chinese tourists in the UK, it reveals a number of influences that determine the nature and significance of the Chinese gaze, including the traditions of pre-modern travel in China, contemporary socio-economic transformations in China, the role of the Chinese state, and technology, the latter suggesting the increasing adoption of a more universal, de-exoticized (photographic) gaze.

Introduction

An intimate relationship has long existed between the modern practices of tourism and photography. In other words, although people have for centuries engaged in various forms of tourism (or, more precisely, travel), Louis Daguerre's invention of photography in 1839 virtually coincided with Thomas Cook's first organized tour in 1841, an event that for many represented the beginning of modern mass tourism (Garlic, 2002). Since then, the histories of tourism and photography have been, as Teymur (1993, p. 6) suggests, so intertwined that it is difficult to separate them; indeed 'the emergence of mass tourism and popular photography owe a great deal to one another' (Belk & Yeh, 2011, p. 345). Consequently, tourism and photography go hand in hand. 'To be a tourist … is to be, almost by necessity, a photographer' (Markwell, 1997, p. 131) or, as Haldrup and Larsen (2003, p. 24) alternatively observe, 'taking photographs is an emblematic tourist practice; it is almost unthinkable to travel for pleasure without bringing the light-weight camera along and returning home without snapshot memories'.

It is not surprising, therefore, that academic attention has long been paid to the phenomenon of tourist photography. Following the publication of Susan Sontag's (1977) seminal text *On Photography*, the practice of tourist photography – that is, why and how tourists take photographs, their relationship with the subjects of their photographs, and the significance of photography to the overall tourist experience – has been considered from a variety of disciplinary perspectives, as have the cultural and technological processes, particularly the advent of mobile technology and the camera phone (Larsen, 2008; Larsen & Sandbye, 2014), that have rendered tourist photography an increasingly dynamic and complex phenomenon. Nevertheless, according to Stylianou-Lambert (2012), two contrasting perspectives dominate the literature.

On the one hand, and taking the lead from Sontag's (1977) argument that tourists exploit the opportunity to appropriate the places and people they visit through their photographs in a voyeuristic yet detached manner, it is suggested by some that tourists typically adopt a passive role when practising photography, seeking out and reproducing images previously encountered in postcards, guidebooks, tourism promotional materials and so on (Jenkins, 2003; Markwell, 1997). In so doing, they verify their anticipated experience, completing, as Albers and James (1988, p. 136) put it, 'a hermeneutic circle, which begins with the photographic appearances that advertise and anticipate a trip … and ends up with travelers certifying and sealing the very same images in their own photographic productions'. More specifically, it has been found that even tourists engaging culturally with the destination take photographs that reveal 'ideologies of Western power and dominance' (Caton & Santos, 2008, p. 7) although other studies suggest more ambivalent evidence of a photographic hermeneutic circle (Garrod, 2008).

On the other hand, building upon Edensor's (2000) notion of tourists as performers, tourist photography has more recently been considered an embodied and proactive activity, as indeed has the consumption of tourism more generally (Bærenholdt, Haldrup, Larsen, & Urry, 2004; Larsen, 2005). That is, rather than being a passive form of consumption manifested in the capture of predetermined, representational images of places and people, it is suggested that, for tourists, photography is increasingly becoming a performance that 'lights up the tourist experience' (Scarles, 2009 p. 465). Tourist photography is now 'less concerned with spectatorship and "consuming places" than with *producing* place myths, social roles, and social relationships' (Larsen, 2005, p. 417, emphasis in original). Moreover, this transformation to a great extent reflects the digital revolution in photography, with smart phone camera technology challenging not only the traditional sociology of photography in general (Bourdieu, 1996; Gye, 2007; Harrison, 2004) but the role and significance of photography in the tourist experience in particular (Larsen & Sandbye, 2014; Lo, McKercher, Lo, Cheung, & Law, 2011).

Irrespective of these debates, however, two points deserve emphasis. First, despite the increasing academic attention paid to the phenomenon of tourist photography, extant research continues to adopt a primarily Western perspective. In other words, the majority of studies are concerned with the photographic practices of Western (principally international) tourists; conversely, more limited attention has been paid to those of non-Western tourists. In particular, although China is now the world's largest source of international tourists, and research has begun to identify the factors that determine the behaviours of Chinese tourists in general (e.g. Arlt, 2006; Nyiri, 2009; Pearce, Wu, & Osmond, 2013), few if any attempts have been made to date to explore how and why they engage in photography in particular during their international travels. This is not to say that photography as both an historical and contemporary cultural phenomenon in China has not benefitted from academic scrutiny (Roberts, 2013). For example, it is understood that, unlike in the West, early photography in China followed a tradition of expression rather than imitation (Moore, 2007); that is, 'even though photography was invented against a background of scientific fever and an urge to capture an absolute reality in Europe, this modernist ambition was mostly lost when it was introduced to China' (Wu & Guoqiang, 2007, p. 32). Partly as a consequence (and partly reflecting the political-economy of China in the late nineteenth and early twentieth centuries) the widespread popular adoption of photography occurred later in China than elsewhere, explaining perhaps the historical and continuing popularity of studio portraits (Hay, 2007; Huang, 2006). However, few if any attempts have been made to consider the photographic practices of contemporary Chinese outbound tourists.

And second, many of these studies reflect or are influenced by the concept of the 'tourist gaze' (Urry, 1990; Urry & Larsen, 2011). As noted above, it is now recognized that tourists both possess and express the ability to decide freely where to travel, what to visit and what to photograph and, indeed, to rewrite the prescribed meanings of tourist places through their photography. Nevertheless, their photographic practices tend to be culturally tailored or directed. That is, Western tourist photography is considered not to be 'natural' but subject to the influence both of tourism institutions and of cultural and technological forces which define it as the socially organized, reflexive and embodied practice of visually consuming what is extraordinary, prevalent and significant in contemporary society (Urry & Larsen, 2011).

The purpose of this paper is to address these two issues. Specifically, drawing on research amongst Chinese tourists in the UK, not only does it analyse their photographic practices in general but also, in particular, it seeks to explore the extent to which the significance and nature of these practices are shaped by an identifiable set of cultural influences or, in other words, by what might be referred to as a Chinese tourist gaze. First, therefore, it is necessary to consider the potential constituents of such a gaze – that is, the cultural and other forces that might influence Chinese tourist photography – as a conceptual framework for the subsequent discussion of the empirical research.

Conceptualizing the Chinese tourist gaze

As discussed above, contemporary research into tourist photography follows two distinct perspectives: as an activity that is either passive and manifested in the reproduction of prescribed images, or proactive, embodied and individualistic. Interestingly, the concept of the (Western) tourist gaze, first proposed by Urry (1990) and subsequently refined over two decades (Urry, 2002: Urry & Larsen, 2011), embraces both. Fundamental to the concept is the socially constructed nature of the gaze. That is, what and how tourists gaze upon is largely influenced by a variety of social and cultural factors which, in a Foucauldian sense, both drive and inform the direction and focus of the gaze (Hollinshead, 1999). The gaze is, therefore, not fixed but dynamic, responding to wider socio-cultural transformations. At the same time, however, the tourist gaze is both reflexive and embodied: 'Places are chosen to be gazed upon because there is anticipation, especially through daydreaming and fantasy, of intense pleasures, either on a different scale or involving different senses from those customarily encountered' (Urry & Larsen, 2011, p. 4). In other words, tourists visit places with which they engage physically and reflexively, interpreting, comparing and evaluating what they gaze upon (Urry & Larsen, 2011, p. 17). Moreover, and of specific relevance to this paper, the tourist gaze has been fashioned by transport, information, communication and, in particular, photographic technologies; the gaze and photography have become enmeshed and transformed through advances in photographic and digital technology, endowing the tourist gaze/photography with dematerialized immediacy (Murray, 2008) and embodied performativity and, perhaps, contributing to the 'de-exoticisation' of the gaze, or a blurring of the tourist and everyday photography (Gye, 2007; Haldrup & Larsen, 2010).

It is not the purpose here to consider the concept of the (Western) tourist gaze in detail (see e.g. Hollinshead & Kuon, 2013). The point is, however, that the (photographic) gaze of non-Western tourists, such as those from China, may be similarly conceptualized; that is, it is logical to suggest that the photographic practices of Chinese tourists are similarly patterned or influenced by a distinctive set of cultural influences whilst displaying the increasingly widespread performativity of contemporary digital/smartphone photography. It has been suggested, for example, that Chinese international tourists typically prefer to gaze on what is distinctive or different, not simply to experience the 'Other' but to cast a comparative 'second gaze' (MacCannell, 2001) back on China as a form of patriotic identity reinforcement (Chan, 2006). More pragmatically, other commentators note that Chinese tourists take photographs to meet traditional family obligations; they capture images of places that they believe absent family members would like to see (Nyiri, 2006, 2009; Yeh, 2009). The question that remains, however, is what are the factors that may potentially define a distinctive Chinese tourist gaze?

A useful starting point in addressing this question is Pearce et al.'s (2013) review of the burgeoning literature into Chinese tourist behaviours. Not surprisingly, as China has emerged as the world's largest international tourist market – more than 127 million Chinese tourists travelled internationally in 2015 (UNWTO, 2016) – increasing academic attention has been paid to the scale, scope and nature of Chinese tourism, both domestic and outbound. Pearce et al. (2013) identify key themes in the literature, such as motivations, social behaviours and activities and interests, before offering a number of underpinning 'broad explanatory forces' (p. 151) which they refer to collectively as a 'Triple-C Gaze'. This comprises Confucianism (to refer to the significance of harmony within traditional Chinese culture); Communism (broadly referring to the role of the contemporary Chinese state); and Capitalism (to refer to the increasing orientation towards consumerism in China).

Whilst highlighting undoubted influences on Chinese tourism, however, the alliterative 'Triple-C' concept is more descriptive than explanatory; that is, it describes three general characteristics of contemporary China yet – perhaps inevitably given its broad perspective – is unable to reveal, for example, why particular sites/sights are gazed upon (and photographed) or how they are understood or interpreted. Nevertheless, it provides a valid framework for proposing more specific influences on the Chinese tourist gaze, as follows.

The pre-modern travel gaze in China

According to Pearce et al. (2013), traditional Chinese culture or, more precisely, Confucianism and its fundamental focus on harmony, continues to be a significant influence on contemporary Chinese tourism (see also Kwek & Lee, 2010). However, the cultural dynamics of the Chinese tourist gaze, or the ways in which Chinese tourists see (and, specifically, take photographs), may in particular be better understood when related to traditional Chinese travel practices and their underpinning philosophies. A detailed consideration of the cultural history of travel in China is beyond the scope of this paper (see Gong, 2001) but, for the present purposes, two points deserve emphasis. First, early travellers in China from the pre-Qin Dynasty era onwards were often motivated by the search for *tao,* a concept that emerged from *Yijing*, the cosmological foundation of Chinese philosophy (Liu, 2006). *Yijing* established the principle that human beings exist in a network comprising other human beings (society), nature (heaven and earth), and the world as a whole, as well as in both the past and the future (Cheng, 1989). As a consequence of this inseparability, people can only learn about themselves and their world through their interactions with it but, at the same time, must learn to act within the bounds of their knowledge of the world. This interdependency between humanity and nature and knowledge and action explains the significance of 'harmony' to Chinese philosophies, whilst the search for *tao*, (tao being the source of all things), becomes a search for the 'truth' as a basis for learning, reflection and self-improvement. Significantly, therefore, prevailing philosophies encouraged pre-modern Chinese travellers not to gaze on nature in the detached, implicitly dominant manner that later came to define nineteenth-century western sightseeing (Adler, 1989) but, rather, to regard themselves as part of nature. Hence, they tended to engage introspectively with what they were gazing upon to generate potential connotative meanings from it to guide their life and cultivate their character (Fung, 1922), an approach that, arguably, has been inherited by modern generations.

Second, travel in ancient China was considered to be a privileged activity undertaken by gods and immortals who would gaze down from elevated positions, such as mountain tops, on the panoramas beneath them. Moreover, legends from the Zhou Dynasty (*c.*1045 BC–256 BC) told of mountain-top encounters between travellers and gods/immortals, encounters which endowed the traveller with immortality. Hence, not only did some mountains come to be seen as sacred places, such as Taishan, or Mount Tai, which remains an important cultural site and tourist attraction in China, but also pre-modern travellers sought out these mountains and began to favour sublime views and panoramas, seeking the experience (and immortality) enjoyed by the gods (Brook, 1998; Gong, 2001). In so doing, they adopted a new way of seeing, a kind of mental travelling and contemplative gazing down from above over extensive panoramas, a visual practice referred to as *guan* (Chang, 2014, p. 40; Sullivan, 1979). Engaging in *guan*, travellers tended to deny their real self, imaginatively creating a picture of fantasy and/or interaction with what was being gazed upon. Consequently, their gazes were transcendent, reciprocal and expressive, characteristics which, again, potentially influence the contemporary Chinese tourist gaze. Indeed, research has revealed that contemporary Chinese tourists may create an imagined mental picture of the places they visit, perhaps reflecting an emphasis on 'decoration, worship [and] aesthetic expression' (Wu & Guoqiang, 2007, p. 32) in early Chinese photography. For example, Chinese backpackers in Tibet often imagine a mysterious, magical and spiritual place; in effect, they tour an imagined rather than a real destination (Shepherd, 2009).

The role of the Chinese state

As noted above, the Western tourist gaze is, to a great extent, culturally determined, influenced by developments in technology and shaped by tourism institutions. In other words, the gaze is, in a sense, 'unregulated'; it responds to transformations in culture, increasingly globalized flows of information, advances in transport and other technologies and the activities of an innovative travel and tourism industry, whilst the role of the state is to facilitate such processes (Urry, 2000). In contrast, the Chinese state arguably remains a dominant force in determining the nature and direction of the Chinese tourist gaze. That is, despite significant political and economic transformations that have witnessed the decentralization of authority and a shift towards a market economy, including within the tourism sector (Ryan & Huang, 2013), the state retains its role in sustaining pre-existing ideologies and constraining social mobilities, especially the movement of ideas, images and sounds, through education, tourism and media (see Nyiri, 2006, 2009; Oakes, 1998). More specifically, tourism (and tourists) in China continue to be exploited by the state to support economic growth, promote cultural hegemony, to create an alternative or 'false' (Oakes, 1998) modernity and to reinforce national identity.

And it does so in two ways. First, through the development of tourist sites and attractions, the state seeks to develop an imagined nostalgic landscape (explicitly linked to pre-modern travel as discussed in the preceding section) as a foundation for building national identity and an alternative modernity (Anagnost, 1993; Kolås, 2004; Oakes, 1998). Specifically, the state promotes the construction of new scenic places and attractions and the rebuilding of existing heritage, typically as theme parks and open air museums (Nyiri, 2006, 2009), which are framed within pre-determined narratives and selectively rewritten cultural and historical meanings (Shepherd, 2009). Such narratives are reinforced through the media whilst state-run tourism bureaus indirectly control travel flows to these destinations. Hence, the gaze of Chinese tourists within China is directed by the state towards the pre-scripted national narratives of tourist sites, inevitably influencing how they then gaze on places during international travel.

Second, that gaze of outbound Chinese tourists is also influenced intentionally by the state, primarily through the manner in which the Chinese diaspora in Western countries are portrayed in the media in China (Arlt, 2006), specifically in Chinese-produced films and TV shows. Not only is international mobility encouraged and considered to be a sign of modernity China (Liu, 1997), but major western cities are portrayed (in the Chinese media) as sites of an emergent global Chinese modernity (Sun, 2002). More specifically, the Chinese diaspora are shown to be contributing not only to their own individual modernization but also, through their sense of connection to their cultural homeland, to the modernization of China too (Nyiri, 2006). Hence, it has been noted that many international Chinese tourists seek to gaze on signs of modernity during their travels (Ong & du Cros, 2012).

Technology

Although traditional film-based photography remained dominant for more than a century, over the last two decades it has become almost non-existent as digital photography has become increasingly widespread (Urry & Larsen, 2011, p. 181). More specifically, the adoption of smartphones as means of taking, instantaneously viewing, sharing and deleting images has fundamentally transformed the practice and meaning of photography (Van House, 2011), not least through the popularity of the

Moreover, as considered shortly, the subsequent discussion of photographs taken during the second study not only revealed significant cultural influences on the gaze of these younger tourists but also suggested, as others have also noted (Chow & Murphy, 2008), increasing commonality between the gaze of Chinese tourists and that of their Western counterparts. Nevertheless, a number of themes emerged across both studies which, as now discussed, reveal the existence of what may be defined as a distinctive Chinese tourist gaze.

The pervasive use of smartphone cameras

The great majority of respondents relied on smartphones for taking photographs; indeed, even those respondents who also carried more complex digital cameras with them on the trip tended to favour their smartphones for capturing images. For many, the attraction of smartphones lay in their ease of use:

> I am … not good at using complicated cameras. After all, photo-taking to me is no more than a means of recording my travel experience. My photos are to be chiefly looked at by myself. I really don't like to edit the photos after a trip and show them off on the social media … A smart phone is enough.

However, the widespread use of smartphones by the respondents for capturing images of their trip also reflected a distinction between two Chinese words: 摄影 ('photography') and 拍照 ('photo-taking'); the former is regarded as the preserve of professional photographers and artists, whilst the latter is associated with recording and subsequent sharing with 'distant others'. Many respondents alluded to the fact that they were not 'professional photographers', whilst one respondent in the second study observed:

> Particularly as my parents have not travelled to the UK before, I wanted to take photos to show my parents when I am back to China – to tell them my travel experiences based around those photos.

Thus, the use of smartphones facilitated, for most respondents, two basic functions of their tourist photography: recording places they have been to and fulfilling traditional family obligations (Yeh, 2009). At the same time, a number of respondents implied that their photography as tourists has become de-exoticized by the use of smartphones; that is, they took similar kinds of photographs whether at home or on holiday. One, for example, indicated she had a passion for flowers and using her smartphone enabled her to take pictures of flowers when and wherever she encountered them. Moreover, and as discussed shortly, for the younger tourists in the second study, the practice of smartphone photography (taking and sharing photographs, particularly of each other) became fundamental to their overall tourist experience. Indeed, the age of respondents was identified in the research as a significant influence on their photographic practices although, as will also be considered, their photographic gaze was not always age-dependent.

Influence of age

The research revealed evident distinctions between the (photographic) gaze adopted by older respondents, or those around 40 years of age and above, and that of younger respondents. For the older tourists, two specific characteristics of their gaze/photo-taking emerged from the photo-elicitation.

Traditional, pre-modern gaze

As considered above, not only did early Chinese travellers regard themselves not separate from but part of nature, but they also gazed on nature introspectively. More specifically, although they engaged in the visual practice of *guan* (Chang, 2014, p. 40) – that is, imagining themselves as immortal in their gaze – natural spectacles such as a flowing river or leaves falling in autumn were, as recognized by Confucius and others (Ku, 1898), symbolic of the passing of time. Hence, early Chinese travelers were reminded of the inevitability of their own aging and mortality.

Interestingly, a similar response was identified amongst some of the older tourists in the first study. For example, one respondent (a 69-year-old university professor) revealed that, on seeing (and photographing) a field of yellow rapeseed, he felt compelled to write in his dairy: 'These yellow blossoms have effective strategies to preserve their stunning beauty ... they dare to face the indifferent and inhuman flowing of time with pride and dignity.' During an interview, he also lamented that:

> You know, I am almost seventy. The time left to me is not too much, as I am getting older and older. I try my best to indulge myself in enjoying every single trip, as each one is an incredible opportunity for me ... a younger person might not totally understand my feelings and my sense of sadness

At the same time, some older respondents implied that they would have welcomed the opportunity to immerse themselves more in nature, but were restricted by a tight schedule:

> There is a saying in China: 'Appreciate the beauty of flowers on the side of road, while on the back of a galloping horse'. I think our appreciation of the views and tourist spots [on the tour] has been more brief and shallower than this.

> I think it is much better if the tour guide could give me the opportunity to walk along the road to that valley over there ... I really want to experience the wind, the soil and the smell in the air ...

In other words, the respondents expressed the desire to gaze in a fashion similar to that adopted by their ancestors. Their experience was, however, often more superficial than they would have liked, their inherent wish to be part of what they were gazing upon and photographing being superseded by structure of the tour.

Gazing on modernity/development

Previous studies have suggested that Chinese tourists are influenced to seek out and gaze upon signs of modernity during their international travels (Nyiri, 2006; Ong & du Cros, 2012). This research similarly found that the older respondents in particular deliberately sought out and photographed representations of modernity or 'advancement', though not to simply celebrate or admire such modernity. Rather, it became evident that they deliberately gazed on aspects of modernity that highlighted what they considered to be problems or challenges in contemporary Chinese society. In other words, their gaze on modernity was, in effect, stimulated by not a patriotic but a critical 'second' gaze (MacCannell, 2001) back on China. For example, one female respondent, travelling with her daughter, took many photographs of and, during her interview, spoke at length about what she considered to be an effective interactive learning opportunity for young visitors to the British Museum:

> What actually impressed me was the collaboration and team spirit within a family. At the same time, it was fabulous to see the kids being encouraged to work out or build something instead of simply learning facts. This is very different from the situation in China, where kids just have the opportunity to view facts on a computer screen ... Also, I saw how a teacher was delivering a history class to the kids. The teacher was dressed in the clothes from specific period in history, and sometimes acted as some political figure ... The class was pretty dynamic ... In contrast, the museums in our country are no more than stores for antiques. It is a shame. They fail to attract the kids back to the museum for another visit

She went on to say that she would upload her photographs onto social media to show her friends how much 'better' she considered the learning experience to be in the UK. More generally, respondents frequently referred to the wide green spaces and traditional villages that they saw and photographed as they travelled past, comparing them favourably to the situation in China where they agreed that land is crowded, traditional built heritage has been destroyed and nature is presented in recreated theme parks, and where rapid economic growth and urbanization has resulted in a variety of environmental and social problems:

> If we compare these two countries, we can see that the English make good use of their land. The landscape is much better. Everywhere, especially the countryside, was beautiful, clean, organised and fantastic. Everywhere, what we see is a great picture.

In short, the gaze of the older Chinese tourists in this study was influenced significantly by what they considered to be negative outcomes of the state's social and economic policies, not least the attempt to direct the gaze of domestic tourists in China towards the prescribed meanings of reconstructed heritage and nature (Shepherd, 2009). Consequently, scenes of 'natural' countryside in the UK featured prominently in the respondents' photographs.

To some extent, the gaze and photographic practices of younger Chinese tourists, particularly those in the second study in this research, were similarly influenced by characteristics of contemporary China. Specifically, their gaze was revealed as a response to the dramatic social and economic transformations that have occurred in recent decades, though their gaze and activities differed from that of the older tourists described above. At the same time, however, they also displayed more 'traditional' behaviour, noted by others (Li, 2008), such as mentally immersing themselves in the landscape. Reference has already been made to the manner in which these younger tourists rejected 'formal' sightseeing at an historic property in favour of enjoying time on the nearby beach; nevertheless, their descriptions during interviews of this experience reveal that traditional element in their behaviour. As one respondent explained:

> ... I became quite excited and wanted to play as soon as my feet touched the sea water ... the movement of water gave me the feeling that I was interacting with nature. From my perspective, dead water is no more than being dead; while sea water is very different – it is living ... I mean, because of the rise and fall of tide, I could feel the movement of sea water. It was strikingly different from dead, still and ordinary water

In a sense, this respondent became alive through her interactions with the sea. However, of greater significance was the explanation that she, and indeed other respondents, gave for wanting to 'play' on the beach; it represented for them an opportunity to escape temporarily from the realities of everyday life. Many of the respondents implied that not only had rapid socio-economic change in China resulted in a sense of alienation, but also they felt under pressure to succeed, to get a good job. Hence, gazing on and immersing themselves in the seascape combined the traditional Chinese way of gazing on nature with the more contemporary purpose of seeking emotional well-being through enjoying 'blue space' (Wheeler, White, Stahl-Timmins, & Depledge, 2012). For example:

> It must be beach. When I saw the extensive and borderless sea, the pressures that I have been suffering recently almost disappeared. I was playing freely and cheerfully on the beach ... You see this photo ...
>
> That's it! Compared to the vast and extensive sea, I was so tiny and trivial that nobody would care about what I was doing or would do. So I was crazily playing, running and taking photos on the beach ... a lot of stuff, you see?

Moreover, for the younger tourists in the second study, taking photographs became central to the experience of the beach and other places they visited. That is, rather than capturing images of the scenery or other sights so that they could later mentally re-visit them or show them to others, they engaged in taking and subsequently sharing images of each other 'at play' (see e.g. Figure 1); taking and sharing their photographs became part of the experience of being there, part of the pleasure of the tourist experience (Haldrup & Larsen, 2010). As one respondent remarked:

> Not only did such kinds of photo-sharing help to develop friendship and communication amongst us, but also it was an extra source of pleasure, particularly when we exchanged our thoughts about photographs with each other .... Someone always joked about these photos ...

As such, the gaze of these younger tourists was subject to a number of influences, both traditional (being part of nature) and modern (contemporary life in China); however, their photographic practices were clearly defined by technology, the opportunities afforded by smartphones allowing them to take and share spontaneous, informal photographs. In this respect, therefore, their photographic gaze was not only de-exoticized but also conformed to the increasingly universal practice of embodied and personalized spontaneous photo-taking and sharing, reflecting the earlier observation that smartphone photography in general, and the taking and sharing of selfies in particular,

Figure 1. Playing on the beach (Respondent's photograph: reproduced with permission).

has become as much an established social practice and cultural norm in China as elsewhere (de Seta & Proksell, 2015; Zhang, 2017).

Stimulated desires – denying the self

Irrespective of the age of respondents, the sense of alienation felt by the younger tourists in the second study and revealed in their photo-taking/sharing was also expressed by other respondents in the research, particularly those on the tour in the first study. However, rather than engaging in the playful activities and photographic practices demonstrated by the tourists in the second study, this sense of alienation was manifested in the significance and discussion of more 'traditional' photographs. More specifically, a number of respondents took photographs of natural, tranquil scenes and, in a manner similar to the transcendent gaze of pre-modern travellers in China, or the visual practice of *guan,* imagined themselves to be part of that scene. For example, in one interview, a female respondent from Shanghai spoke about her discomfort living in a crowded, fast-moving city and how, in a photograph she took of Loch Lomond in Scotland (see Figure 2), she imagined herself into the picture:

> As for this photograph of Loch Lomond, I really like it because the lake offered [me] a sense of isolation and tranquillity. When I looked at it, I felt like I had become the boat on the lake ... All the stress and bad feeling went away. I don't have to think about anything. Just to be there ... a fantasy world ... Yes, I like to be quiet, and living in a busy city sometimes makes me feel uncomfortable, shallow and superficial.

In other words, this respondent had the desire to consume something or somewhere that represented the counterpoint to her everyday life, something that would compensate for her discomfort and feelings of alienation. And to satisfy this need, she imagined herself as the boat in the tranquil

Figure 2. The boat on Loch Lomond (Respondent's photograph: reproduced with permission).

scene of Loch Lomond, in so doing denying her 'self' and enjoying an extraordinary, tranquil though fleeting experience.

Others expressed similar feelings. For example:

> Sometimes, when I have gazed upon a cloud hovering over an extensive wild area, I have felt very lonely and isolated, particularly when there was nobody else but me there. This feeling of isolation occasionally even brought me a hint of sadness. I think we, modern people, are becoming increasingly hollow and lonely … I live in a metropolitan city, which is full of change. Everything is changing so swiftly … .

This respondent, however, imagined himself through his photographs to be living a more tranquil life in a provincial English city (Bath) that was on the tour itinerary; again, he denied his 'self' to become part of what he described or imagined as the 'happy world' of everyday life in that city. Hence, the everyday home life of these respondents directed them into a transcendent gaze; their photographs, therefore, were not simply records of what they had gazed upon, but desired images into which they could again place themselves.

Imagined authenticity

A final theme that emerged from the research was what is referred to here as the 'imagined authenticity' that was evident in the Chinese tourists' gaze and photography. Again as a manifestation of the traditional transcendent gaze of pre-modern Chinese travelers, in the interviews many respondents discussed how they mentally transformed places they gazed upon and photographed to fit a predetermined, imagined reality or authenticity. In other words, they were less concerned with seeing, understanding and capturing images of the reality of what they were gazing upon than with mentally transporting the scene into an imaged (perhaps idealized) authenticity.

For example, following a visit to Cambridge, one respondent (female, 45 years old) stated in an interview that: 'Cambridge made an incredible impression on me ... The academic atmosphere at Cambridge University was very enriching and dynamic ... a lot of photos of mine were taken there,' adding that 'I saw the style of the buildings in the University was unique. Red bricks, aren't they? So I assumed that the students in this University studied very hard.' In other words, based solely on the style of the university buildings she had gazed upon (and extensively photographed) she imagined the authenticity of the university, in so doing thinking about her own student days: 'Since I hadn't studied hard while I was at University, I felt much regret while I explored ... '

In a similar vein, another respondent, when discussing a photograph she had taken in Edinburgh, said:

> You see, dark clouds covered the sky. It looked like demons would come to this world ... to this place. Yes, I like Edinburgh very much, especially when dark clouds covered it. How to say? I like this sort of tranquillity before the storm. This view took me back to the seventeenth Century.

Despite all the evidence of modernity, such as shops and cars, and the presence of many other tourists, through gazing down a historic street in Edinburgh this respondent sensed an Edinburgh of the past, an Edinburgh facing uncertain times 'before the storm'. And this was an imagined authenticity that both excited and satisfied her, and she which attempted (albeit unsuccessfully) to recreate in a photograph:

> If this bus disappeared, it [this photo] is really equivalent to a picture of Edinburgh in the seventeenth Century. This photo should have been much better. But this bus ...

The extent to which this imagined authenticity was stimulated by the characteristics of the tour – that is, given the tight and rigid schedule, the tourists did not have time to immerse themselves in the reality of the destination – is unclear. However, even when they were given the opportunity to spend more time to wander freely, they tended not to do so. In other words, they displayed the 'typical' behaviour of Chinese tourists (at least, those on organized tours) recognized by others, preferring to travel and remain in groups rather than exploring more independently (Fu et al., 2012). Hence, the imagined authenticity element of their gaze may be related to this cultural characteristic.

Overall, then, the research revealed a number of distinctive characteristics with regards to the gazed and photographic practices of the Chinese tourists participating in the studies. The question remains, however, as to what extent these characteristics are representative of a Chinese tourist (photographic) gaze.

Conclusion: towards a Chinese tourist gaze?

The purpose of this paper has been twofold; first, to address the lack of attention paid to the practices of non-Western, specifically Chinese, tourists in the tourism photography literature and second, in so doing, to consider the extent to which those practices reveal a distinctive set of influences that distinguish the Chinese tourist gaze from its Western counterpart. As discussed earlier in the paper, the (Western) tourist gaze is, according to Urry and Larsen (2011), the outcome of two broad sets of influences. On the one hand, it is socially and culturally fashioned; what and how tourists gaze upon (and photograph) is determined by their dynamic socio-cultural home environment which, it is controversially argued, encourages tourists to seek out and gaze upon the extraordinary or the 'exotic'. And on the other hand, it is becoming increasingly embodied and reflexive, not least through the opportunities afforded by digital technology and, in particular, the pervasive use of smartphones.

From the research discussed in this paper, it is evident that the gaze and, in particular, photographic practices, of Chinese tourists are similarly fashioned; that is, they share some of the characteristics of the (Western) gaze as originally conceptualized. Certainly, Chinese tourists experience, consume, make sense of and, though their photography, collect meaningful images of the people, places and cultures they gaze upon and, moreover, the manner in which they do so is shaped by a variety of social, cultural, institutional and technological forces. For the Chinese, tourism has become, as Urry (1990) observed in the case of Western tourism, democratized, and their gaze reflects transformations in social structures and tastes, the influence of social and political institutions and the dominant role of transport, information and communication technologies.

Nevertheless, the outcomes of the research suggest that the Chinese gaze differs in a number of ways (see Figure 3). First, it is evident that contemporary Chinese tourists have to some extent inherited the cultural gaze of pre-modern travellers in China. That is, in comparison to the values underpinning western 'sightseeing' (Adler, 1989) which, in effect, separate people from what they are gazing upon, Chinese tourists continue to not only place themselves in nature, in the world they are visiting, but also to engage in a variety of practices reminiscent of those of their forebears. For example, many respondents were found to favour a form of mental travel, engaging with the places and sights they encountered and photographed in a transcendental and interactive manner, constructing and consuming an imagined authenticity. And more specifically, some of the older respondents visualized the passing of time in the scenes they encountered, further and distinctively rooting their gaze in the philosophical traditions of Chinese travel. At the same time, however, some recognized elements of contemporary Chinese culture were also evident in their gazes and photographic practices, not least recording images to fulfil family obligations.

Second, it is apparent from the research that the gaze of many respondents was fashioned by recent rapid transformations in socio-economic conditions in China. For some, particularly the younger tourists but also some of the older respondents, a sense of alienation and expressed feelings of loneliness and shallowness encouraged them to gaze on and immerse themselves in places and sights that enabled them to escape temporarily from such feelings. In the second study, this was manifested most notably in the respondents' playful photo-taking sharing – a practice that contrasts remarkably with more conventional concepts of the purpose and practice of tourist photography – whilst others compensated for the pressures of their contemporary life by engaging in the traditional practice of denying the self, imagining themselves into the scenes they gazed upon and

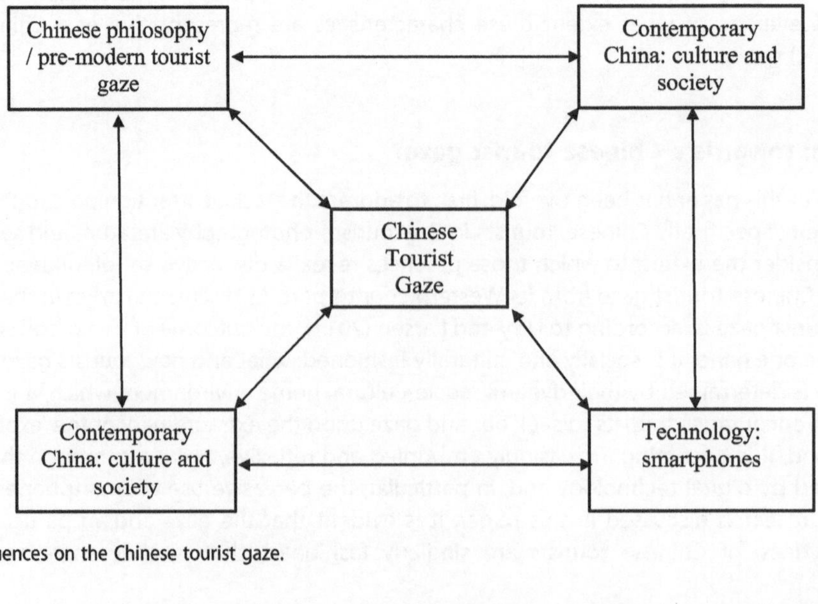

Figure 3. Influences on the Chinese tourist gaze.

photographed. More pragmatically, some respondents also delighted in gazing upon natural scenes untainted by pollution or environmental damage, or on aspects of life that considered more 'advanced' that in China. In all cases, however, it could be argued that, unlike the alleged focus of the western gaze on the extraordinary or exotic, Chinese tourists seek to gaze on or experience the 'ordinary' or 'normal' and, moreover, that their gaze is balanced by a negative 'second gaze' back towards China.

Third, and related, the role of the state remains dominant in shaping the Chinese tourist gaze. Not only have the economic policies of recent decades in China that have brought about rapid economic growth and urbanization encourage Chinese tourists to adopt the gazes/photographic practices described above but, as discussed earlier in the paper, the Chinese state and its relevant institutions continue to employ tourism and the media as a means of achieving cultural hegemony, of normalizing Chinese tourists' gaze and manipulating what they desire to see. However, in recognizing the state-driven adaptation and commodification of natural and built heritage in China, Chinese tourist defy such attempts to achieve cultural hegemony; they gaze with pleasure on what they consider to be the natural, the traditional, the uncommodified places and sights encountered during their travels. In this sense, the Chinese gaze is a manifestation of resistance to the state.

To conclude, then, this paper suggests that, from the evidence of this research at least, there exists a distinctive Chinese tourist (photographic) gaze. That is, a number of influences upon can be identified that endow the gaze of Chinese tourists with a distinctive character. The extent to which this can be considered universal amongst all Chinese tourists will, of course, be dependent on further research; not only there is a need to explore the (photographic) gaze of Chinese tourists in different contexts (and, indeed, of other non-Western tourists), but also developments in the nature of Chinese outbound tourism, particularly the more widespread adoption of independent as opposed to organized travel, will enhance the dynamic nature of the gaze.

Nevertheless, two clear points emerge from the research. First, an evident relationship exists between an inherited traditional Chinese gaze and the gaze as a response to the contemporary socio-economic conditions in China. Indeed, the outcomes of this research reveal as much about the respondents' attitudes to socio-economic change in China as they do about the nature of their tourist gaze and photographic practices and, as such, they make a contribution to the related anthropology/photography literature (e.g. Pinney, 2011). And second, the distinctions between the needs and activities of older and younger Chinese tourists, particularly the preference amongst younger tourists for playful, performative photo-taking and sharing in which they themselves, rather than the places visited become the object of the photographic gaze, begin to challenge conventional perspectives on the practice and significance of tourist photography.

Finally, although this paper has adopted a primarily conceptual perspective, these points in turn suggest a number of practical implications that arise from the research. For older tourists, the tour experience may be more satisfying if more time is allowed for them to gaze upon and immerse themselves in natural scenery and traditional built heritage rather than simply to 'collect' well-known sites; whilst capturing images to show to absent others remains an obligation, it is the culturally influenced traditional gaze that adds meaning to Chinese tourists' photographic practices. The same may be said for younger tourists although their behaviours revealed by the research suggests that, over time, formal organized tours will become less attractive to the Chinese tourist market. That is, independent travel will inevitably become more popular amongst Chinese tourists. Of greatest significance, however, the outcomes of this study imply that the continuing growth in outbound Chinese tourism may eventually contribute to demands for political change in China itself, although only time and further research will determine whether this will indeed the case.

Disclosure statement

No potential conflict of interest was reported by the authors.

References

Adler, J. (1989). Origins of sightseeing. *Annals of Tourism Research, 16*(1), 7–29.

Albers, P., & James, W. (1988). Travel photography: A methodological approach. *Annals of Tourism Research, 15*(1), 134–158.

Anagnost, A. (1993). The nationscape: Movement in the field of vision. *Positions, 1*(3), 585–606.

Arlt, W. (2006). *China's outbound tourism*. Abingdon: Routledge.

Bærenholdt, O. J., Haldrup, M., Larsen, J., & Urry, J. (2004). *Performing tourist places*. Aldershot: Ashgate.

Belk, R., & Yeh, J. (2011). Tourist photographs: Signs of self. *International Journal of Culture, Tourism and Hospitality Research, 5*(4), 345–353.

Bourdieu, P. (1996). *Photography: A middle-brow art*. Cambridge: Polity.

Brook, T. (1998). *The confusions of pleasure: Commerce and culture in Ming China*. Berkeley: University of California Press.

Caton, K., & Santos, C. (2008). Closing the hermeneutic circle? Photographic encounters with the other. *Annals of Tourism Research, 35*(1), 7–26.

Cederholm, E. (2012). Photo-elicitation and the construction of tourist experiences: Photographs as mediators in interviews. In T. Rakić & D. Chambers (Eds.), *An introduction to visual research methods in tourism* (pp. 92–107). Bristol: Channel View Publications.

Chan, Y. (2006). Coming of age of the Chinese tourists: The emergence of non-Western tourism and host-guest interactions in Vietnam's border tourism. *Tourist Studies, 6*(3), 187–213.

Chang, M. (2014). Envisioning the spectacle of Emperor Qianlong's tours of southern China: An exercise in historical imagination. In J. Cook, J. Goldstein, M. Johnson, & S. Schmalzer (Eds.), *Visualizing modern China: Image, history and memory 1750-present* (pp. 25–45). London: Lexington Books.

Cheng, C.-Y. (1989). Chinese metaphysics as non-metaphysics: Confucian and Daoist insights into the nature of reality. In R. Allison (Ed.), *Understanding the Chinese mind: The philosophical roots* (pp. 167–208). Oxford: Oxford University Press.

Chow, I, & Murphy, P. (2008). Travel activity preferences of Chinese outbound tourists for overseas destinations. *Journal of Hospitality Marketing & Management, 16*(1), 61–80.

de Seta, G., & Proksell, M. (2015). The aesthetics of *zipai*: From WeChat selfies to self-representation in contemporary Chinese art and photography. *Networking Knowledge: Journal of the MeCCSA Postgraduate Network, 8*(6), 1–27.

Edensor, T. (2000). Staging tourism: Tourists as performers. *Annals of Tourism Research, 27*(2), 322–344.

Feifer, M. (1985). *Going places*. London: Macmillan.

Fu, X., Lehto, X. Y., & Cai, L. A. (2012). Culture-based interpretation of vacation consumption. *Journal of China Tourism Research, 8*(3), 320–333.

Fung, Y. L. (1922). Why China has no science: An interpretation of the history and consequences of Chinese philosophy. *The International Journal of Ethics, 32*(3), 237–263.

Garlic, S. (2002). Revealing the unseen: Tourism, art and photography. *Cultural Studies, 16*(2), 289–305.

Garrod, B. (2008). Understanding the relationship between tourism destination imagery and tourist photography. *Journal of Travel Research, 47*(3), 346–358.

Gong, P. C. (2001). *The history of travelling spirit and culture in China*. Shijiangzhuang: Hebei Education Press.

Gye, L. (2007). Picture this: The impact of mobile camera phones on personal photographic practices. *Continuum: Journal of Media & Cultural Studies, 21*(2), 279–288.

Haldrup, M., & Larsen, J. (2003). The family gaze. *Tourist Studies, 3*(1), 23–46.

Haldrup, M., & Larsen, J. (2010). *Tourism, performance and the everyday: Consuming the orient*. Abingdon: Routledge.

Harper, D. (2002). Talking about pictures: A case for photo elicitation. *Visual Studies, 17*(1), 13–26.

Harrison, B. (2004). Snap happy: Toward a sociology of 'everyday' photography. In C. Pole (Ed.), *Seeing is believing? Approaches to visual research*, Studies in qualitative methodology (Vol. 7, pp. 23–39). Oxford: Elsevier.

Hay, J. (2007). Notes on Chinese photography and advertising in late nineteenth-century Shanghai. In J. Kuo (Ed.), *Visual culture in Shanghai, 1850s–1930s* (pp. 95–119). Washington, DC: New Academia Publishing.

Holiday, S., Lewis, M. J., Nielsen, R., Anderson, H. D., & Elinzano, M. (2016). The selfie study: Archetypes and motivations in modern self-photography. *Visual Communication Quarterly, 23*(3), 175–187.

Hollinshead, K. (1999). Surveillance of the worlds of tourism: Foucault and the eye-of-power. *Tourism Management, 20*(1), 7–23.

Hollinshead, K., & Kuon V. (2013). The scopic drive of tourism: Foucault and eye dialectic. In O. Moufakkir & Y. Reisinger (Eds.), *The host gaze in global tourism* (pp. 1–18). Wallingford: CABI.

Huang, X. (2006). Performing gender: Nostalgic wedding photography in contemporary China. *Ethnologies, 28*, 81–111.

Hurworth, R. (2004). Photo-interviewing. *Qualitative Research Journal, 4*(1), 73–79.

Jenkins, O. (2003). Photography and travel brochures: The circle of representation. *Tourism Geographies, 5*(3), 305–328.

Kolås, Å. (2004). Tourism and the making of place in Shangri-La. *Tourism Geographies, 6*(3), 262–278.

Ku, H.-M. (1898). *The discourses and sayings of Confucius*. Hong Kong: Kelly and Walsh.

Kwek, A., & Lee, Y. S. (2010). Chinese tourists and Confucianism. *Asia Pacific Journal of Tourism Research, 15*(2), 129–141.

Larsen, J. (2005). Families seen sightseeing: Performativity of tourist photography. *Space and Culture, 8*(4), 416–434.

Larsen, J. (2008). Practices and flows of digital photography: An ethnographic framework. *Mobilities, 3*(1), 141–160.

Larsen, J., & Sandbye, M. (2014). *Digital snaps: The new face of photography*. London: I.B. Taurus.

Li, F.-M. (2008). Culture as a major determinant in tourism development of China. *Current Issues in Tourism, 11*(6), 492–513.

Liu, X. (1997). Space, mobility and flexibility: Chinese villagers and scholars negotiate power at home and abroad. In A. Ong & D. Nonini (Eds.), *Ungrounded empires: The cultural politics of modern Chinese transnationalism* (pp. 91–114). London: Routledge.

Liu, J. (2006). *An introduction to Chinese philosophy: From ancient philosophy to Chinese Buddhism*. Oxford: Blackwell.

Lo, I. S., McKercher, B., Lo, A., Cheung, C., & Law, R. (2011). Tourism and online photography. *Tourism Management, 32*(4), 725–731.

MacCannell, D. (2001). Tourist agency. *Tourist Studies, 1*(1), 23–37.

Markwell, K. (1997). Dimensions of photography in a nature-based tour. *Annals of Tourism Research, 24*(1), 131–155.

Moore, O. (2007). Photography in China: A global medium locally appropriated. *IIAS Newsletter, 44*, 6–7.

Murray, S. (2008). Digital images, photo-sharing, and our shifting notions of everyday aesthetics. *Journal of Visual Culture, 7*(2), 147–163.

Nyiri, P. (2006). *Scenic spots: Chinese tourism, cultural authority and the state*. Seattle: University of Washington Press.

Nyiri, P. (2009). Between encouragement and control: Tourism, modernity and discipline in China. In T. Winter, P. Teo, & T. Chang (Eds.), *Asia on tour: Exploring the rise of Asian tourism* (pp. 153–170). Abingdon: Routledge.

Oakes, T. (1998). *Tourism and modernity in China*. London: Routledge.

Oliffe, J., & Botterff, J. (2007). Further than the eye can see? Photo elicitation and research with men. *Qualitative Health Research, 17*(6), 850–858.

Ong, C.-E., & du Cros, H. (2012). The post-Mao gazes: Chinese backpackers in Macau. *Annals of Tourism Research, 39*(2), 735–754.

Pearce, P., Wu, M.-Y., & Osmond, A. (2013). Puzzles in understanding Chinese tourist behaviour: Towards a triple-C gaze. *Tourism Recreation Research, 38*(2), 145–157.

Pinney, C. (2011). *Photography and anthropology*. London: Reaktion Books.

Roberts, C. (2013). *Photography and China*. London: Reaktion Books.

Ryan, C., & Huang, S. (2013). Planning for tourism places. In C. Ryan & S. Huang (Eds.), *Tourism in China: Destinations, planning and experiences* (pp. 227–236). Bristol: Channel View.

Scarles, C. (2009). Becoming tourist: Renegotiating the visual in the tourist experience. *Environment and Planning D: Society and Space, 27*(3), 465–488.

Shepherd, R. (2009). Cultural presertvation, tourism and 'donkey travel' on China's frontier. In *Asia on tour: Exploring the rise of Asian tourism* (pp. 256–263). London: Routledge.

Sontag, S. (1977). *On photography*. New York: Delta Books.

Statista. (2017). Number of smartphone users in China from 2013 to 2021. Retrieved from www.statista.com/statistics/467160/forecast-of-smartphone-users-in-china/

Stylianou-Lambert, T. (2012). Tourists with cameras: Reproducing or producing? *Annals of Tourism Research, 39*(4), 1817–1838.

Sullivan, M. (1979). *Symbols of eternity: Art of landscape painting in China*. Oxford: Oxford University Press.

Sun, W.-N. (2002). *Leaving China: Media, migration and transnational imagination*. New York, NY: Rowman & Littlefield.

Teymur, N. (1993). Phototourism – or the epistemology of photography in tourism. *Tourism in Focus, 6*, 6–16.

UNWTO. (2016). Tourism highlights 2016 edition. Retrieved from www.e-unwto.org/doi/pdf/10.18111/9789284418145

Urry, J. (1990). *The tourist gaze: Leisure and travel in contemporary societies*. London: Sage.

Urry, J. (2000). *Sociology beyond societies. Mobilities for the 21st century*. London: Routledge.

Urry, J. (2002). *The tourist gaze* (2nd ed.). London: Sage.

Urry, J., & Larsen, J. (2011). *The tourist gaze 3.0*. London: Sage.

Van House, N. (2011). Personal photography, digital technologies and the uses of the visual. *Visual Studies, 26*(2), 125–134.

Warren, S. (2005). Photography and voice in critical qualitative management research. *Accounting, Auditing & Accountability Journal, 18*(6), 861–882.

Wheeler, B., White, M., Stahl-Timmins, W., & Depledge, M. (2012). Does living by the coast improve health and well-being? *Health and Place, 18*(5), 1198–1201.

Wu, J., & Guoqiang, Y. (2007). Beyond propaganda, aestheticism and commercialism: The coming of age of documentary photography in China. *Javnost – The Public, 14*(3), 31–48.

Yeh, J. (2009). Still vision and mobile youth: Tourist photos, travel narratives and Taiwanese modernity. In T. Winter, P. Teo, & T. Chang (Eds.), *Asia on tour: Exploring the rise of Asian tourism* (pp. 302–314). Abingdon: Routledge.

Zhang, W. (2017). Smartphone photography in urban China. *International Journal of Social, Behavioral, Educational, Economic, Business and Industrial Engineering, 11*(1), 2312–2239.

Exploring qualitative applications of social media data for place-based assessments in destination planning

Allie McCreary, Erin Seekamp, Mae Davenport and Jordan W. Smith

ABSTRACT

Understanding the physical attributes of a destination valued by visitors allows tourism managers to consider those components when planning, managing, and marketing destinations. However, determining the key components of visitors' destination image can be difficult. This study utilizes social media data to explore visitors' destination image of a nature-based tourism locale. Specifically, we examined user-generated content by visitors to the tourism region adjacent to Lake Superior in Minnesota USA during the summer 2015 tourism season. Content analysis was utilized to determine the major themes of visitors' destination image. Our findings are similar to those of existing studies: natural resources, built resources, and human subjects are the most prevalent components of destination image. However, our method of ascertaining these components of destination image – specifically, thematically analyzing social media data – present a technique that may be more easily accessible to tourism providers, as it can be a lower cost and time investment. We discuss how these findings can inform the development of recreation opportunities and marketing materials. This qualitative analysis of social media data can be applied in other locales as a relatively efficient and real-time method to inform place-based management and marketing.

Introduction

Place-based management and marketing depends on understanding visitors' perceptions of a destination (e.g. Swinney, Lang, & Runyan, 2012; Yung, Freimund, & Belsky, 2003). Destination image (Stylidis, Sit, & Biran, 2014) – also known as place form (Beckley, Stedman, Wallace, & Ambard, 2007) and place image (Kavaratzis & Hatch, 2013) – refers to the physical characteristics of a geographic space. Previous research has established a relationship between destination image and place satisfaction (Chen & Tsai, 2007; Lee, Lee, & Lee, 2014), intention to re-visit a place (Assaker & Hallak, 2013), and destination loyalty (Prayag & Ryan, 2012). Understanding visitor perceptions of a destination's physical characteristics can enhance place-based management efforts (Lin, Morais, Kerstetter, & Hou, 2007; Stedman, 2003), as physical attributes are a common pull factor for tourism-based economies (e.g. Cocolas, Walters, & Ruhanen, 2016). Understanding which specific attributes of place are important to visitors can help tourism providers understand key resources (Cahyanto, Pennington-Gray, & Thapa, 2013) and better brand destinations to ensure sustained tourism flows (Lew, 2017).

In this study, we explore destination image using social media data. While social media data is of growing interest and has been analyzed quantitatively (e.g. to understand visitor flows, Wood, Guerry, Silver, & Lacayo, 2013), we were interested in utilizing a qualitative content analysis of social media posts to better understand visitors' destination image. To achieve this goal, we established two research questions:

1. What are the dominant themes captured in images of a nature-based tourism destination that visitors share (via social media) with family and friends?
2. How can photographic social media data inform place-based management and marketing for tourism destinations?

We utilized a modified visitor-employed photography (VEP) approach to explore the content of photographs that visitors to a nature-based tourism destination shared with family and friends via the Instagram social media platform. Photographs that were posted to Instagram were analyzed thematically (e.g. Richards & Friess, 2015) to reveal the key themes of destination image. Such user-generated content (UGC) can help tourism providers determine the dominant characteristics of destination image as conceptualized by visitors and, based on this information, strategically set place-based management and marketing goals. While the findings presented here are destination-specific, our method of using UGC to assess place-based characteristics and inform place-based management and marketing is transferable to other destinations.

Background

Understanding destination image

The concept of 'place' has been used in various disciplines (sociology, geography, environmental studies, psychology, planning, and human health) to understand how individuals' residence or the outdoor recreation settings they visit influences various aspects of their lives (Seamon & Sowers, 2008). The earliest research regarding place established how this concept is defined, distinct of space, by three main components: (1) physical attributes; (2) the meanings individuals and groups assign; and (3) the activities which occur within those spaces (Relph, 1976). Other research has built on this foundational understanding of place by exploring how 'sense of place', or one's relational and affective operationalization of space, is manifested and maintained (Yi-Fu, 1974).

The prevailing trajectory of place research has explored the affective component of place: attachment; identity; and dependence on a place (Altman & Low, 1992; Proshansky, Faban, & Kaminof, 1983; Stokols & Shumaker, 1981; Williams & Roggenbuck, 1989). More recently, place meanings were posited as the functional, descriptive foundations of affective ties (Davenport & Anderson, 2005). Place meanings are the values individuals assign to geographic spaces and can range from the abstract (e.g. a place of spiritual significance) to the tangible (e.g. one's business location is dependent on a particular place). Place meanings have been reliably measured through quantitative survey items, which further operationalize place meanings into various dimensions such as individual identity, family identity, community identity, self-expression, ecological integrity, economic dependency, and place dependency (Kil, Holland, & Stein, 2014; Smith, Anderson, & Moore, 2012). Additionally, qualitative studies have used personal narratives to explore place meanings specific to a geographic space, such as a river valued for providing sustenance to residents living along a waterway (Davenport & Anderson, 2005). Collectively, research has demonstrated that place meanings are related to physical characteristics of the landscape (Stedman, 2003). However, there remains limited research on methodologies used to determine individuals' destination image.

VEP is a common method used to ascertain individuals' visual preferences or in other cases where understanding physical characteristics are a key component of the research design (e.g. Amsden, Stedman, & Kruger, 2010; Kim, Chang, & Shelby, 2003; Michaelidou, Siamagka, Moraes, & Micevski,

2013; Taylor, Czarnowski, Sexton, & Flick, 1995). VEP allows visitors to capture their own images (often on disposable cameras provided by a researcher) and then discuss those images with a member of the research team (e.g. Dorwart, Moore, & Leung, 2009). The VEP method allows for the generation of rich data; however, the technique is time-intensive as researchers must first distribute cameras and instructions and then reconnect with participants to administer post-photography surveys or interviews.

Social media data have the potential to inform tourism science in new and meaningful ways, including understanding social values, enhancing place-based management and branding, and informing resource conservation (Di Minin, Tenkanen, & Toivonen, 2015; Metaxas & Mustafaraj, 2014). Recently, UGC has been used to quantify landscape values at a continental scale (van Zanten et al., 2016). UGC has also been used to assess sense of place through analysis of the tags that accompany Flickr photographs from urban areas throughout the United States (Feick & Robertson, 2015). UGC joins conventional mechanisms such as stories, art, film, and music as examples of media 'through which landscapes are created, recreated, and redefined' (Greider & Garkovich, 1994, p. 18). UGC allows individuals to define the meaning of a landscape and share that meaning with others. UGC can also generate more generalizable research, as it enables participation by individuals with varying levels of literacy, eligibility, and availability (Sancar & Severcan, 2010).

Studying visitors' photographs alone may not reveal the complex and dynamic meanings individuals assign to the landscape. However, analyzing visitors' photographs can reveal the key tourism resources to which visitors assign these complex and dynamic values. Landscape characteristics are important as 'they underpin both place attachment and satisfaction' (Stedman, 2003, p. 682) and allow tourism providers to better understand the favorable components of a destination (Cahyanto et al., 2013; Lew, 2017). Tourism providers cannot directly manage for the many, unique values of each individual visitor; however, tourism providers can seek to understand the key elements of a setting that facilitate the realization of visitors' place meanings.

Informing place-based management and marketing

Exploring issues in tourism management from a place-based perspective allows for diverse views and consequently diverse management strategies to be considered by managers (Yung et al., 2003). For example, public participation geographic information systems (PPGIS) can be used to identify areas that visitors believe are important; these data can then be used to inform place-based planning processes (Brown & Weber, 2011; Lowery & Morse, 2013). Interviews are often used to elicit narratives that can inform theory regarding how and why places take on certain meanings for individuals (e.g. Gustafson, 2001). Qualitative data from focus groups, meetings, and workshops have also been used to identify support for natural resource management and intentions to participate in planning process (Jacobs & Buijs, 2011; Kil et al., 2014).

A more recent line of place research extends the idea of place-based management to include the concept of place branding or marketing. Generally defined, place-based management is 'a coordinated, area-based, multi-stakeholder approach to improve locations, harnessing the skills, experiences, and resources of those in the private, public, and voluntary sectors' (IPM, 2016, n.p.). Place-based marketing is defined through three components: visual; verbal; and behavioural expressions of place (Kavaratzis & Hatch, 2013), or 'the link between identity, experience, and image' (Govers & Go, 2009, p. 23). Both of these definitions suggest an important component of place-based management and marketing is the visual image of a place. In some cases, the image may transcend verbal marketing cues that can be misinterpreted by individuals with different socio-cultural backgrounds (Govers & Go, 2009; Kavaratzis & Hatch, 2013). As such, we contend that UGC posted to social media platforms is an emerging and readily available data source for informing place-based management and marketing.

Methods

We qualitatively analyzed social media data to understand destination image as conceptualized by visitors. Specifically, we analyzed photographs and textual information posted to the Instagram social media platform and affiliated with Lake Superior's 'North Shore' during the summer of 2015. The North Shore region is a tourism dependent, natural-resource rich region in northeastern Minnesota, USA. The destination is characterized by small towns, locally-owned businesses, and nature-based recreation opportunities, with the height of annual tourism occurring in the summer during which time hiking, boating, camping, and fishing are popular recreational activities (Davenport, Schneider, Date, & Filter, 2011). Eight Minnesota state parks, along with private tourism operators (lodges, outfitters, and guides), provide access and infrastructure to visitors seeking to experience the destination's forests, coast, and waterbodies.

Research design

We used a phenomenological research design using qualitative thematic analysis to explore how summer visitors to the North Shore conceptualized key attributes of the region's destination image. Specifically, visitors' photographs, which were posted to Instagram with the hashtag 'MyNorthShore' (hereafter, '#MyNorthShore'), were thematically coded using interpretive content analysis to reveal key themes in the destination image of the North Shore tourism region (Drisko & Maschi, 2015). The coding strategy was developed to ascertain and illustrate the specific and tangible resources captured and shared by North Shore visitors on social media.

Sampling

The initial sample consisted of individuals visiting the North Shore between 15 July and 3 August 2015, which corresponded with the sampling period for on-site visitor survey. We determined that the on-site visitor survey (see Hestetune et al., 2018) was an ideal opportunity to prompt visitors to participate in the photographic component of the research project. Visitors, as characterized by data resulting from the summer 2015 on-site survey ($n = 4{,}298$; 57% response rate), are typically in-state (71%) visitors who make an average of two trips to the North Shore per summer season (Table 1). Common activities of North Shore visitors consist of scenic driving, hiking, and visiting cultural and historic sites.

We did not set a target sample size for the content analysis of the UGC, as the research design was predominantly exploratory. Additionally, we did not make any explicit efforts to reduce sampling error within the target population. Rather, data generation was intended to occur organically via visitors' exposure to #MyNorthShore through either primary or secondary prompts. A primary prompt was made when visitors were intercepted by a research assistant during the on-site survey. Specifically, #MyNorthShore stickers were handed out to visitors (primary prompt), regardless of their eligibility to participate in the survey, with directions on how to use the hashtag and a prompt to post images from their current trip to social media using the hashtag (Figure 1). Remaining stickers and directions were available for visitors to freely take from sampling locations after on-site survey sampling concluded (secondary prompt). Stickers were selected as a prompt due to their popularity among visitors and the low cost of the stickers as a data-generation mechanism. Two thousand stickers were distributed at various recreation sites, although we cannot verify how many visitors took stickers, as any particular visitor may have taken no stickers, one sticker, or multiple stickers. Additional prompting occurred when individuals saw other individuals (friends/family/individuals they 'follow' on social media) using the hashtag on social media and also elected to use the hashtag in their own social media posts. The sticker design was based on the popular white oval decals with black 'place' initials (originally devised in Europe as a way to determine a vehicle's country of origin) with '#MyNorthShore' text. While the directions (see Figure 1) referenced a variety of social media sites, review of content on multiple sites revealed that, after the data generation period concluded, the majority of #MyNorthShore posts were made to Instagram. For this reason, Instagram was the platform from which we collected data.

Table 1. Descriptive statistics emerging from the summer 2015 on-site survey of North Shore visitors.

Survey measure	North Shore summer visitors
Gender	56% female
Age	75% over the age of 35
Income	29% make more than $100,000 annually (2015 USD)
Education	66% hold at least a Bachelor's degree
Trip purpose	82% are primarily recreating at North Shore sites
Origin	71% from Minnesota
Years making visits to the North Shore	mean = 16 years (Std. Dev. = 15.82)
Trip length	mean = 4 nights (Std. Dev. = 4.01)
Party size	mean = 3 people (Std. Dev. = 2.08)
No. of current trips per summer season	mean = 2 trips per summer season (Std. Dev. = 1.60)
Activity participation:	
Scenic driving	89%
Hiking	81%
Visiting cultural/historic sites	65%
Swimming	51%
Picnicking	46%
Wildlife viewing	46%
Rock collecting	38%
Camping	36%
Bicycling	15%
Fishing	15%
Non-motorized boating (inland lakes)	10%
Creating art	9%
Non-motorized boating (Lake Superior)	9%
Motorized boating (Lake Superior)	6%
Gathering wild plants (foods)	6%
Motorized boating (inland lakes)	6%
ATV riding	2%
Horseback riding	1%
Hunting	1%

|Note: 1,298 total usable survey resulted from on-site summer sampling (a 57% response rate).

Data gathering

A Python script, enabled by an Instagram Application Program Interface (API) token, was used to collect the content associated with #MyNorthShore from the Instagram website. Python is a computer programming language that enables crawling (or searching the internet) and gathering data from the internet where content is retrievable through a permanent Uniform Resource Locator (URL, or web address) (Metaxas & Mustafaraj, 2014). The resulting data were organized into an Excel spreadsheet. Data in the spreadsheet included the individuals' username and user ID, time and date of the post, a URL of the image, any tags and captions posted with the image, type of filter used on the image, and number of likes and comments on the image. A geo-tagged location was available for some posts, depending on individual users' Instagram settings and voluntary choice to designate the location of the photograph in their social media post (see Figure 2 for spatial distribution of those images which were geo-tagged). We used photographs posted between July 15 (when sticker distribution/data prompting began) and 7 September 2015 (Labor Day, a typical date considered to be the end of a summer season within the United States) for analysis.

Analysis

Analysis was completed within QSR International N*Vivo (v. 10) software; the software enabled us to create and assign parent category codes and themes to individuals' posts (both images and caption text), query data, and export results. We created an initial codebook based on existing place-related literature (e.g. commonly reported attributes of place, particularly in coastal and/or nature-based tourism destinations). We analyzed Instagram posts using interpretive content analysis, which focuses on identifying characteristics (the subjects or objects) of an image or text rather than

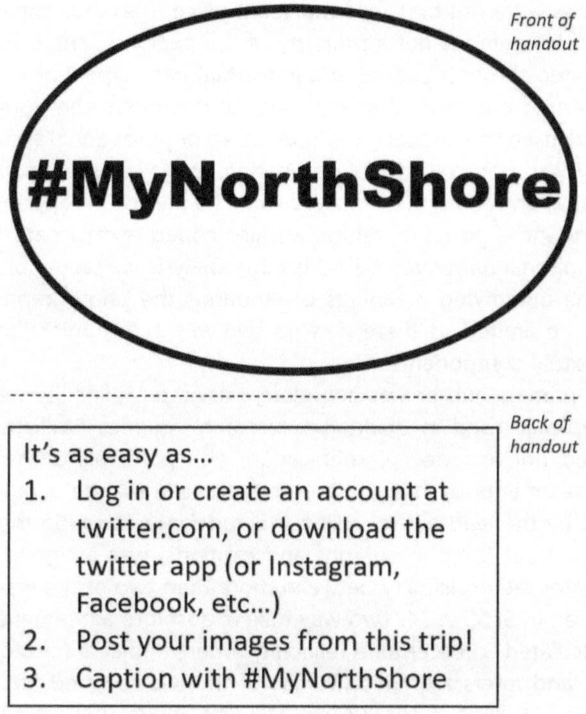

Figure 1. Directional handout provided to visitors for data collection of North Shore destination image study.

quantitative descriptions (i.e. words counts) (Drisko & Maschi, 2015). We coded photographs iteratively; that is, as new themes emerged and were added to the codebook, we reanalyzed previously coded photographs. Coding was not restrictive; meaning multiple themes could be applied to a single photograph or its accompanying text.

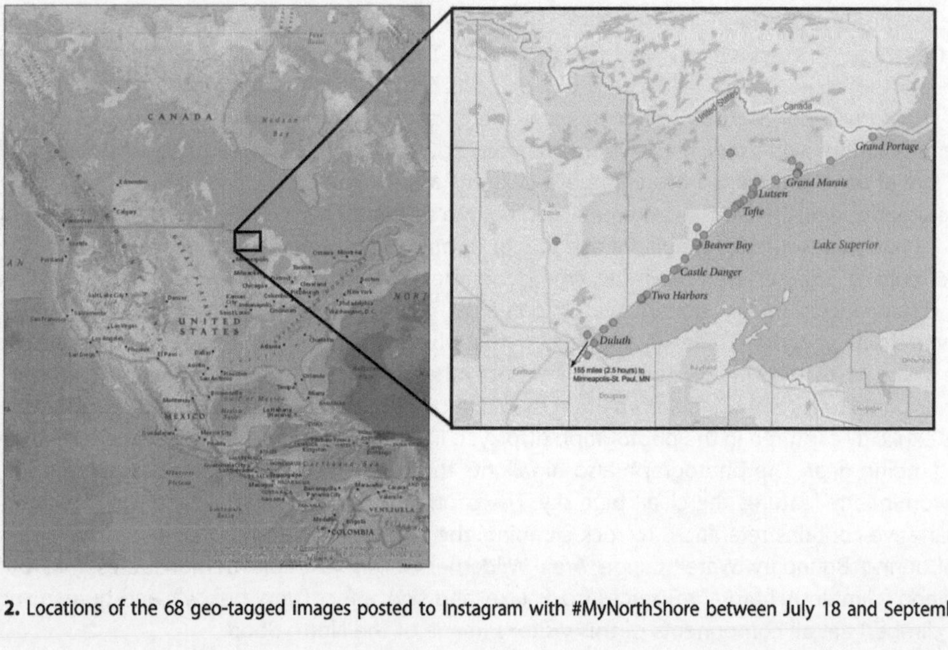

Figure 2. Locations of the 68 geo-tagged images posted to Instagram with #MyNorthShore between July 18 and September 7, 2015.

Categories and themes were not meant to represent place meanings explicitly, as meanings are socially constructed and 'meaning is not a property of the person or the object, but a relationship between the two mediated through culture and individual past experience' (Williams & Patterson, 2007, p. 936). As the desired outcome of analysis was to document the dominant components of a destination's image, we used the codebook to provide an organizational system for understanding the most dominant physical characteristics of destination image included in visitors' UGC. Although coding focused on physical attributes, the concept of affect emerged in captions and additional hashtags related to #MyNorthShore posts; therefore, we also coded textual data for instances of these positive or negative emotional narratives. Based on this analysis we were able to make exploratory inferences regarding the underlying meanings or emotions the photographs may contain. These inferences or analyses are limited as there was no two-way communication with participants to verify our analysis of textual components.

After coding by the primary author was complete, inter-rater reliability was assessed. Although member checks are typically used in qualitative research inquiries (Williams & Patterson, 2007), this research design did not include communication with participants. As such, we conducted inter-rater reliability assessments by having three researchers analyze a random selection of the images/captions coded by the lead author, using the codebook to guide their analysis. A random sample of 10% of the dataset (18 photographs and captions) was assigned to each of the three researchers. To assess inter-rater reliability between more than two of the researchers, an intraclass correlation was computed in SPSS v. 24. Two-way mixed, absolute agreement intraclass correlation (ICC) statistics were calculated to determine reliability where subjects (photographs and captions) are randomly assigned and raters are fixed. Results of the ICC revealed high reliability of coding among the coders. For all 18 images, the ICC statistic was above .792 (mean = .819). This exceeds the commonly accepted threshold of .6 as an acceptable level of reliability among raters (Bruton, Conway, & Holgate, 2000).

Results

We collected a total of 194 photographs posted to Instagram with #MyNorthShore between 18 July and 7 September 2015. A total of 8 photographs were removed from the dataset due to irrelevance (e.g. advertisements for e-cigarettes that contained #MyNorthShore, among other tags related to Lake Superior and Minnesota). A total of 53 unique users posted the remaining 186 photographs for an average of 3.6 posts per user (range = 1–33). The majority ($n = 37$) of users posted three or fewer photographs. A few avid users ($n = 6$) posted more than three photographs. These avid users individually contributed 5% or more of the dataset (8 or more photographs). To explore the influence of more avid users, results were analyzed both including and excluding avid users to determine if avid users influenced the emergent patterns and themes.

We organized results from the thematic coding into six parent categories (Table 2). These categories were: (1) natural resources; (2) built infrastructure; (3) human subjects; (4) outdoor recreation; (5) affect; and (6) culture. Themes are codes that emerged under each of these six categories. There were not large discrepancies in the dominance of particular categories or themes when avid users were removed from analysis (Table 2), suggesting that the avid Instagram users' conceptualization of North Shore destination image does not differ from other users. Many photographs and corresponding captions contained components from more than one of these categories; for example, outdoor recreation is explicitly captured in the photograph displayed in Figure 3 as it portrays an active recreationist in rock climbing gear. The photograph also highlights the destination's rock formations, Lake Superior, and prominently features the clear, blue sky. The caption supports the application of these codes as the narrative contains references to rock climbing, the type of rock, and the lake itself (as well as the neighbouring Boundary Water Canoe Area Wilderness). This example demonstrates that outdoor recreation (climbing), natural resources (rock, lake, and sky), affect ('very heady'), and human subjects (rock climber) are all components of this visitor's image of the North Shore.

Table 2. Definitions and density of categories and themes used to analyze #MyNorthShore Instagram posts.

Category	Definition	Full (Post Count)	Full (% of Posts)	Avid removed (Post Count)	Avid removed (% of Posts)
I. Natural Resources	Photo includes imagery of natural resources; text refers to a natural resource listed below.	205	45%	117	43%
Themes include: Wildlife, weather, waterfalls, rocks, plants, rivers, Lake Superior, inland lake, landscapes, insects, forests, coastline.					
II. Built infrastructure	Photos includes, or text references, an element of built infrastructure	70	15%	40	15%
Themes include: Roads (waysides, transportation corridors, streetscapes), recreation infrastructure (e.g. trails, boat ramps, bridges), historic sites, buildings (e.g. lighthouses, retail shops, restaurants) hotels, cabin or lodge.					
III. Human Subjects	Image or text containing or referring to a person.	66	14%	44	16%
Themes include: Self, recreation provider, pets, number of subjects in photo, family and friends, youth, mixed generation, adults					
IV. Outdoor Recreation	Photos depict and/or text includes a reference to participation in a recreation activity listed below.	51	11%	29	11%
Themes include: Visiting cultural and historic sites, swimming, scenic driving, rock collecting, recreation constraints, recreation substitution, picnicking, leisure at home, interpretation, visitor's centre, signs, museums, hunting, hiking, gathering wild plants (foods), fishing, creating art, contemplation, rock climbing, camping, campfires, boating, bicycling, ATV use.					
V. Affect	Text containing, 'awe', 'epic', 'best', etc. or 'horrible', 'worst', 'lame' etc.	39	9%	18	7%
Themes include: Include both references to positive and negative emotional responses.					
VI. Culture	Photo includes images and/or text references a cultural component listed below.	29	6%	22	8%
Themes include: Ojibwe (native American symbols or cultural references), food and drink, art					

Natural resources was the most dominant category present in the posts (45% of images included some natural resource). The most common themes under the natural resources parent category were water (one-half of all photographs coded with the natural resources category were also coded with 'open water', most often Lake Superior), forests, and rocks. Landscape-level photographs were also prominent, as well as photographs of waterfalls. The example photograph displayed in Figure 4 illustrates the themes of water (Lake Superior) and rocks. Agate stones pebbling the shoreline, as well as the large stone bluffs surrounding Lake Superior emerged from our analysis as key components of the North Shore destination image. Additionally, the post shown in Figure 4 captures a less dominant theme under the natural resources parent category, 'weather'.

Built infrastructure was the second most dominant category present in the posts (15% of images included built infrastructure). The most common themes of this category include buildings, lighthouses, and roads. Recreation infrastructure (e.g. trails, boat ramps, signs, camp sites or cabins) were less dominant, indicating that while these are known components of visitors' trips (the infrastructure upon which tourism depends), they are not dominant aspects of the North Shore destination image as conceptualized by the sample population. It is important to note that photographs that contained built infrastructure, which also may be consider a cultural resource (such as the lighthouse), were coded as both 'built infrastructure' and 'culture'; that is, codes for these or any other categories are not mutually exclusive. A photograph that captures an iconic North Shore lighthouse (built infrastructure and cultural resource) along the shoreline is presented in Figure 5.

A similar proportion (14%) of photographs included images with people and/or animals. These human subject photographs had captions that most frequently contained references to family and friends. Additionally, adults were more prominent than youth (and pets less prominent than actual humans), and photographs containing one human subject was most common (no groups of five or more captured in the photographs). An example of human subjects as a component of North Shore destination image is displayed in Figure 6. The caption of this photograph refers to watching the waves break on the shore; however, the photograph does not simply contain the water and shoreline but also includes two individuals occupying this space. The photograph demonstrates, that for many visitors, travel companions are an integral component of the destination image.

Figure 3. Example of photograph and caption with multiple themes present.

Overall, the category of outdoor recreation was not prominent in the posts (11% of images explicitly included or mentioned outdoor recreation). This could be due to the difficult nature of capturing actual recreation participation (e.g. it is difficult, and potentially risky, to take photographs while

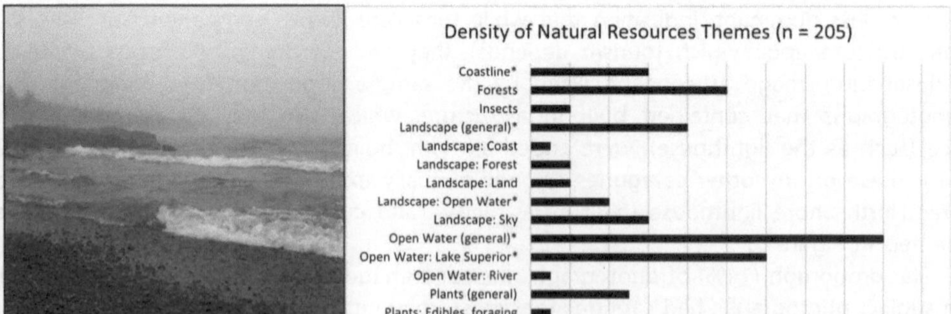

Figure 4. Example of natural resources photograph and caption, with chart of the natural resources category theme density. An * denotes natural resources themes that were applied to the example photograph on the left.

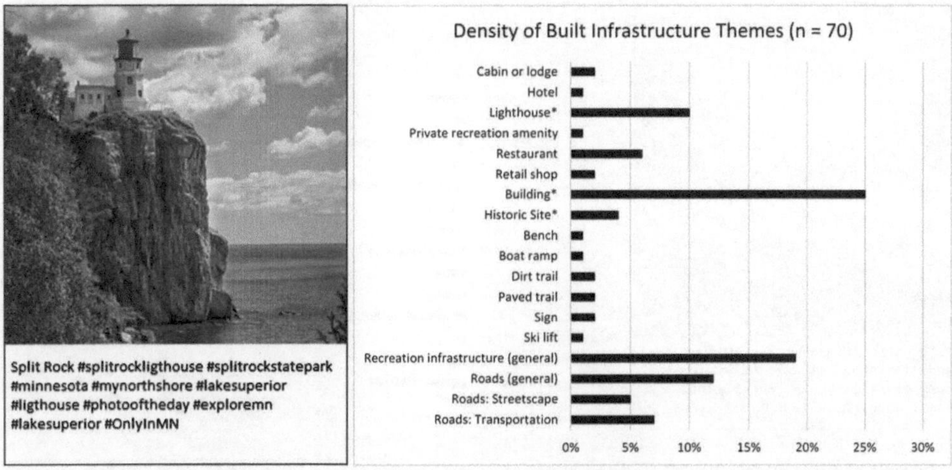

Figure 5. Example of built infrastructure photograph and caption and chart of built infrastructure theme density. An * denotes built infrastructure themes that were applied to the example photograph on the left.

fishing). For those photographs and captions that included elements of outdoor recreation, the most common themes were hiking and boating (particularly non-motorized boating such as canoeing and kayaking) and many of the photographs within the outdoor recreation category were representations of these pursuits. For example, an individual shared a photograph of a companion paddling a non-motorized boat (Figure 7). This photograph demonstrates that many photographs and captions coded as outdoor recreation were also coded with a human subject category; most often the photograph of outdoor recreation is not focused on the gear or outdoor recreation resource, but rather individuals sharing in the recreation experience.

Culture was a less dominant category (6% of Instagram posts included images or references to cultural components). Culture was conceptualized and operationalized through themes of art, food, and drink, and Native American culture. Among these components, photographs of food and drink were most common, such as the example photograph displayed in Figure 8. North Shore visitors' destination image includes the art, food, and culture of the place they visit; however, these themes were less dominant than physical attributes of place. Photographs, such as the one displayed in Figure 8, demonstrate that visitors' images of the destination's food culture

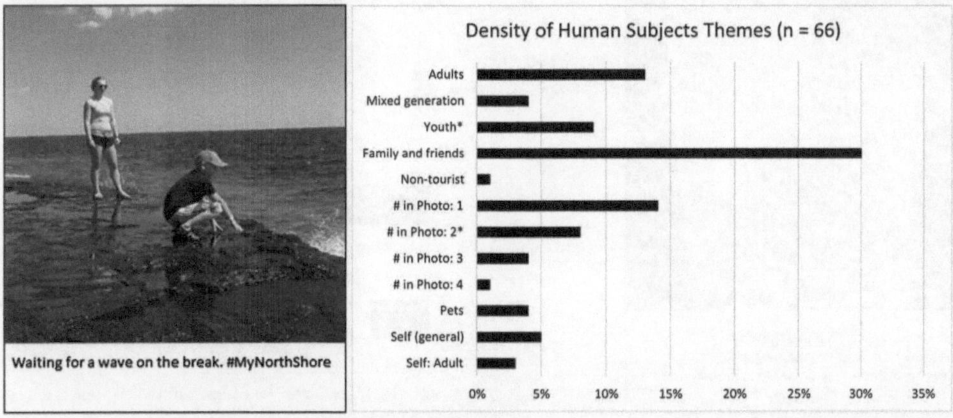

Figure 6. Example of human subjects photograph and caption and chart of human subject theme density. An * denotes human subject themes that were applied to the example photograph on the left.

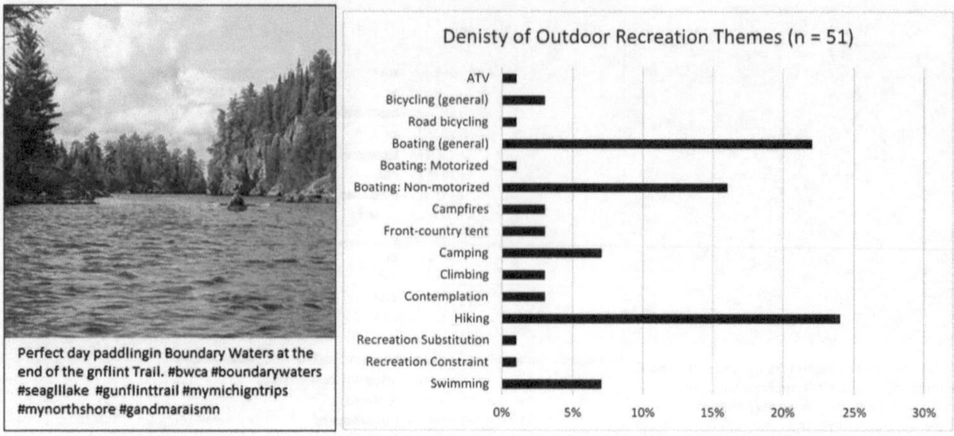

Figure 7. Example of outdoor recreation photograph and caption and chart of theme prominence in the outdoor recreation category. The photograph on the left was coded using the parent category outdoor recreation, however, none of the themes were applied. Other themes were applied including natural resources and human subject themes.

are focused on local, unique offerings. Captions of images classified as cultural namely included referencing the food or drink item (e.g. #worldsBestDonuts or #cheeseCurds).

Affect was also less prominent (9% of posts were coded for instances of positive or negative emotion); it was captured through captions and additional hashtags associated with the photographs visitors posted to Instagram. Positive affect was much more dominant than negative affect (with less than 5 captions being coded as negative affect, such as the latter part of this quote: 'We had a beautiful hike to the north end of the Superior Hiking Trail today. The view was 270°, but hazy. Hot as hell'.). For example, positive affect is demonstrated in the caption to the photograph displayed in Figure 9, which reads 'certainly a good way to start off my day!' Other examples of positive affect included: 'Idyllic weekend #mynorthshore!' 'Within this place–this whole lakeshore as far as I have explored–there is a special kind of serenity. Anxieties, concerns and worries fade into nothingness … ' and 'Grand Marais is incredible! There's so much beauty in our own backyard, and I will surely be back soon'. Such captions indicate that destination image inherently includes the emotional ties to a space, which emerge even from a study grounded in understating the physical attributes of place.

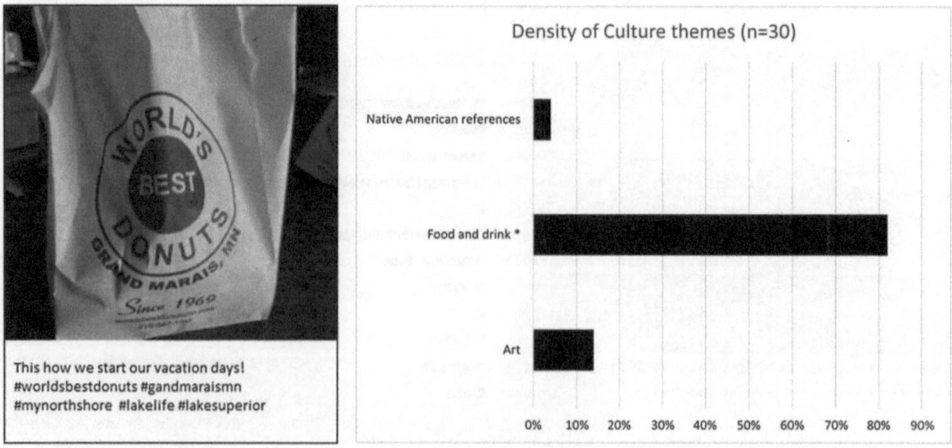

Figure 8. Example of a culture photograph and caption and chart of culture theme density. An * denotes culture themes that were applied to the example photograph on the left.

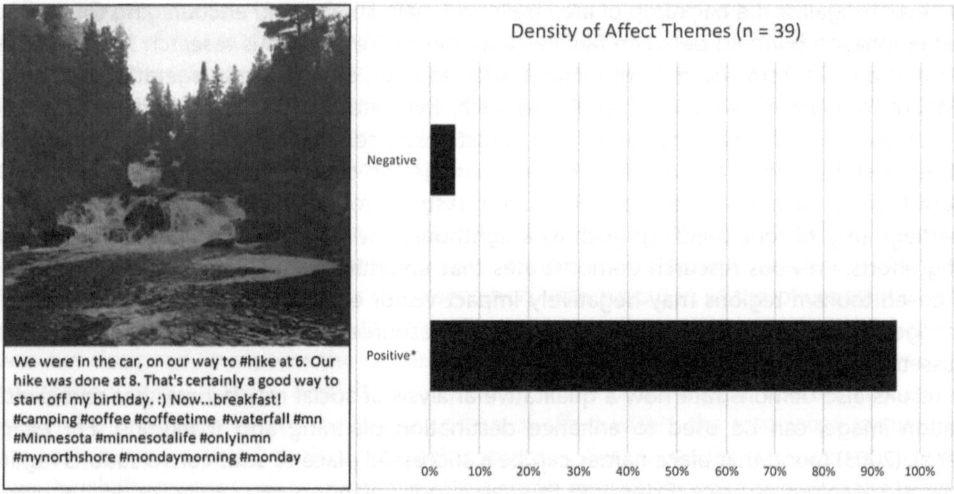

Figure 9. Example of a caption demonstrating affect and chart affect theme density. An * denotes the affect theme that was applied to the example photograph on the left.

Discussion

This study demonstrates the feasibility of qualitatively utilizing social media data as a rather rapid and low-cost process to determine the key components of North Shore visitors' destination image and subsequently inform place-based management and marketing. Other studies, such as Taylor et al. (1995), that couple VEP data with more in-depth qualitative content (e.g. photo logs) have illustrated that a physical characteristic's persistence in photographic data may be a more reliable indicator of unconscious elements of visitors' destination image. As such, our study expands current methodological approaches to understanding important place-based characteristics in regional tourism destinations by accessing and analyzing readily available UGC posted to social media platforms.

Related to our first research objective of *determining the key components of destination image,* the key components of North Shore visitors' destination image include natural resources, built infrastructure, and human subjects. The dominant categories and themes identified in our study are consistent with existing research that demonstrates natural resources are an important component of destination image (Lin et al., 2007). In particular, the abundance of water-based features (i.e. waterfalls, rivers, Lake Superior, inland lakes, coastlines) has also been found in other studies using VEP to understand visitor preferences (Taylor et al., 1995). A somewhat novel component of water features was the emergence of weather as theme in North Shore visitors' photographs. The wet, misty weather along the North Shore in summer, when many other locales are experiencing hot and dry conditions, is characteristic of the region and its destination image. The prevalence of coupled characteristics, such as natural and built resources, can be important to destination image (and the underlying place meanings assigned to these spaces) as these are the settings where interaction and engagement occur (Amsden et al., 2010). Additionally, human subjects have also been found to be a primary component of destination image (Michaelidou et al., 2013). Michaelidou et al. (2013) determined that the common appearance of human subjects in visitor photographs is likely related to visitors' desire to capture participation and interaction during a recreation experience.

Regarding the second objective of *informing place-based management and marketing,* it is important to note that while the methodology of assessing visitors' destination image to inform place-based management and marketing may be transferable, the specific findings or recommendations are likely to be destination specific. For example, our findings suggest that destination management organizations in the North Shore region can enhance visitors' exposure to the physical attributes of the region by promoting opportunities for both land and water-based recreation experiences, themes of destination image using social media posts, which can inform place-based management and marketing.

Acknowledgements

The authors would also like to thank Derek Van Berkel and Boris van Zanten for their assistance with data collection.

Disclosure statement

No potential conflict of interest was reported by the authors.

Funding

This work was prepared by Allie McCreary, Erin Seekamp, Mae Daveport, and Jordan Smith using federal funds under award NA14OAR4170080 from Minnesota Sea Grant, National Sea Grant College Program, National Oceanic and Atmospheric Administration, U.S. Department of Commerce. The statements, findings, conclusions, and recommendations are those of the author(s) and do not necessarily reflect the views of NOAA, the Sea Grant College Program, or the U.S. Department of Commerce.

References

Allen, S. D., Wickwar, D. A., Clark, F. P., Dow, R. R., Potts, R., Snyder, S. A., & Snyder, S. A. (2009). *Values, beliefs, and attitudes technical guide for forest service land and resource management, planning, and decisionmaking*. Portland, OR: US Department of Agriculture, Forest Service, Pacific Northwest Research Station.

Amsden, B. L., Stedman, R. C., & Kruger, L. E. (2010). The creation and maintenance of sense of place in a tourism-dependent community. *Leisure Sciences, 33*(1), 32–51.

Antoniou, V., Morley, J., & Haklay, M. (2010). Web 2.0 geotagged photos: Assessing the spatial dimension of the phenomenon. *Geomatica, 64*(1), 99–110.

Assaker, G., & Hallak, R. (2013). Moderating effects of tourists' noelty-seeking tendencies on destination image, visitor satisfaction, and short- and long-term revisit intentions. *Journal of Travel Research, 52*(5), 600–613.

Beckley, T. M., Stedman, R. C., Wallace, S. M., & Ambard, M. (2007). Snapshots of what matters most: Using residents-employed photography to articulate attachment to place. *Society & Natural Resources, 20*(10), 913–929.

Brando, C., & Bucher, B. (2010). Quality in user generated spatial content: A matter of specifications. *13th AGILE International Conference of Geographic Information Science*, 1–8.

Brown, G., & Weber, D. (2011). Public participation GIS: A new method for national park planning. *Landscape and Urban Planning, 102*(1), 1–15.

Bruton, A., Conway, J. H., & Holgate, S. T. (2000). Reliability: What is it and how is it measured? *Physiotherapy, 86*(2), 94–99.

Bullock, S. D., & Lawson, S. R. (2007). Examining the potential effects of management actions on visitor experiences on the summit of Cadillac Mountain, Acadia National Park. *Research in Human Ecology, 14*(2), 140–157.

Cahyanto, I., Pennington-Gray, L., & Thapa, B. (2013). Tourist-resident interfaces: Using reflexive photography to develop responsible rural tourism in Indonesia. *Journal of Sustainable Tourism, 21*(5), 732–749.

Chen, C. F., & Tsai, D. (2007). How destination image and evaluative factors affect behavioral intentions? *Tourism Management, 28*, 1115–1122.

Chin, C. L. M., Moore, S. A., Wallington, T. J., & Dowling, R. K. (2000). Ecotourism in Bako National Park, Borneo: Visitors' perspective on enviornmental impacts and their management. *Journal of Sustainable Tourism, 8*(1), 20–35.

Cocolas, N., Walters, G., & Ruhanen, L. (2016). Behavioural adaptation to climate change among winter alpine tourists: An analysis of tourist motivations and leisure substitutability. *Journal of Sustainable Tourism, 24*(6), 846–865.

Davenport, M. A., & Anderson, D. H. (2005). Getting from sense of place to place-based management: An nterpretive investigation of place meanings and perceptions of landscape change. *Society & Natural Resources, 18*, 625–641.

Davenport, M. A., Schneider, I., Date, A., & Filter, L. (2011). *Minnesota's network of parks & trails: Northeast region profile*. Retrieved from http://ccl.design.umn.edu/mnpat.html

Di Minin, E., Tenkanen, H., & Toivonen, T. (2015). Prospects and challenges for social media data in conservation science. *Frontiers in Environmental Science, 3*(September), 1–6.

Dorwart, C. E., Moore, R. L., & Leung, Y. F. (2009). Visitors' perceptions of a trail environment and effects on experiences: A model for nature-based recreation experiences. *Leisure Sciences, 32*(1), 33–54.

Drisko, J., & Maschi, T. (2015). *Content analysis*. New York, USA: Oxford University Press.

Environmental Protection Agency (EPA). (2002). Community Culture and the Environment: A Guide to Understanding a Sense of Place, US Environmental ProtectionAgency (EPA), available at: www.epa.gov/care/library/community_culture.pdf (accessed 10 January 2015).

Feick, R., & Robertson, C. (2015). Identifying locally- and globally-distinctive urban place descriptors from heterogeneous user-generated content. *Advances in Geographic Information Science, 19*, 51–63.

Govers, R., & Go, F. (2009). *Place branding: Glocal, virtual and physical identities, constrcuted, imagined and experienced*. Basingstoke: Palgrave Macmillan.

Greider, T., & Garkovich, L. (1994). Landscapes: The social construction of nature and the environment. *Rural Sociology, 59* (1), 1–24.

Gustafson, P. (2001). Meanings of place: Everyday experience and theoretical conceptualizations. *Journal of Environmental Psychology, 21*, 5–16.

Hestetune, A., McCreary, A., Holmberg, K., Wilson, B., Seekamp, E., Davenport, M. A., & Smith, J. W. (2018). Research note: Climate change and the demand for summer tourism on Minnesota's North Shore. *Journal of Outdoor Recreation and Tourism, 24*(October), 21–25.

Higginbotham, N., Connor, L., Albrecht, G., Freeman, S., & Agho, K. (2007). Validation of an environmental distress scale. *EcoHealth, 3*, 245–254.

IPM. (2016). *Institute of place management*. Retrieved from http://placemanagement.org/

Jacobs, M. H., & Buijs, A. E. (2011). Understanding stakeholders' attitudes toward water management interventions: Role of place meanings. *Water Resources Research, 47*(1), 1–11.

Kavaratzis, M., & Hatch, M. J. (2013). The dynamic of place brands: An identity-based approach to place branding theory. *Marketing Theory, 13*(1), 69–86.

Kerstetter, D. L., & Bricker, K. (2009). Exploring Fijian's sense of place after exposure to tourism development. *Journal of Sustainable Tourism, 17*(6), 691–708.

Kil, N., Holland, S., & Stein, T. (2014). Place meanings and participatory planning intentions. *Society & Natural Resources, 27* (5), 475–491.

Kim, S. O., Chang, H. L., & Shelby, B. (2003). Utilization of photographs for determining impact indicators for trail management. *Environmental Management, 32*(2), 282–289.

Lee, B., Lee, C. K., & Lee, J. (2014). Dynamic nature of destination image and influence of tourist overall satisfaction on image modification. *Journal of Travel Research, 53*(2), 239–251.

Lew, A. A. (2017). Tourism planning and place making: place-making or placemaking. *Tourism Geographies, 19*(3), 448–466.

Lin, C. H., Morais, D. B., Kerstetter, D. L., & Hou, J. S. (2007). Examining the role of cognitive and affective image in predicting choice across natural, developed, and theme-park destinations. *Journal of Travel Research, 46*(2), 183–194.

Low, S. M., & Altman, I. (1992). Place attachment. In *Place attachment* (pp. 1–12). Boston, MA: Springer.

Lowery, D. R., & Morse, W. C. (2013). A qualitative method for collecting spatial data on important places for recreation, livelihoods, and ecological meanings: Integrating focus groups with public particpation geographic information systems. *Society & Natural Resources, 26*, 1422–1437.

Metaxas, P., & Mustafaraj, E. (2014). Sifting the sand on the river bank: Social media as a source for research data. *IT - Information Technology, 56*(5), 230–239.

Michaelidou, N., Siamagka, N. T., Moraes, C., & Micevski, M. (2013). Do marketers use visual representations of destinations that tourists value? Comparing visitors' image of a destination with marketer-controlled images online. *Journal of Travel Research, 52*(6), 789–804.

Pew. (2016). *Instagram user demographics*. Retrieved from http://www.pewinternet.org/2015/08/19/mobile-messaging-and-social-media-2015/2015-08-19_social-media-update_09/

Prayag, G., & Ryan, C. (2012). Antecedents of tourists' loyalty to Mauritius: The role and influence of destination image, place attachment, personal involvement, and satisfaction. *Journal of Travel Research, 51*(3), 342–356.

Proshansky, H. M., Faban, A. K., & Kaminof, R. (1983). Place identity: Physical world socialization of the self. *Journal of Environmental Psychology, 3*, 57–83.

Relph, E. (1976). *Place and placelessness* (Vol. 1). Pion.

Richards, D. R., & Friess, D. A. (2015). A rapid indicator of cultural ecosystem service usage at a fine spatial scale: Content analysis of social media photographs. *Ecological Indicators, 53*, 187–195.

Sancar, F. H., & Severcan, Y. C. (2010). Children's places: Rural-urban comparisons using particpatory photography in the Bodrum Peninsula, Turkey. *Journal of Urban Design, 15*(3), 293–324.

Seamon, D, & Sowers, J. (2008). Place and placelessness (1976): Edward relph. *Key texts in human geography*, 43–52.

Smith, J. W., Anderson, D. H., & Moore, R. L. (2012). Social capital, place meanings, and perceived resilience to climate change. *Rural Sociology, 77*(3), 380–407.

Stedman, R. C. (2003). Is it really just a social construction?: The contribution of the physical environment to sense of place. *Society & Natural Resources, 16*(8), 671–685.

Stokols, D., & Shumaker, S. A. (1981). People in places: A transactional view of settings. In J. Harvey (Ed.), *Cognition, social behavior, and the environment* (pp. 441–488). Hillsdale, NJ: Erlbaum.

Stylidis, D., Sit, J., & Biran, A. (2014). An exploratory study of residents' perception of place image: The case of Kavala. *Journal of Travel Research, 55*(5), 659–674.

Swinney, J., Lang, C., & Runyan, R. (2012). An exploration of rural community branding efforts from the perspective of community residents. *International Journal of Rural Management, 8*(1&2), 35–47.

Taylor, J. G., Czarnowski, K. J., Sexton, N. R., & Flick, S. (1995). The importance of water to Rocky Mountain National Park visitors: An adaptation of visitor-employed photography to natural resources management. *Journal of Applied Recreation Research, 20*(1), 61–85.

van Zanten, B. T., van Berkel, D. B., Meetemeyer, R. K., Smith, J. W., Tieskens, K. F., & Vergurg, P. H. (2016). *Continental scale quatification of landscape values using social media data*. In *Proceedings of the National Academy of Sciences* (pp. 1–7).

White, E., Bowker, J. M., Askew, A. E., Langner, L. L., Arnold, J. R., & English, D. B. (2016). *Federal outdoor recreation trends: Effects on economic opportunities*. (Gen. Tech. Rep. PNW-GTR-945). Portland, OR: US Department of Agriculture, Forest Service, Pacific Northwest Station, 46 p., 945.

Williams, D. R., & Patterson, M. E. (2007). Snapshots of what, exactly? A comment on methodological experimentation and conceptual foundations in place research. *Society & Natural Resources, 20*(10), 931–937.

Williams, D. R., & Roggenbuck, J. W. (1989). *Measuring place attachment: Some preliminary results*. In *paper presented at the session on outdoor planning and management NRPA symposium on leisure research* (pp. 1689–1699), San Antonio, TX. https://doi.org/10.1017/CBO9781107415324.004

Wood, S. A., Guerry, A. D., Silver, J. M., & Lacayo, M. (2013). Using social media to quantify nature-based tourism and recreation. *Scientific Reports, 3*(2976), 2976.

Yi-Fu, T. (1974). *Topophilia: A Study of Enviromental Perception, Attitude, and Values*. Prentice Hall.

Yung, L., Freimund, W. A., & Belsky, J. M. (2003). The politics of place: Understanding meaning, common ground, and political differnce on the Rocky Mountain Front. *Forest Science, 49*(6), 855.

Index